The theory of approximate methods and their application to the numerical solution of singular integral equations

Monographs and textbooks on mechanics of solids and fluids
editor-in-chief: G. Æ. Oravas

Mechanics: Analysis
editor: V. J. Mizel

1. M. A. Krasnoselskii *et al.*
   Integral operators in spaces of summable functions

2. V. V. Ivanov
   The theory of approximate methods and their application to the numerical solution of singular integral equations

3. A. Kufner *et al.*
   Function spaces

The theory of approximate methods and their application to the numerical solution of singular integral equations

V. V. Ivanov

Cybernetics Institute
Ukrainian Academy of Sciences
Kiev, U.S.S.R.

Translated by

A. Ideh

Edited by

R. S. Anderssen

Computer Centre
Australian National University, Canberra

D. Elliott

Mathematics Department
University of Tasmania, Hobart

NOORDHOFF INTERNATIONAL PUBLISHING
LEYDEN

Softcover reprint of the hardcover 1st edition 1976
A division of A. W. Sijthoff International Publishing Company B.V.,
Leyden, The Netherlands

ISBN 978-90-481-8462-0

Original title:
'Teoria priblizshennykh metodov i ee primenenie k chislennomu resheniju singuljarnykh integral'nykh uravnenii'
published in 1968 in Russian by Naukova Dumka Press, Kiev

We heard what she said, but we knew what she meant.

The Two Ronnies

CONTENTS

CHAPTER 4

NUMERICAL SOLUTION OF SINGULAR INTEGRAL EQUATIONS OF CONVOLUTION TYPE

CHAPTER 5

NUMERICAL SOLUTION OF OTHER FORMS OF SINGULAR EQUATIONS

AUTHOR'S PREFACE

The present monograph summarizes the author's investigations into the theory, application, and basic structure of methods and algorithms for the approximate solution of singular integral equations. The author has attempted to explain all the significant advances connected with the given theme, and therefore has cited well known results without proof, but with appropriate references.

To read the book, a knowledge is required of computational methods, elements of functional analysis, theory of functions of a complex variable, the constructive theory of functions and the theory of singular integral equations. Relevant texts on these subjects are by Berezin and Zhidkov [8], Lyusternik and Sobolev [96], Lavrentiev and Shabat [90], Natanson [109], Muskhelishvili [108] and Gakhov [33].

In attempting to make the book available to a very wide group of specialists, the author has merely stated necessary auxiliary information about functional analysis, the theory of singular integral equations and the constructive theory of functions. Additional material, available for the reader with an advanced mathematical training, is set with a vertical line in the margin within the text.

The primary stimulus for the author was the profound work of N.I. Muskhelishvili on singular integral equations (see [108]). Subsequently the author has been consistently encouraged by F.D. Gakhov and I.N. Vekua. The author thanks them for their frankness.

TRANSLATION EDITORS' PREFACE

Professor Ivanov's book was published in 1968; it continues to be the only book so far written that deals extensively with the approximate solution of singular integral equations. Although it is seven years since it was published, no comprehensive text of a similar nature has yet appeared in the English language. The editors therefore considered that it was appropriate to translate Professor Ivanov's book into English. The first part of the book (Chapter 1) contains much on the approximate solution of operator equations that has not previously been available to readers of English. The remaining four Chapters deal with singular integral equations.

A first translation was done by Mrs. A. Ideh under the direction of the Canberra editor (R.S.A.). In revising the first draft translation, the Hobart editor (D.E.) took the opportunity to correct many of the misprints that appeared in the original book. The editors would like to express their thanks to Professor V.V. Ivanov whose assistance in the correction of the major errors is much appreciated.

Thanks are also due to the Australian Research Grants Committee for a grant that enabled this project to get started, and to the Computer Centre, Australian National University, for assistance which helped to take it through to its completion. One of us (D.E.) would like to thank Drs. W.R. Bloom and T.P. Wojtowicz, and Messrs. M.L. Dow and G.S. Kelly for their assistance with this translation. The other of us (R.S.A.) would like to thank Mrs. Judy McHugh for her help with the typing of the first draft of the manuscript. Both editors would like to thank Mrs. A. Ideh for agreeing to do the initial translation, and Mrs. L. Brown whose excellent typing has made the publication of this translation possible.

Canberra
Hobart May 1975

INTRODUCTION

Lavrentiev's work [89] was the first in which questions about the solution of singular integral equations with a Cauchy kernel were examined with full mathematical rigour without the use of regularization (i.e., without reduction to a second kind Fredholm equation). In this work, two numerical methods were proposed and justified: the first was based on a polygonal approximation to the solution; the second anticipated, by means of a particular example, the general Ritz-Galerkin method.

In the preface to [89], Academician S.A. Chaplygin wrote, *It is with pleasure that, with the appearance of Professor M.A. Lavrentiev's work, one expresses the hope that the study will continue and that an attempt will be made to simplify the application of the method and increase its convergence. For the solution of this problem has many possibilities, and to me it is conceivable that the results will repay the expended effort.*

It is interesting that at about the same time, there appeared independently the work of Enskog [157], in which a particular form of singular integral equation was solved by a modified least squares method; and the work of Wiener and Hopf [171], in which the now well known Wiener-Hopf equations of the first and second kind were discussed and solved (without relating them to singular integral equations with a Cauchy kernel).

The rapid development of the general theory of singular integral equations in the following years, their connection with boundary value problems, and multidimensional applications in the theory of elasticity, hydrodynamics, and many parts of mathematical physics, has directed the attention of mathematicians to the approximate solution of singular integral equations. This led chiefly to the necessity to derive solutions for singular integral equations arising in applications, although the same theory proved to be important in the investigation of constructive methods.

Muskhelishvili [108], when summarizing the many years of investigations into the subject of singular integral equations and their applications asserted that, *Future development of this (M.A. Lavrentiev's) and analogous methods for the approximate solution of singular integral equations is, in my opinion, one of the most important problems in the theory of equations.*

The present work is devoted to a basic description of numerical schemes, the rigorous foundation and comparison of a series of approximate methods and algorithms, and their application to the numerical solution of singular integral equations.

By *foundation* is meant an examination of the stability of solutions, the proof of the convergence of approximate solutions to the exact, an examination of the rate of convergence and methods of best convergence, the determination of practical methods of error estimation, and the examination of effective methods and algorithms from the point of view of their computer implementation.

Firstly, there are the general approximate methods for the solution of linear operator equations in Hilbert and Banach spaces. Linear singular integral equations represent a particular class of such equations. It is therefore natural to examine initially these general methods in some detail together with their connection with numerical algorithms, and thereby answer the question about what specific difficulties arise as a result of the application of these general schemes to the approximate solution of singular integral equations.

Secondly, it is well known that there exist various approximate methods for the solution of Fredholm equations of the second kind, and therefore it is natural to examine the question about which of these methods can be applied to singular integral equations without regularization.

There is a small group of methods which were specifically designed for the approximate solution of singular integral equations. The basis for these methods are closed form solutions for certain classes of singular integral equations.

Finally, some generalizations of well known methods are examined. In addition, a series of non-linear programming methods, connected with the more general problem of the solution of non-linear operator equations with additional constraints on the solution, is studied.

Questions connected with the widely known applications of singular integral equations to the theory of elasticity and hydrodynamics, are hardly touched upon in the book. We have examined in greater detail the comparatively little known applications of singular integral equations to the solution of problems in the analysis and synthesis of systems in the theory of automatic control, and to some problems in theoretical physics. In the book, results for the computational implementation and solution of a series of particular problems from these fields are considered.

In conclusion, we emphasize some principal aspects of the book:

1. A distinctive aspect of the numerical solution of singular integral equations is the under-determined (or over-determined) nature of the corresponding approximate equations. Therefore, particular attention in the book has been concentrated on the construction and verification of algorithms which are effective in these situations. As is shown in the book, there are algorithms which are useful even if the given equation does not have a unique solution.

2. It is known that intermediate calculations and results cannot be reproduced with an accuracy which is superior to the accuracy of the given data. In the analysis of different algorithms for the method of least squares and a series of other methods, it is shown that the accuracy of the approximation, the number of iterations, and certain other parameters of the algorithms, must be closely related to the accuracy of the original data. An attempt is made to define the optimal values of these parameters and it is shown that the use of a high accuracy method for low accuracy data is not only inefficient, but also harmful since it can lead to a considerable loss of accuracy.

3. At many places throughout the book, the author develops ideas on the solution of the largest number of various problems by means of a minimum set of algorithms, which are of great importance in increasing the effectiveness of computational implementations, see [100]. Such collections of algorithms for the majority of problems are: the construction of Hermite-Lagrange interpolation formulas (for the solution of problems of approximation and best convergence), algorithms for the calculation of $y_k = Ae_k$ for a given class of operators A and elements e_k; the summing of products (for the calculation of scalar products); algorithms for simple iteration, coordinate-wise descent and steepest descent (for the solution of linear equations in a Hilbert space), and optimization (for the solution of more complex problems).

Chapter 1

METHODS AND ALGORITHMS FOR THE APPROXIMATE SOLUTION OF GENERAL OPERATOR EQUATIONS

§1 RELEVANT ASPECTS OF FUNCTIONAL ANALYSIS[1]

§1.1 Abstract Spaces

A set X is called a *metric space*, if, for every pair of its elements x and y, there exists a non-negative number $\rho_X(x,y)$ which satisfies the conditions:

1) $\rho_X(x,y) = 0$ if and only if $x = y$ (identity axiom);

2) $\rho_X(x,y) = \rho_X(y,x)$ (symmetry axiom);

3) $\rho_X(x,y) + \rho_X(y,z) \geq \rho_X(x,z)$ for any $z \in X$ (triangle inequality).

The number $\rho_X(x,y)$ is called the *distance* between the elements x and y. If there is no ambiguity as to which metric space X is being examined, then $\rho(x,y)$ is written instead of $\rho_X(x,y)$. The elements of a metric space are called its *points*.

Every subset Y of X, examined with respect to the distance in X, is itself a metric space and is called a *subspace* of X.

An element x of X is called a *limit* of the sequence of the elements $x_1, x_2, \ldots, x_n, \ldots$ of X if $\rho(x_n, x) \to 0$ as $n \to \infty$. In this case, we write $x_n \to x$ or $\lim\limits_{n\to\infty} x_n = x$.

The set of points $x \in X$ which satisfy the inequality $\rho(x, a) < r$ (or the inequality $\rho(x,a) \leq r$) for a fixed $a \in X$ and non-negative real number r define the *open sphere* $S(a, r)$ (or the *closed sphere* $\tilde{S}(a,r)$) with centre a and radius r. A *neighbourhood* of $x \in X$ is any open sphere with centre at this point. The set of points Y is called *bounded* if there exists an open sphere containing Y.

Many of the more important concepts of the theory of point sets lying on a straight line can be introduced into a metric space. Thus, for a given set $M \subset X$, a point $a \in X$ is called a *boundary (or limit) point* of M if every neighbourhood of a contains at least one point in the set $M - \{a\}$ i.e. the set of points of M not including a. The set consisting of M and all its boundary

---

(1) The material of this section is taken from [96].

points is called the *closure* of M and is denoted by $\bar{M}$. The set M is said to be closed if $\bar{M} = M$, and *open* if its complement $X - M$ [the set of points of X which are not in M] is closed. M is said to be *dense* in a set G if $G \subseteq \bar{M}$. In particular M is said to be *everywhere dense*, if $\bar{M} = X$. Finally, M is said to be *nowhere dense* in X if every open sphere in X contains an open sphere S such that $S \cap M$ is empty.

A sequence $\{x_n\}$ of elements in a metric space X is called a *Cauchy sequence* or a *fundamental sequence*, if, for a given $\epsilon > 0$, there exists an $n_0(\epsilon)$ such that $\rho(x_m, x_n) < \epsilon$ for $n,m \geq n_0(\epsilon)$. If every Cauchy sequence in X converges to an element of X, then X is said to be *complete*.

A space X is *separable* if it contains a countable subset which is everywhere dense in it.

Some examples of metric spaces are:

1) E_n, the n dimensional Euclidean Space of all the ordered arrays of n real numbers with

$$\rho(x,y) = \left\{ \sum_{i=1}^{n} (x_i - y_i)^2 \right\}^{1/2}, \quad x = (x_1, x_2, \ldots, x_n), \quad y = (y_1, y_2, \ldots, y_n). \tag{1.1}$$

2) $C(E)$ - the space of all bounded continuous complex-valued functions defined on a set E with the Chebyshev metric

$$\rho(x,y) = \sup_{t \in E} |x(t) - y(t)| . \tag{1.2}$$

3) $L_p(\Gamma)$ - the space of all pth integrable functions defined on the rectifiable curve Γ with

$$\rho(x,y) = \left\{ \int_\Gamma |x(t) - y(t)|^p |dt| \right\}^{1/p} . \tag{1.3}$$

4) m - the space of all bounded numerical sequences with

$$\rho(x,y) = \sup_i |x_i - y_i| , \quad x = (x_1, x_2, \ldots), \quad y = (y_1, y_2, \ldots), \tag{1.4}$$

$$|x_i| \leq C_x, \quad |y_i| \leq C_y ,$$

where C_x and C_y denote constants which do not depend on i.

5) ℓ_p - the space of pth summable numerical sequences with

$$\rho(x,y) = \left\{ \sum_{i=1}^{\infty} |x_i - y_i|^p \right\}^{1/p} , \quad \sum_{i=1}^{\infty} |x_i|^p < \infty, \quad \sum_{i=1}^{\infty} |y_i|^p < \infty . \tag{1.5}$$

6) S - the space of all numerical sequences with

$$(1.6) \qquad \rho(x,y) = \sum_{i=1}^{\infty} \frac{|x_i - y_i|}{2^i \{1 + |x_i - y_i|\}}$$

From an examination of many actual spaces, it is found that the elements of these spaces {functions, numerical sequences, etc.} can be added to each other or multiplied by a scalar to yield elements of the same spaces.

Any space E which satisfies the following axioms is called a *linear space* or a *vector space*:

I. E is an Abelian group with respect to the binary operation of addition. This means that the sum $x+y$ of any two elements $x,y \in E$ is also an element of E and

1) $x+y = y+x$ (commutative axiom);
2) $x+(y+z) = (x+y)+z$ for any $z \in E$ (associative axiom);
3) there exists a uniquely defined element $0 \in E$ such that $x+0 = x$ for all $x \in E$;
4) for every $x \in E$ there exists a uniquely defined element $(-x) \in E$ such that $x+(-x) = 0$. The element 0 is called the *zero* or *null element* and $(-x)$ is referred to as the *negative* of x. It is usual to write $x-y$ instead of $x+(-y)$.

II. The multiplication of the elements $x,y,z,\ldots$ of E by real (or complex) scalars $\lambda, \mu, \nu,\ldots$, is defined with $\mu x, \mu y,\ldots \in E$ and the following conditions are satisfied:

1) $\lambda(\mu x) = (\lambda\mu)x$ (associative rule for scalar multiplication);
2) $\lambda(x+y) = \lambda x + \lambda y$, $(\lambda+\mu)x = \lambda x + \mu x$ (the two distributive rules for scalar multiplication);
3) $1.x = x$.

Depending on whether the scalars are real or complex, E is a real or complex linear space.

In a linear space, the concepts of linear independence and dependence can be introduced. The elements $x_1, x_2,\ldots,x_n$ of a linear space are *linearly independent* if the only values of the scalars $\lambda_i (i = 1,2,\ldots,n)$ for which

$$(1.7) \qquad \lambda_1 x_1 + \lambda_2 x_2 + \ldots + \lambda_n x_n = 0$$

is valid, are $\lambda_i = 0$ $(i = 1,2,\ldots,n)$. If, on the other hand, there exist

$\lambda_i (i = 1,2,\ldots,n)$ not all zero for which (1.7) is valid, then the elements $x_i (i = 1,2,\ldots,n)$ are *linearly dependent*. For example, let $\lambda_n \neq 0$ when the $x_i (i = 1,2,\ldots,n)$ are linearly dependent. Then $x_n = \alpha_1 x_1 + \alpha_2 x_2 + \ldots + \alpha_{n-1} x_{n-1}$, where $\alpha_i = - \lambda_i / \lambda_n$. In this case, the element x_n is said to be a *linear combination* of the elements x_i $(i = 1,2,\ldots,n-1)$.

The set of elements of E which satisfy $y = tx$, $x \in E$, $x \neq 0$, $-\infty < t < \infty$, is called the *straight line* defined with respect to x, and the set

$$\{y : y = (1-t)x_1 + tx_2, \text{ with } x_1, x_2 \in E \text{ and } 0 \leq t \leq 1\},$$

is called the *interval* which joins the points x_1 and x_2. A set K of E is *convex* if the interval which joins any two points of K lies wholly in K.

A non-empty set L of the linear space E is called a *linear manifold* if, together with the elements x_i $(i = 1,2,\ldots n)$, L contains every linear combination $\alpha_1 x_1 + \alpha_2 x_2 + \ldots + \alpha_n x_n$ of these elements.
The totality of all possible sums $\sum_{i=1}^{k} \alpha_i x_i$ is called the linear manifold *resulting from* the given elements, or the linear manifold *defined over* these elements.

Let E be a linear space such that $L_1, L_2, \ldots, L_n$ are linear manifolds in E. If every element $x \in E$ has a unique representation of the form

$$x = x_1 + x_2 + \ldots + x_n, \; x_i \in L_i, \; i = 1,2,\ldots,n , \tag{1.8}$$

then E is the *direct sum* of the linear manifolds $L_1, L_2, \ldots, L_n$.
We write

$$E = \bigoplus_{i=1}^{n} L_i . \tag{1.9}$$

If a linear space is also a metric space, then it is called a *linear metric space*. A particular type of linear metric space is the normed linear space.

A set E is called a *normed linear space* if:

1) E is a linear space with respect to multiplication by real or complex scalars, and

2) To every $x \in E$ there exists a real number, called the *norm* of x and denoted by $\|x\|$, which satisfies the following conditions:

1) $\|x\| \geq 0$, with $\|x\| = 0$ iff $x = 0$;
2) $\|x+y\| \leq \|x\| + \|y\|$ for all $x,y \in E$;
3) $\|\lambda x\| = |\lambda| \, \|x\|$ for any scalar λ.

In a normed linear space, distance (metric) can be defined by

(1.10) $$\rho(x,y) = \|x-y\| .$$

Convergence with respect to this metric is called *convergence in norm* or *strong convergence.*

If a normed linear space is complete with respect to convergence in norm, then it is called a *Banach space* or *a space of type B.*

Examples of spaces of type B are E_n, C, L_p, m and ℓ_p with, in each case,

(1.11) $$\|x\| = \rho(x,0) .$$

The Banach space $C^k[0,1]$ is the space of functions defined on $[0,1]$ which have continuous derivatives of order k with

(1.12) $$\|x\| = \max \{\max_t \; |x(t)|, \max_t |x'(t)|, \ldots, \max_t \; |x^{(k)}(t)|\} \quad .$$

A system of elements $e_1, e_2, \ldots, e_n, \ldots$ in E is called *complete* in E if, for given $\epsilon > 0$ and $x \in E$, there exists an n and constants

$\alpha_1, \alpha_2, \ldots, \alpha_n$ such that $\|x - \sum_{k=1}^{n} \alpha_k e_k\| \leq \epsilon$.

An example of a complete system is the sequence of monomials t^n $(n = 0,1,2,\ldots)$ in the space $C[0,1]$.

A system of elements $e_1, e_2, \ldots, e_n, \ldots$ in E is called a *basis* for E if, for arbitrary $x \in E$, there exists a unique representation of the form

(1.13) $$x = \sum_{i=1}^{\infty} \lambda_i e_i \quad .$$

In an n-dimensional real (complex) vector space E_n the operation of the scalar product of two vectors can be defined. The *scalar product* (x,y) of the two vectors $x = (x_1, \ldots, x_n)$ and $y = (y_1, \ldots, y_n)$ is defined by

(1.14) $$(x,y) = \sum_{i=1}^{n} x_i \bar{y}_i \; .$$

Using the scalar product, the norm or length of a vector x is defined by

(1.15) $$\|x\| = \sqrt{(x,x)} \ .$$

In analysis, the scalar product has wide application. It is therefore natural to examine the class of linear spaces H in which the operation of scalar product is defined. Such spaces are called *Hilbert spaces* and are defined by the following axioms:

1) H is a linear space.
2) For every pair of elements x and y in H, there exists a complex number (x,y) which is called their *scalar product* and satisfies:
   (a) $(x,y) = \overline{(y,x)}$;
   (b) $(x_1+x_2,y) = (x_1,y) + (x_2,y)$;
   (c) $(\lambda x,y) = \lambda(x,y)$ for any complex scalar λ;
   (d) $(x,x) \geq 0$ with $(x,x) = 0$ if and only if $x = 0$;

   The norm of x becomes $\|x\| = \sqrt{(x,x)}$.
3) H is complete with respect to the metric $\rho(x,y) = \|x-y\|$.

The complex space ℓ_2 is a Hilbert space if, for every two elements $x = (x_1,x_2,\ldots,x_n,\ldots)$ and $y = (y_1,y_2,\ldots,y_n,\ldots)$, we define

(1.16) $$(x,y) = \sum_{i=1}^{\infty} x_i \bar{y}_i \ .$$

Another example of a Hilbert space is $L_2(\Gamma,\rho)$, the space of complex functions which are defined on the rectifiable curve Γ and which satisfy

(1.17) $$\int_\Gamma \rho(t)\ |x(t)|^2\ |dt| < \infty \ ,$$

where $\rho(t) > 0$ is called the *weight function*. The scalar product in this space is defined by

(1.18) $$(x,y) = \frac{1}{L} \int_\Gamma \rho(t)\ x(t)\ \overline{y(t)}\ |dt| \ ,$$

where L is the length of Γ. In particular, if $\rho(t) = 1$, we obtain the Hilbert space $L_2(\Gamma)$.

Two elements x and y in H are said to be *orthogonal* (we write $x \perp y$) if $(x,y) = 0$. An element x is said to be orthogonal to the subspace $L \subset H$, if x is orthogonal to every element $y \in L$. In this case, we write $x \perp L$.

Theorem 1.1 *If $x \in H$ and L is a subspace of H, then the decomposition*

$$x = y + z \ , \tag{1.19}$$

where $y \in L$, and $z \perp L$, is unique.

In this decomposition, the element y is called the *projection* of x on L.

The set M of all elements orthogonal to the subspace L is also a subspace which is called the *orthogonal complement* of L, and is denoted by $H \ominus L$. We say that H is the *orthogonal sum* of the spaces L and M, and we write $H = L \oplus M$.

A system $\{e_i\} = \{e_1, e_2,\ldots,e_n,\ldots\}$ of elements of H is called an *orthonormal system* if

$$(e_i, e_j) = \delta_{ij} = \begin{cases} 1 & \text{if } i = j \ , \\ 0 & \text{if } i \neq j \ . \end{cases} \tag{1.20}$$

In the space $L_2([0,1])$, the functions $exp(2\pi int)$, where $n = 0,\pm1,\pm2,\ldots$ give an example of an orthonormal system.

The numbers $c_i = (x, e_i)$ are called the *Fourier coefficients* of the element x with respect to the orthonormal system $\{e_i\}$.

§1.2 Operators and Functionals

Consider two spaces X and Y, and let M be a subset in X. If, for every element $x \in M$, there is assigned a unique element $y \in Y$, then we say that an *operator* or *abstract function* f is defined on M, and we write $y = fx = f(x)$. The set M is called the *domain* of f and is usually written as $D(f)$. The set $f(M)$ in Y is called the *range* of f and is usually written as $R(f)$.

If Y is the real line, then the operator or abstract function is known as a *functional*.

Let X and Y be metric spaces. The operator f is said to be *continuous* at $x_0 \in D(f)$ if, for every $\epsilon > 0$, there exists a $\delta > 0$ such that $\rho_Y(f(x), f(x_0)) < \epsilon$ for all $x \in D(f)$ which satisfy $\rho_X(x, x_0) < \delta$.

An operator A is said to be *linear* if

1) $D(A)$ is a linear space;
2) the operator is *additive*; i.e., for all x_1 and x_2 in $D(A)$,

$$A(x_1 + x_2) = Ax_1 + Ax_2 \quad ; \tag{1.21}$$

3) the operator is *homogeneous*; i.e., for all x in $D(A)$ and all scalars λ,

$$A(\lambda x) = \lambda Ax \quad . \tag{1.22}$$

An example of a linear operator is a rectangular matrix A which maps vectors $x_n \in E_n$ to vectors $y_m \in E_m$.

Let E_X and E_Y be two normed linear spaces and let A be a linear operator mapping E_X into E_Y. The operator A is *bounded* if there exists a constant K such that

$$\|Ax\| \leq K\|x\| \text{ for all } x \in E_X \; . \tag{1.23}$$

The least value of K for which (1.23) is satisfied is called the *norm of the operator* A, and is denoted by $\|A\|$.

Below, we shall denote the set of all operators mapping from E_X into E_Y by $(E_X \to E_Y)$.

Let $A_1, A_2, \ldots, A_n, \ldots$ be a sequence of linear operators in $(E_X \to E_Y)$. If there exists an operator $A \in (E_X \to E_Y)$ such that

$$\|A_n - A\| \to 0 \quad (n \to \infty) \; , \tag{1.24}$$

then this sequence of operators is said to *converge in norm* to A.

If for each fixed x

$$\|A_n x - Ax\| \to 0 \quad (n \to \infty) \; , \tag{1.25}$$

then the sequence of operators is said to *converge pointwise* to A.

If $A \in (E_X \to E_Y)$ and $A^{-1} \in (E_Y \to E_X)$ are such that

$$A^{-1}Ax = A^{-1}(Ax) = x \tag{1.26}$$

for every $x \in E_X$ and

$$A\,A^{-1}y = A(A^{-1}y) = y \tag{1.27}$$

for every $y \in E_Y$, then the operators A and A^{-1} are said to be *mutually inverse* operators. If A^{-1} satisfies only one of these conditions, then it is called a *left* or *right inverse* of A depending on whether (1.26) or (1.27) respectively is satisfied.

The operator $I \in (E_X \to E_X)$ which satisfies the condition $Ix = x$ for all $x \in E_X$ is called the *identity operator*.

Equations of the form

(1.28) $$Ax - \lambda x = y \text{ or } (A - \lambda I)x = y \quad ,$$

where A is a linear operator and λ is a parameter, are often encountered in various branches of mathematics. Along with these equations, we can examine the corresponding *homogeneous equations*

(1.29) $$Ax - \lambda x = 0 \text{ or } (A - \lambda I)x = 0 \ .$$

These equations always have the solution $x = 0$, which is referred to as the *trivial solution.*

For a given value of λ, suppose the inverse of $A - \lambda I$ is defined, and let $(A - \lambda I)^{-1} = R_\lambda$. The operator R_λ is called the *resolvent operator,* and the values of λ for which R_λ is bounded is called the *resolvent set* of the operator A.

If, for a given λ, the homogeneous equation has a solution other than the trivial solution, this value of λ is called an *eigenvalue* (or *characteristic value*) of A, and the non-trivial solution is called an *eigenvector* of A corresponding to the eigenvalue λ.

The values of λ which are not in the resolvent set form the *spectrum* of A.

A linear operator which maps into the real line is called a *linear functional.*

The set of all linear functionals defined on a normed linear space E forms a Banach space E^*, which is called the *conjugate space* of E.

If, for any linear functional $f \in E^*$, $f(x_n) \to f(x_0)$ as $n \to \infty$, then we say that the sequence $\{x_n\} \in E$ *converges weakly* to $x_0 \in E$.

Consider a bounded linear operator $A \in (E_X \to E_Y)$. Let ϕ be a linear functional defined on E_Y. Then there exists a linear functional f on E_X which is defined by

(1.30) $$\phi(y) = \phi(Ax) = f(x) \quad ,$$

where $x \in E_X$ and $y = Ax$. In this way, we observe that there exists a correspondence between functionals $\phi \in E_Y^*$ and $f \in E_X^*$. This defines an operator A^* which maps from E_Y^* to E_X^* and is called the *adjoint* of A.

If E_X and E_Y are Hilbert spaces, then $\phi(y) = (z_\phi, y)$, $f(x) = (z_f, x)$, where $z_\phi \in E_Y$, $z_f \in E_X$. The adjoint operator A^* is defined by the correspondence

(1.31) $$(z_\phi, Ax) = (A^*z_\phi, x) \quad ,$$

which is valid for all $x \in E_X$.

We illustrate the above concepts by examining the integral operator

(1.32) $$Ax(t) = \int_0^1 k(t, s)\, x(s) ds \quad .$$

If $k(t, s)$ is a continuous function on the square $0 \le t, s \le 1$, then A is a bounded linear operator which maps $C([0,1])$ into itself with

(1.33) $$\|A\| = \max_t \int_0^1 |k(t, s)| ds \quad .$$

If $k(t, s)$ is square integrable on $0 \le t, s \le 1$, i.e., if

(1.34) $$\int_0^1 \int_0^1 |k(t, s)|^2 \, ds\, dt < \infty \quad ,$$

then A is a bounded linear operator which maps $L_2([0,1])$ into itself, and

(1.35) $$\|A\|^2 = \int_0^1 \int_0^1 |k(t, s)|^2 \, ds\, dt \quad .$$

In both cases, the spectrum of A consists only of a finite or countable number of points, where there are no non-zero points of the spectrum of A except for the eigenvalues of A. All the eigenvalues lie in the disc $|\lambda| \le \|A\|$, and in the case of a countable spectrum there exists a unique limit point, $\lambda = 0$.

In the case when A is defined on the Hilbert space $L_2([0,1])$, the adjoint operator is defined by

(1.36) $$A^*\phi(t) = \int_0^1 \overline{k(s,t)}\, \phi(s)\, ds \quad .$$

§1.3 Compact Sets

A set K contained in a metric space X is called *compact* if every infinite subset of the set K contains a convergent sequence. If these limits belong to K, then K is called *compact in itself*, or a *closed compact set*.† If these limits lie in X, but not necessarily in K, then K is said to be *compact in X*, or *compact with respect to X*. If, in particular, every infinite subset of

† Modern usage has *sequentially compact* for *compact in itself*, and *relatively sequentially compact* for *compact in X*. We shall keep to the author's terminology. (Eds.)

X contains a sequence converging to an element of X, then the space X is called *compact*.

To illustrate the very important role played by the theory of compact sets in the examination of the structure of approximate methods, we cite some basic theorems which depend on this definition.

Theorem 1.2. *Let K be a closed compact set in X, and f be a continuous functional defined on this set. Then*

1) *f is bounded on K;*
2) *f attains its upper and lower bounds on K.*

A set N in a metric space X is called an *ϵ-net* of the set $M \subseteq X$ if, for every point $x \in M$, there exists a point $x_\epsilon \in N$ such that $\rho(x, x_\epsilon) < \epsilon$.

Theorem 1.3 (Hausdorff). *In order that a set K in a metric space X be compact, it is necessary that for every $\epsilon > 0$, there exists a finite ϵ-net for K. If the space X is complete, then this condition is also sufficient.*

The functions of a set $M \subset C([0,1])$ are said to be *uniformly bounded*, if there exists a constant C such that $|x(t)| \leq C$ for all $x \in M$ and for arbitrary $t \in [0,1]$. They are *equicontinuous* if, for given $\epsilon > 0$, there exists a $\delta > 0$, which depends only on ϵ, such that for every t_1 and t_2 in $[0,1]$ which satisfy $|t_1-t_2| < \delta$, $|x(t_1)-x(t_2)| < \epsilon$ for all x in M.

Theorem 1.4 (Arzelà). *A necessary and sufficient condition for a set M, contained in $C([0,1])$, to be compact is that the functions $x \in M$ are uniformly bounded and equicontinuous.*

Theorem 1.5 (Riesz). *For a set K contained in $L_p([0,1])$ to be compact, it is necessary and sufficient that the x contained in K be uniformly bounded in norm and equicontinuous in norm. That is, there exists a constant C and, for every $\epsilon > 0$, there exists a $\delta(\epsilon) > 0$ such that, for all x in K,*

$$(1.37) \qquad \int_0^1 |x(t)|^p\, dt \leq C^p \text{ and } \int_0^1 |x(t+h)-x(t)|^p dt < \epsilon^p .$$

where $0 < h < \delta(\epsilon)$.

Theorem 1.6 (Gel'fand). *For a set K in a separable Banach Space E to be compact, it is necessary and sufficient that, for any sequence of functionals $\{f_n\}$ which converge weakly to zero, the limit relationship $f_n(x) \to 0$ as $n \to \infty$ holds uniformly for all x in K. That is, given ϵ there exists n_ϵ such that $|f_n(x)| < \epsilon$ for $n > n_\epsilon$ and arbitrary x in K.*

A set M, contained in a normed linear space E, is called *weakly compact*, if every sequence $\{x_n\}$ of M contains a weakly convergent subsequence $\{x_{n_k}\}$. That is, there exists an $x_0 \in E$ such that $f(x_{n_k}) \to f(x_0)$ as $k \to \infty$ for arbitrary f in E^*. If all the x_0 are contained in M, then M is called a *closed weakly compact set*.

Theorem 1.7. *If the space E is separable, then every sphere in the conjugate space E^* forms a weakly compact set. In particular, every sphere in a separable Hilbert space forms a weakly compact set.*

§2 THE GENERAL THEORY OF APPROXIMATE METHODS

§2.1 Linear Problems in a Hilbert Space

Let H_1 and H_2 be given Hilbert spaces, and consider the operator equation

(2.1) $$Ax = y\ ,\quad y \in H_2,$$

where A is a bounded linear operator[2] with domain $D(A) \subseteq H_1$ and range $R(A) \subseteq H_2$. We denote by $\overline{D(A)}$ and $\overline{R(A)}$ the closure of $D(A)$ and $R(A)$ respectively, and by A^* the adjoint of A. It is well known that if A is bounded, then it can be assumed without loss of generality that $\overline{D(A)} = D(A) = H_1$, and $D(A^*) = H_2$. We denote by $L(A)$ the null space of A and by $D(A) \ominus L(A)$ the orthogonal complement of $L(A)$ with respect to $D(A)$.

If $\overline{R(A)} = R(A)$, then A is said to be *normally solvable*[3]. In this case, since A defines a 1-1 mapping from $D(A) \ominus L(A)$ onto $R(A)$, it follows from a theorem of Banach that there exists a bounded operator A^{-1} such that

$$D(A^{-1}) = R(A)\ ,\ R(A^{-1}) = D(A) \ominus L(A)\ .$$

If $\overline{R(A)} \neq R(A)$, then A^{-1} is unbounded (see [48]).

We assume that A and y are known with an accuracy of ϵ; i.e., there exists a bounded linear operator A_ϵ and an element y_ϵ of H_2 such that

(2.2) $$\|A - A_\epsilon\| \leq \epsilon\ ,\ \|y - y_\epsilon\| \leq \epsilon$$
$$D(A_\epsilon) = D(A)\ .$$

Following the method of regularization [129], we introduce the quadratic functional

(2.3) $$I^{\alpha,\epsilon}(x) = \|A_\epsilon x - y_\epsilon\|^2 + \alpha\|x\|^2\ ,$$

in which α is a non-negative parameter which depends upon ϵ.[4]

---

(2) The case of an unbounded operator is considered at the end of this section.

(3) From [123; §185], if A is normally solvable then a necessary and sufficient condition for $Ax = y$ to be solvable is that y is orthogonal to all solutions of $A^*u = 0$.

(4) In certain cases, it is more useful to replace $\alpha\|x\|^2$ in (2.3) with the value at x of some functional f. In particular, if H_1 is a function space and the solution of (2.1) is differentiable r times, then it is possible to set

$$f(x) = \sum_{\nu=0}^{r} c_\nu \|x^{(\nu)}\|^2\ ,\ c_\nu > 0\ .$$

Theorem 2.1. *For $\alpha > 0$, there exists a unique element $\overset{*}{x}^{\alpha,\epsilon}$ which is the solution of the Euler (Euler-Lagrange) equation corresponding to the functional $I^{\alpha,\epsilon}(x)$. That is*

$$(2.4) \qquad (\alpha I + A_\epsilon^* A_\epsilon)x^{\alpha,\epsilon} = A_\epsilon^* y_\epsilon ,$$

and

$$(2.5) \qquad I^{\alpha,\epsilon}(\overset{*}{x}^{\alpha,\epsilon}) = \inf_{x \in D(A_\epsilon)} \{I^{\alpha,\epsilon}(x)\} .$$

Proof. It is not difficult to construct the operator $(\alpha I + A_\epsilon^* A_\epsilon)^{-1}$, the inverse of $(\alpha I + A_\epsilon^* A_\epsilon)$, explicitly. In fact, expanding the function $1/(\alpha+z)$ as a Taylor series about $z = a$, we obtain

$$\frac{1}{\alpha+z} = \sum_{k=0}^{\infty} \frac{(a-z)^k}{(a+\alpha)^{k+1}} , \qquad 1 = (\alpha+z) \sum_{k=0}^{\infty} \frac{(a-z)^k}{(a+\alpha)^{k+1}} .$$

In a similar way, we obtain formally

$$(2.6) \qquad (\alpha I + A_\epsilon^* A_\epsilon)^{-1} = \sum_{k=0}^{\infty} \frac{(aI - A_\epsilon^* A_\epsilon)^k}{(a+\alpha)^{k+1}} , \; I = (\alpha I + A_\epsilon^* A_\epsilon) \sum_{k=0}^{\infty} \frac{(aI - A_\epsilon^* A_\epsilon)^k}{(a+\alpha)^{k+1}} .$$

We now show that for $a \geq \|A_\epsilon\|^2/2$, the series $\sum_{k=0}^{\infty} (aI - A_\epsilon^* A_\epsilon)^k/(a+\alpha)^{k+1}$ converges in norm to the operator $(\alpha I + A_\epsilon^* A_\epsilon)^{-1}$. From

$$\|(aI - A_\epsilon^* A_\epsilon)^k\| \leq \|aI - A_\epsilon^* A_\epsilon\|^k , \text{ and}$$

$$\|(aI - A_\epsilon^* A_\epsilon)x\|^2 = a^2\|x\|^2 - 2a\|A_\epsilon x\|^2 + \|A_\epsilon^* A_\epsilon x\|^2$$

$$\leq a^2\|x\|^2 + (-2a + \|A_\epsilon\|^2)\|A_\epsilon x\|^2 \leq a^2\|x\|^2 ,$$

it follows that

$$(2.7) \qquad \|(aI - A_\epsilon^* A_\epsilon)^k\| \leq a^k ,$$

which proves the result.

It remains to show that the solution of (2.4), viz. $\overset{*}{x}^{\alpha,\epsilon}$, is the point at which $I^{\alpha,\epsilon}(x)$ attains its minimum.

Putting $x = \overset{*}{x}^{\alpha,\epsilon} + u$, we obtain from (2.4) that

$$(2.8)\quad I^{\alpha,\epsilon}(\overset{*}{x}^{\alpha,\epsilon} + u) = I^{\alpha,\epsilon}(\overset{*}{x}^{\alpha,\epsilon}) + 2\mathrm{Re}(\alpha\overset{*}{x}^{\alpha,\epsilon} + A_\epsilon^* A_\epsilon \overset{*}{x}^{\alpha,\epsilon} - A_\epsilon^* y_\epsilon, u) + \|A_\epsilon u\|^2 + \alpha\|u\|^2$$

$$= I^{\alpha,\epsilon}(\overset{*}{x}^{\alpha,\epsilon}) + \|A_\epsilon u\|^2 + \alpha\|u\|^2 \geq I^{\alpha,\epsilon}(\overset{*}{x}^{\alpha,\epsilon}).$$

Hence the theorem is proved. #

Theorem 2.2.[5] *If A is normally solvable, then*

$$(2.9)\qquad A^*Ax = A^*y$$

is solvable for arbitrary $y \in H_2$, and conversely. Furthermore, every solution of (2.9) *minimizes*

$$(2.10)\qquad I(x) = \|Ax - y\| .$$

Proof. We write y in the form $y = y^{(1)} + y^{(2)}$ where $y^{(1)} \in R(A)$ and $y^{(2)} \perp R(A)$. Since $(A^*y^{(2)}, x) = (y^{(2)}, Ax) = 0$ for arbitrary x in $D(A)$, $A^*y^{(2)} = 0$ and $A^*y = A^*y^{(1)}$. The solvability of (2.9) is an immediate consequence of this result.

For the proof of the converse, we assume that the contrary holds. Then there exists a y_* which is orthogonal to all solutions of $A^*u = 0$, and such that $Ax = y_*$ is not solvable. Together with this, there exists an $\overset{*}{x}$ such that $A^*(A\overset{*}{x} - y_*) = 0$. This implies that $A\overset{*}{x} - y_* = u_* \neq 0$, $A^*u_* = 0$. But $(y_*, u_*) = (y_* - A\overset{*}{x}, u_*) + (A\overset{*}{x}, u_*) = -(u_*, u_*) \neq 0$, which is not possible.

The last part of the theorem follows easily from (2.8) by putting α and $\epsilon = 0$. #

We shall write

$$(2.11)\qquad I^{(\alpha)}(x) = \|Ax - y\|^2 + \alpha\|x\|^2 .$$

Theorem 2.3 *Let $\overset{*}{x}^{\alpha,\epsilon}$ and $\overset{*}{x}^{(\alpha)}$ be the elements for which the functionals* (2.3) *and* (2.11) *attain their minimum. Then*

$$(2.12)\qquad \|\overset{*}{x}^{(\alpha)} - \overset{*}{x}^{\alpha,\epsilon}\| \leq \frac{\epsilon}{\alpha}\{\|y\| + \|A_\epsilon\| + (\|A_\epsilon\| + \|A\|)\|\overset{*}{x}^{\alpha,\epsilon}\|\}$$

and

$$(2.13)\qquad \|\overset{*}{x}^{(\alpha)} - \overset{*}{x}^{\alpha,\epsilon}\| \leq \frac{\epsilon}{\alpha}\{\|y\| + \|A_\epsilon\| + (\|A_\epsilon\| + \|A\|)\|\overset{*}{x}^{(\alpha)}\|\} .$$

---

(5) For a special case, see [103].

Proof. From Theorem 2.1, we have

$$(\alpha I + A_\epsilon^* A_\epsilon)\overset{*}{x}^{\alpha,\epsilon} = A_\epsilon^* y_\epsilon, \quad (\alpha I + A^*A)\overset{*}{x}^{(\alpha)} = A^*y .$$

Subtracting, we obtain

$$(\alpha I + A^*A)(\overset{*}{x}^{(\alpha)} - \overset{*}{x}^{\alpha,\epsilon}) = A^*y - A_\epsilon^* y_\epsilon + (A_\epsilon^* A_\epsilon - A^*A)\overset{*}{x}^{\alpha,\epsilon} ,$$

$$(\alpha I + A_\epsilon^* A_\epsilon)(\overset{*}{x}^{\alpha,\epsilon} - \overset{*}{x}^{(\alpha)}) = A_\epsilon^* y_\epsilon - A^*y + (A^*A - A_\epsilon^* A_\epsilon)\overset{*}{x}^{(\alpha)} .$$

Now, from (2.7) and (2.6), it follows that

$$\|(\alpha I + A_\epsilon^* A_\epsilon)^{-1}\| \le \sum_{k=0}^{\infty} a^k/[(a + \alpha)^{k+1}] = 1/\alpha .$$

Similarly, we obtain $\|(\alpha I + A^*A)^{-1}\| \le 1/\alpha$. Consequently

$$\text{(2.14)} \quad \begin{aligned} \|\overset{*}{x}^{(\alpha)} - \overset{*}{x}^{\alpha,\epsilon}\| &\le [\|A^*y - A_\epsilon^* y_\epsilon\| + \|(A_\epsilon^* A_\epsilon - A^*A)\overset{*}{x}^{\alpha,\epsilon}\|]/\alpha , \\ \|\overset{*}{x}^{(\alpha)} - \overset{*}{x}^{\alpha,\epsilon}\| &\le [\|A^*y - A_\epsilon^* y_\epsilon\| + \|(A_\epsilon^* A_\epsilon - A^*A)\overset{*}{x}^{(\alpha)}\|]/\alpha . \end{aligned}$$

It remains to note that (2.12) and (2.13) now follow from (2.14) together with (2.2), and the obvious condition $\|A_\epsilon^* - A^*\| = \|A_\epsilon - A\| \le \epsilon$. #

Note. If, for a given operator A, there exists a sequence $\{x_n\}$ such that $\|Ax_n\| = c_n\|x_n\|$, with $c_n \to 0$ as $n \to \infty$, then $\|(\alpha I + A^*A)^{-1}\| = 1/\alpha$. In fact

$$\begin{aligned} \|(\alpha I + A^*A)x_n\|^2 &= \alpha^2\|x_n\|^2 + 2\alpha\|Ax_n\|^2 + \|A^*A\, x_n\|^2 \\ &\le (\alpha^2 + 2\alpha c_n{}^2 + \|A\|^2 c_n{}^2)\|x_n\|^2 , \end{aligned}$$

and thus, as $n \to \infty$,

$$\begin{aligned} \|(\alpha I + A^*A)^{-1}\|^2 &\ge \|x_n\|^2/\{\|(\alpha I + A^*A)x_n\|^2\} \\ &\ge 1/\{\alpha^2 + 2\alpha c_n + \|A\|^2 c_n{}^2\} \to 1/\alpha^2 . \end{aligned}$$

Theorem 2.4. *Let $A^*Ax = A^*y$ be solvable, and $\overset{*}{x}$ its unique solution which is orthogonal to the null space of A. Then*

$$\text{(2.15)} \quad \|\overset{*}{x} - \overset{*}{x}^{(\alpha)}\| \le \{\delta_R + (\delta_R^2 + \alpha R^2)^{1/2}\}/2 ,$$

where

$$\text{(2.16)} \quad \delta_R = \inf_{\|w\| \le R} \|\overset{*}{x} - A^*w\|$$

and, in addition, $\|\overset{*}{x} - \overset{*}{x}^{(\alpha)}\| \to 0$ *as* $\alpha \to 0$.

Proof. From (2.8) with $\epsilon = 0$ and $u = \overset{*}{x} - \overset{*}{x}^{(\alpha)}$,

$$I^{(\alpha)}(\overset{*}{x}) = I^{(\alpha)}(\overset{*}{x}^{(\alpha)}) + \|A(\overset{*}{x} - \overset{*}{x}^{(\alpha)})\|^2 + \alpha\|\overset{*}{x} - \overset{*}{x}^{(\alpha)}\|^2 \tag{2.17}$$

$$= \|A\overset{*}{x} - y\|^2 + \alpha\|\overset{*}{x}\|^2 \quad .$$

Writing y in the form $y = y^{(1)} + y^{(2)}$, $y^{(1)} \in R(A)$, $y^{(2)} \perp R(A)$, and noting that

$$\|A\overset{*}{x} - y\|^2 = \|y^{(2)}\|^2 \quad ,$$

$$\|A\overset{*}{x}^{(\alpha)} - y\|^2 = \|A(\overset{*}{x} - \overset{*}{x}^{(\alpha)})\|^2 + \|y^{(2)}\|^2 \; ,$$

we obtain from (2.17) that

$$2\|A(\overset{*}{x} - \overset{*}{x}^{(\alpha)})\|^2 + \alpha\|\overset{*}{x}^{(\alpha)}\|^2 + \alpha\|\overset{*}{x} - \overset{*}{x}^{(\alpha)}\|^2 = \alpha\|\overset{*}{x}\|^2 \; , \tag{2.18}$$

from which we find

$$\|A(\overset{*}{x} - \overset{*}{x}^{(\alpha)})\|^2 = \frac{\alpha}{2}\,(\|\overset{*}{x}\|^2 - \|\overset{*}{x}^{(\alpha)}\|^2 - \|\overset{*}{x} - \overset{*}{x}^{(\alpha)}\|^2) \le \frac{\alpha}{2}\,\|\overset{*}{x}\|^2 \quad , \tag{2.19}$$

and

$$\|\overset{*}{x}^{(\alpha)}\|^2 = \|\overset{*}{x}\|^2 - \|\overset{*}{x} - \overset{*}{x}^{(\alpha)}\|^2 - \frac{2}{\alpha}\|A(\overset{*}{x} - \overset{*}{x}^{(\alpha)})\|^2 \le \|\overset{*}{x}\|^2 . \tag{2.20}$$

On the other hand, it follows from the identity

$$\|\overset{*}{x} - \overset{*}{x}^{(\alpha)}\|^2 = \|\overset{*}{x}^{(\alpha)}\|^2 - \|\overset{*}{x}\|^2 + 2\mathrm{Re}(\overset{*}{x} - \overset{*}{x}^{(\alpha)}, \overset{*}{x}) \tag{2.21}$$

that

$$\|\overset{*}{x}\|^2 - \|\overset{*}{x}^{(\alpha)}\|^2 - \|\overset{*}{x} - \overset{*}{x}^{(\alpha)}\|^2 = 2\mathrm{Re}(\overset{*}{x} - \overset{*}{x}^{(\alpha)}, \overset{*}{x}) - 2\|\overset{*}{x} - \overset{*}{x}^{(\alpha)}\|^2 \; . \tag{2.22}$$

The substitution of (2.22) into (2.19) yields, on using (2.16),

$$\begin{aligned} \|A(\overset{*}{x} - \overset{*}{x}^{(\alpha)})\|^2 &= \alpha\mathrm{Re}(\overset{*}{x} - \overset{*}{x}^{(\alpha)}, \overset{*}{x}) - \alpha\|\overset{*}{x} - \overset{*}{x}^{(\alpha)}\|^2 \\ &\le \alpha\|A(\overset{*}{x} - \overset{*}{x}^{(\alpha)})\|R + \alpha\|\overset{*}{x} - \overset{*}{x}^{(\alpha)}\|\delta_R - \alpha\|\overset{*}{x} - \overset{*}{x}^{(\alpha)}\|^2 \; , \end{aligned} \tag{2.23}$$

whence

$$\begin{aligned} (\|A(\overset{*}{x} - \overset{*}{x}^{(\alpha)})\| - \alpha R/2)^2 + \alpha(\|\overset{*}{x} - \overset{*}{x}^{(\alpha)}\| - \delta_R/2)^2 \\ \le \alpha^2 R^2/4 + \alpha\delta_R^2/4 \; , \end{aligned} \tag{2.24}$$

and

$$\|\overset{*}{x} - \overset{*}{x}^{(\alpha)}\| \le \{\delta_R + (\delta_R^2 + \alpha R^2)^{\frac{1}{2}}\}/2 \quad .$$

We shall now show that $\delta_R \to 0$ as $R \to \infty$. Let us put $R_1 = \overline{R(A^*)}$, $R_2 = H_1 \ominus R_1$ and write $\overset{*}{x}$ as $\overset{*}{x} = \overset{*}{x}^{(1)} + \overset{*}{x}^{(2)}$ where $\overset{*}{x}^{(1)} \in R_1$ and $\overset{*}{x}^{(2)} \in R_2$. We find that $(\overset{*}{x}^{(2)}, A^*w) = 0$ for all $w \in H_2$. In particular, with $w = A\overset{*}{x}^{(2)}$, we find that $(\overset{*}{x}^{(2)}, A^*A\overset{*}{x}^{(2)}) = \|A\overset{*}{x}^{(2)}\|^2 = 0$. This implies that $\overset{*}{x}^{(2)}$ is in the null space of A. Since $\overset{*}{x}$ is assumed to be orthogonal to the null space of A, it follows that $\|\overset{*}{x}^{(2)}\|^2 = (\overset{*}{x}^{(2)}, \overset{*}{x} - \overset{*}{x}^{(1)}) = 0$. Consequently $\overset{*}{x} = \overset{*}{x}^{(1)} \in R_1$.

To complete the proof of the theorem, for a given arbitrarily small $\epsilon > 0$, we choose R so that $\delta_R \le \epsilon/2$. Then with this value of R fixed, we choose α so that

$$\|\overset{*}{x} - \overset{*}{x}^{(\alpha)}\| \le \frac{\epsilon}{4} + \{(\frac{\epsilon}{4})^2 + \alpha(\frac{R}{2})^2\}^{\frac{1}{2}} < \epsilon.$$

Since ϵ is arbitrary, convergence has been proved. #

We make a number of observations:

1. If $\overset{*}{x} = A^*\overset{*}{w}$, then from (2.15) we obtain $\|\overset{*}{x} - \overset{*}{x}^{(\alpha)}\| \le \alpha^{\frac{1}{2}}\|w\|/2$. More precisely, $\|\overset{*}{x} - \overset{*}{x}^{(\alpha)}\| = \alpha\|\overset{*}{w}^{(\alpha)}\|$, where $\alpha\overset{*}{w}^{(\alpha)} + A^*A\overset{*}{w}^{(\alpha)} = \overset{*}{x}$.
2. It follows from (2.18) that

$$\|\overset{*}{x} - \overset{*}{x}^{(\alpha)}\|^2 = \|\overset{*}{x}\|^2 - \|\overset{*}{x}^{(\alpha)}\|^2 - 2\|A(\overset{*}{x} - \overset{*}{x}^{(\alpha)})\|^2/\alpha$$

$$\le \|\overset{*}{x} - \overset{*}{x}^{(\alpha)}\|^2 + 2\|\overset{*}{x} - \overset{*}{x}^{(\alpha)}\| \, \|\overset{*}{x}^{(\alpha)}\| - 2\|A(\overset{*}{x} - \overset{*}{x}^{(\alpha)})\|^2/\alpha ,$$

   whence

$$\|\overset{*}{x} - \overset{*}{x}^{(\alpha)}\| \ge \|A(\overset{*}{x} - \overset{*}{x}^{(\alpha)})\|^2/\{\alpha\|\overset{*}{x}^{(\alpha)}\|\} .$$

3. Noting that $\|\overset{*}{x}\| = \lim_{\alpha\to 0} \|\overset{*}{x}^{(\alpha)}\|$, we obtain

$$\|\overset{*}{x} - \overset{*}{x}^{(\alpha)}\|^2 = \lim_{\alpha\to 0} \{\|\overset{*}{x}^{(\alpha)}\|^2\} - \|\overset{*}{x}^{(\alpha)}\|^2 - 2\|A(\overset{*}{x} - \overset{*}{x}^{(\alpha)})\|^2/\alpha .$$

   The last relationship can be applied for the estimation of $\|\overset{*}{x} - \overset{*}{x}^{(\alpha)}\|$.

4. The behaviour of quantities of the type δ_R have been studied in [68] and [69] in great detail.

We now assume that instead of $\overset{*}{x}^{\alpha,\epsilon}$, we have to find a minimizing sequence $\{x_n^{\alpha,\epsilon}\}$ such that, as $n \to \infty$

$$I^{\alpha,\epsilon}(x_n^{\alpha,\epsilon}) \to I^{\alpha,\epsilon}(\overset{*}{x}^{\alpha,\epsilon}) .$$

<u>Theorem 2.5</u>. *The following estimate*

$$(2.25)\qquad \|\overset{*}{x}{}^{\alpha,\epsilon} - x_n^{\alpha,\epsilon}\|^2 \le \{I^{\alpha,\epsilon}(x_n^{\alpha,\epsilon}) - I^{\alpha,\epsilon}(\overset{*}{x}{}^{\alpha,\epsilon})\}/\alpha \quad ,$$

is valid.

The proof is an immediate consequence of (2.8) on putting $x_n^{\alpha,\epsilon} - \overset{*}{x}{}^{\alpha,\epsilon}$ instead of u.

As a corollary of Theorems 2.3-2.5, we obtain

<u>Theorem 2.6</u>. *If* α, ϵ/α *and* $\{I^{\alpha,\epsilon}(x_n^{\alpha,\epsilon}) - I^{\alpha,\epsilon}(\overset{*}{x}{}^{\alpha,\epsilon})\}/\alpha$ *tend uniformly to zero, then* $\|x_n^{\alpha,\epsilon} - \overset{*}{x}\| \to 0$.

From the estimates (2.13), (2.15), (2.20) and (2.25), we obtain:

<u>Theorem 2.7</u>. *If* $\alpha = O(\epsilon^{\frac{1}{2}})$ *and*

$$I^{\alpha,\epsilon}(x_n^{\alpha,\epsilon}) - I^{\alpha,\epsilon}(\overset{*}{x}{}^{\alpha,\epsilon}) = O(\epsilon^{3/2}), \text{ then}$$

$$\|x_n^{\alpha,\epsilon} - \overset{*}{x}\| = O(\epsilon^{\frac{1}{2}}) \quad ,$$

provided that $\overset{*}{x} = A^*\overset{*}{w}$, *for some* $\overset{*}{w} \in H_2$.

In particular, when $R(A) = H_2$, $R(A_\epsilon) \subseteq H_2$, we let

$$(2.26)\qquad I_\epsilon(x) = \|A_\epsilon x - y_\epsilon\| \quad ,$$

$\overset{*}{x}_\epsilon$ be the element for which $I_\epsilon(\overset{*}{x}_\epsilon) = \inf\limits_{x\in D(A_\epsilon)} \{I_\epsilon(x)\}$, $\overset{*}{x}$ the solution of $Ax = y$ orthogonal to $L(A)$ and A^{-1} the inverse operator of A (restricted to $D(A) \ominus L(A)$) which maps $R(A)$ onto $D(A) \ominus L(A)$.

<u>Theorem 2.8</u>. *If* $\epsilon\|A^{-1}\| < 1$ *then*

1) *the operator*

$$(2.27)\qquad A_\epsilon^{-1} = A^{-1} + A^{-1}(A-A_\epsilon)A^{-1} + A^{-1}[(A-A_\epsilon)A^{-1}]^2 + \ldots$$

defines the inverse of A_ϵ *(restricted to* $D(A) \ominus L(A)$*) and maps* $R(A_\epsilon) = H_2$ *onto* $D(A) \ominus L(A)$, *and*

2) *for any element* $x_n^{(\epsilon)} \in D(A) \ominus L(A)$ *the following estimate*

$$(2.28)\qquad \|x_n^{(\epsilon)} - \overset{*}{x}\| \le \|A^{-1}\| \; \frac{I_\epsilon(x_n^{(\epsilon)}) + \epsilon(1 + \|A^{-1}\| \, \|y\|)}{1 - \epsilon\|A^{-1}\|} \quad ,$$

is valid.

Proof. It is easily seen that, for $\epsilon\|A^{-1}\| < 1$, the series (2.27) converges in norm and

$$\|A_\epsilon^{-1}\| < \|A^{-1}\|/\{1-\epsilon\,\|A^{-1}\|\}.$$

An immediate check confirms the correctness of the condition on $A_\epsilon : A_\epsilon A_\epsilon^{-1}\, y = y,\ y \in H_2$ and $A_\epsilon^{-1}\, A_\epsilon\, x = x,\ x \in D(A) \ominus L(A)$.

For the proof of 2), we note that

$$\begin{aligned}\|\overset{*}{x} - \overset{*}{x}_\epsilon\| &= \|A^{-1}y - A_\epsilon^{-1}\, y_\epsilon\|, \\ &\le \|A_\epsilon^{-1} - A^{-1}\|\,\|y_\epsilon\| + \|A^{-1}\|\,\|y_\epsilon - y\|, \\ &\le \frac{\|A^{-1}\|^2\,\epsilon\|y_\epsilon\|}{1-\epsilon\|A^{-1}\|} + \|A^{-1}\|\epsilon \le \frac{\|A^{-1}\|^2\,\|y\| + \|A^{-1}\|}{1-\epsilon\|A^{-1}\|}\,\epsilon,\end{aligned}$$

and

$$\begin{aligned}\|x_n^{(\epsilon)} - \overset{*}{x}_\epsilon\| &= \|A_\epsilon^{-1}\, A_\epsilon(x_n^{(\epsilon)} - \overset{*}{x}_\epsilon)\| \\ &\le \|A_\epsilon^{-1}\|\,\|A_\epsilon\, x_n^{(\epsilon)} - y_\epsilon\| \le \|A^{-1}\|\,\frac{I_\epsilon(x_n^{(\epsilon)})}{1-\epsilon\|A^{-1}\|}.\end{aligned}$$

Combining these two inequalities, we obtain the required result. #

Changing the roles of A and A_ϵ and retaining the notation and sense of Theorem 2.8, we can prove the following theorem in an analogous way.[6]

Theorem 2.9. *If A_ϵ^{-1} exists and $\epsilon\|A_\epsilon^{-1}\| < 1$, then*

1) *A^{-1} also exists where*

$$A^{-1} = A_\epsilon^{-1} + A_\epsilon^{-1}(A_\epsilon - A)A_\epsilon^{-1} + A_\epsilon^{-1}\,[(A_\epsilon - A)A_\epsilon^{-1}]^2 + \ldots \tag{2.29}$$

and

2) *the following estimate*

$$\|x_n^{(\epsilon)} - \overset{*}{x}\| \le \|A_\epsilon^{-1}\|\,\frac{I_\epsilon(x_n^{(\epsilon)}) + \epsilon(1 + \|A_\epsilon^{-1}\|\,\|y_\epsilon\|)}{1-\epsilon\|A_\epsilon^{-1}\|}, \tag{2.30}$$

is valid.

Unlike (2.28), the estimate (2.30) can, in practice, be effective.

We can now summarize the above analysis. To do so, we need to introduce the following definitions.

(6) Theorems 2.8 and 2.9 are consequences of Kantorovich's general theory of approximation, [74].

Definition 1. *Any element* x_δ *in* $D(A)$ *for which*

$$\|Ax_\delta - y\| \leq \inf_{x\varepsilon D(A)} \|Ax - y\| + \delta \quad ,$$

is called a β*-approximation to the solution of* $Ax = y$ *with accuracy* δ.

Definition 2. *We assume that there exists an* $\overset{o}{x} \varepsilon D(A)$ *such that*

$$\|A\overset{o}{x} - y\| = \inf_{x\varepsilon D(A)} \|Ax - y\| \quad .$$

Any element x_δ *in* $D(A)$ *for which* $\|\overset{o}{x} - x_\delta\| \leq \delta$ *is called an* α*-approximation to the solution of* $Ax = y$ *with accuracy* δ.

It is clear that in the case of a bounded operator A, α-convergence implies β-convergence.

Theorem 2.6 implies that the method of regularization, combined with a method for the construction of the minimizing sequence $x_n^{\alpha,\epsilon}$, defines for any bounded linear operator A an α-approximation for the solution of $Ax = y$ with arbitrary accuracy δ. It is essential to stress that, in the general case, effective *a priori* estimates for α,ϵ and n as a function of δ, which guarantee an α-approximation with a pre-determined accuracy δ, are not known. Particular results of this nature are contained in [131] and [132].

For the normally solvable operators of Theorems 2.8 and 2.9, it follows that an α-approximation with arbitrary accuracy can be constructed without the help of regularization. If $\|A_\epsilon^{-1}\|$ is known, Theorem 2.9 yields an effective procedure for the determination of δ for given ϵ and n.

We examine the case when A is a closed linear operator which is not necessarily bounded.(7) Theorem 2.1 can now be replaced by Theorem 2.10.

Theorem 2.10. *For* $\alpha > 0$, *there exists a unique* $\overset{*}{x}^{(\alpha)} = A^*\overset{*}{w}^{(\alpha)}$, *where* $\overset{*}{w}^{(\alpha)}$ *is the solution of*

(2.31) $$\alpha w + AA^*w = y \ ,$$

such that $$I^{(\alpha)}(\overset{*}{x}^{(\alpha)}) = \inf_{x\varepsilon D(A)} \{I^{(\alpha)}(x)\} \quad .$$

---

(7) An operator A is said to be *closed* if $x_n \to x$ ($x_n \varepsilon D(A)$) and $Ax_n \to y$ imply that $x \varepsilon D(A)$ and $Ax = y$. Differentiation is an example of a closed linear operator which is unbounded, see [3].

Proof. It is known {see [3]} that (2.31) is uniquely solvable for arbitrary y in H_2. Putting $\overset{*}{x}^{(\alpha)} = A^*\overset{*}{w}^{(\alpha)}$, where $\overset{*}{w}^{(\alpha)}$ is the solution of (2.31), we obtain for arbitrary $u \in D(A)$ that

$$(A\overset{*}{x}^{(\alpha)} - y, Au) + \alpha(\overset{*}{x}^{(\alpha)}, u) = (AA^*\overset{*}{w}^{(\alpha)} - y + \alpha\overset{*}{w}^{(\alpha)}, Au) = 0 .$$

Noting that

$$I^{(\alpha)}(x+u) = I^{(\alpha)}(x) + \alpha\|u\|^2 + \|Au\|^2 + 2\mathrm{Re}[(Ax - y, Au) + \alpha(x,u)] \quad , \tag{2.32}$$

it is easy to establish that

$$I^{(\alpha)}(\overset{*}{x}^{(\alpha)} + u) = I^{(\alpha)}(\overset{*}{x}^{(\alpha)}) + \alpha\|u\|^2 + \|Au\|^2 \geq I(\overset{*}{x}^{(\alpha)}) . \tag{2.33}$$

This implies that $I^{(\alpha)}(\overset{*}{x}^{(\alpha)}) = \inf_{x \in D(A)} \{I^{(\alpha)}(x)\}$. If we assume that there exists an $\overset{*}{x}^{(\alpha)} + u$ for which $I^{(\alpha)}(\overset{*}{x}^{(\alpha)} + \overset{*}{u}) = I^{(\alpha)}(\overset{*}{x}^{(\alpha)})$ it follows from (2.33) that we must have $\alpha\|\overset{*}{u}\|^2 = 0$. Since $\alpha > 0$, $\|\overset{*}{u}\| = 0$. This proves the theorem. #

Without loss of generality, we assume A_ϵ and y_ϵ satisfy the following conditions: A_ϵ is a closed operator, and

$$\|y - y_\epsilon\| \leq \epsilon, \ \|(AA^* - A_\epsilon A_\epsilon^*)x\| \leq \epsilon(x)\ \|x\| \ , \ \lim_{\epsilon \to 0} \epsilon(x) = 0 \tag{2.34}$$

for every x.

Theorem 2.11. *For $\epsilon \to 0$, $\overset{*}{w}^{\alpha,\epsilon}$ converges strongly to $\overset{*}{w}^{(\alpha)}$.*

Proof. By analogy with the proof of Theorem 2.3, we obtain

$$(\alpha I + AA^*)(\overset{*}{w}^{(\alpha)} - \overset{*}{w}^{\alpha,\epsilon}) = y - y_\epsilon + (A_\epsilon A_\epsilon^* - AA^*)\overset{*}{w}^{\alpha,\epsilon} \quad ,$$

$$(\alpha I + A_\epsilon A_\epsilon^*)(\overset{*}{w}^{\alpha,\epsilon} - \overset{*}{w}^{(\alpha)}) = y_\epsilon - y + (AA^* - A_\epsilon A_\epsilon^*)\overset{*}{w}^{(\alpha)} \ .$$

From

$$((\alpha I + AA^*)u, u) = \alpha\|u\|^2 + \|Au\|^2 \leq \|(\alpha I + AA^*)u\| \ \|u\| \ ,$$

it follows that

$$\|u\| = \|(\alpha I + AA^*)^{-1}(\alpha I + AA^*)u\| \leq \|(\alpha I + AA^*)u\|/\alpha \quad ,$$

whence

$$\|(\alpha I + AA^*)^{-1}\| \leq 1/\alpha \quad .$$

In a similar way,

$$\|(\alpha I + A_\epsilon A_\epsilon^*)^{-1}\| \leq 1/\alpha \quad .$$

Hence

$$(2.35) \qquad \|\overset{*}{w}^{\alpha,\epsilon} - \overset{*}{w}^{(\alpha)}\| \leq [\epsilon + \epsilon(\overset{*}{w}^{(\alpha)})]/\alpha \to 0 \quad \text{as} \quad \epsilon \to 0 \ ,$$

which establishes the theorem. #

From the proof of Theorem 2.11 and equation (2.14), it follows that the convergence of $\overset{*}{x}^{\alpha,\epsilon}$ to $\overset{*}{x}^{(\alpha)}$ when A and A_ϵ are bounded, guarantees the more general condition:

$$(2.36) \qquad \|A^*y - A_\epsilon^*y_\epsilon\| \to 0 \ , \quad \|(A_\epsilon^*A_\epsilon - A^*A)\overset{*}{x}^{(\alpha)}\| \to 0$$

as $\epsilon \to 0$.

In the general case when A and A_ϵ are closed operators, where A_ϵ^* converges pointwise to A^* as $\epsilon \to 0$, and A_ϵ^* is bounded, taking $x^{\alpha,\epsilon_1} = A_\epsilon^*\overset{*}{w}^{\alpha,\epsilon_1}$ we obtain

$$(2.37) \qquad \|x^{\alpha,\epsilon_1} - \overset{*}{x}^{(\alpha)}\| = \|A_\epsilon^*\overset{*}{w}^{\alpha,\epsilon_1} - A^*\overset{*}{w}^{(\alpha)}\|$$

$$\leq \|(A^* - A_\epsilon^*)\overset{*}{w}^{(\alpha)}\| + \|A_\epsilon^*(\overset{*}{w}^{\alpha,\epsilon_1} - \overset{*}{w}^{(\alpha)})\| \to 0 \text{ as } \epsilon \to 0$$

for sufficiently small $\epsilon_1 \leq \epsilon$.

Theorem 2.12. *If* $A_\epsilon = A$, *then*

$$(2.38) \qquad \|\overset{*}{x}^{\alpha,\epsilon} - \overset{*}{x}^{(\alpha)}\| \leq \epsilon/(2\alpha)^{\frac{1}{2}} \quad .$$

Proof. From (2.8) and (2.33),

$$2\alpha\|\overset{*}{x}^{\alpha,\epsilon} - \overset{*}{x}^{(\alpha)}\|^2 + 2\|A(\overset{*}{x}^{\alpha,\epsilon} - \overset{*}{x}^{(\alpha)})\|^2$$

$$= I^{\alpha,\epsilon}(\overset{*}{x}^{(\alpha)}) - I^{\alpha,\epsilon}(\overset{*}{x}^{\alpha,\epsilon}) + I^{(\alpha)}(\overset{*}{x}^{\alpha,\epsilon}) - I^{(\alpha)}(\overset{*}{x}^{(\alpha)})$$

$$= \|A\overset{*}{x}^{(\alpha)} - y_\epsilon\|^2 - \|A\overset{*}{x}^{\alpha,\epsilon} - y_\epsilon\|^2 + \alpha\|\overset{*}{x}^{(\alpha)}\|^2 - \alpha\|\overset{*}{x}^{\alpha,\epsilon}\|^2 +$$

$$+ \|A\overset{*}{x}^{\alpha,\epsilon} - y\|^2 + \alpha\|\overset{*}{x}^{\alpha,\epsilon}\|^2 - \|A\overset{*}{x}^{(\alpha)} - y\|^2 - \alpha\|\overset{*}{x}^{(\alpha)}\|^2$$

$$= 2\mathrm{Re}(A(\overset{*}{x}^{(\alpha)} - \overset{*}{x}^{\alpha,\epsilon}), y - y_\epsilon) \leq 2\epsilon\|A(\overset{*}{x}^{\alpha,\epsilon} - \overset{*}{x}^{(\alpha)})\| \ .$$

It now readily follows that $\|A(\overset{*}{x}^{\alpha,\epsilon} - \overset{*}{x}^{(\alpha)})\| \leq \epsilon$.

Assuming that $\|A(\overset{*}{x}^{\alpha,\epsilon} - \overset{*}{x}^{(\alpha)})\|^2$ is larger (smaller) than $\alpha\|\overset{*}{x}^{\alpha,\epsilon} - \overset{*}{x}^{(\alpha)}\|^2$, we obtain the required inequality (2.38). #

We now generalize Theorem 2.4, introducing instead of (2.16),

$$\delta_n = \|\overset{*}{x} - A^* w_n\| \quad , \ w_n \in D(A^*) \quad . \tag{2.39}$$

Theorem 2.13. *Let* $y = y^{(1)} + y^{(2)}$, $y^{(1)} \in R(A)$ *and* $y^{(2)} \perp R(A)$, *and let* $\overset{*}{x}$ *be the unique solution of* $Ax = y^{(1)}$ *which is orthogonal to* $L(A)$. *For the convergence of* $\overset{*}{x}{}^{(\alpha)}$ *to* $\overset{*}{x}$, *it is necessary and sufficient that* $\overset{*}{x}$ *be approximated arbitrarily closely by elements of* $R(A^*)$. *In addition, we obtain, using the notation of* (2.39),

$$\|\overset{*}{x} - \overset{*}{x}{}^{(\alpha)}\| \le \delta_n + \{\delta_n^2 + \alpha\|w_n\|^2\}^{1/2}/2 \quad , \tag{2.40}$$

$$\|A(\overset{*}{x} - \overset{*}{x}{}^{(\alpha)})\| \le \alpha\|w_n\| + \{\alpha^2\|w_n\|^2 + \alpha\delta_n^2\}^{1/2}/2 \quad .$$

Proof. The necessary condition is a trivial consequence of the fact that $\overset{*}{x}{}^{(\alpha)} = A^*\overset{*}{w}{}^{(\alpha)}$.

For the sufficiency condition, we note that (2.17)-(2.22) remain valid under the given conditions. Instead of (2.23) and (2.24), we find

$$\begin{aligned}\|A(\overset{*}{x} - \overset{*}{x}{}^{(\alpha)})\|^2 &= \alpha \mathrm{Re}(\overset{*}{x} - \overset{*}{x}{}^{(\alpha)}, \overset{*}{x}) - \alpha\|\overset{*}{x} - \overset{*}{x}{}^{(\alpha)}\|^2 \\ &\le \alpha\|A(\overset{*}{x} - \overset{*}{x}{}^{(\alpha)})\| \, \|w_n\| + \alpha\|\overset{*}{x} - \overset{*}{x}{}^{(\alpha)}\|\delta_n - \alpha\|\overset{*}{x} - \overset{*}{x}{}^{(\alpha)}\|^2 \quad ,\end{aligned} \tag{2.41}$$

and

$$\begin{aligned}\{\|A(\overset{*}{x} - \overset{*}{x}{}^{(\alpha)})\| - \alpha\|w_n\|/2\}^2 + \alpha\{\|\overset{*}{x} - \overset{*}{x}{}^{(\alpha)}\| - \delta_n/2\}^2 \\ \le \alpha^2\|w_n\|^2/4 + \alpha\delta_n^2/4 \quad .\end{aligned} \tag{2.42}$$

The results of (2.40), and hence the theorem, are now a simple consequence of (2.42). #

§2.2 Linear Problems in a Banach Space

Let E_1 and E_2 be given Banach spaces and consider the equation

$$Ax = y, \quad y \in E_2, \tag{2.43}$$

where A is a bounded linear operator with domain $D(A) = E_1$ and range $R(A) \subseteq E_2$.

Usually, approximate methods for the solution of (2.43) start with an alternative equation

(2.44) $$\tilde{A}\,\tilde{x} = \tilde{y} \quad ,$$

in which $\tilde{A}$ and $\tilde{y}$ approximate A and y in some sense. In the case when (2.43) is a Fredholm integral equation of the second kind, most of the known approximate methods choose $\tilde{A}$ to be close to A in the norm. Under the assumption that A^{-1} is bounded, the proof of theorems about the convergence of $\tilde{x}$ to x as $\tilde{A} \to A$ and $\tilde{y} \to y$ can in such cases be readily given.(8)

For singular integral equations and equations of the first kind, A^{-1} is generally unbounded, and in many approximate methods $\tilde{A} \to A$ only in the sense of pointwise convergence of the operators, i.e. in the sense $\|\tilde{A}x - Ax\| \to 0$ for every x in $D(A)$, $(D(\tilde{A}) = D(A))$.

Let us assume that the operator A has an inverse. In general, it is not possible to assert that the operator $\tilde{A}$ also has an inverse which converges to the inverse of A. As an example, suppose $A = I$, $Ix = x$ and $\tilde{A} = P_n$ (P_n is the projection operator which maps every x from E_1 to its best approximation by the first n elements of a (given) complete system $e_1, e_2, \ldots, e_n, \ldots$). It is clear that $\|P_n x - x\| \to 0$ for every x, but the equation $P_n x_n = y_n$, where $x_n = \sum_{k=1}^{n} \alpha_k e_k$, can either not have a solution (when y_n is not an element of the form $\sum_{k=1}^{n} \alpha_k e_k$) or, can have a non-unique solution with respect to α_k (when $e_1, e_2, \ldots, e_n$ are linearly dependent).

An important concept for the approximation of operators, which is different from that for approximation with respect to a norm, and which can be used to prove convergence of approximate methods, has been introduced in [124].

Definition 1. *In a Banach space X, a set*†† *of operators A_ϵ is uniformly completely continuous*† *on $M \subset X$, if it maps every bounded set $\Phi \subset M$ into a compact set; that is $\bigcup_\epsilon A_\epsilon \Phi$ is compact.*

Definition 2. *If, in a Banach space X, the operators A_ϵ converge pointwise to A and are uniformly completely continuous on $M \subset X$, we say that the A_ϵ approximate A on M regularly.*

Theorem 2.14. (S.L. Sobolev). *If there exists a sequence $\{\psi_\epsilon\} \subset M \subset X$ such that*

$$\|\psi_\epsilon\| \geq h > 0, \quad \lim_{\epsilon \to 0} \|(I - A_\epsilon)\,\psi_\epsilon\| = 0 \ ,$$

where A_ϵ approximate A on M regularly, then the operator $I-A$ does not have an inverse.

---

(8) Many of these theorems are corollaries of Theorems 2.8 and 2.9 which clearly are valid for Banach spaces.

† Modern usage is *collectively compact*. We shall keep to the author's terminology. (Eds.)

†† By "a set of operators A_ϵ", we mean a sequence of operators $A_{\epsilon_1}, A_{\epsilon_2}, \ldots, A_{\epsilon_n}, \ldots$ with $\epsilon_1 > \epsilon_2 > \ldots > \epsilon_n > \ldots$ such that $\lim_{n\to\infty} \epsilon_n = 0$. (Eds.)

Theorem 2.15. *On a compact set M in a Banach space X, any operators A_ϵ which converge pointwise to A, approximate A regularly.*

Proof. Let $\{x_\epsilon\}$ be an arbitrary sequence in M. Due to the compactness of M, $\{x_\epsilon\}$ must converge to an element $\zeta_0 \in X$. We show that $A_\epsilon x_\epsilon \to A\zeta_0$. In fact, we have

$$\|A_\epsilon x_\epsilon - A\zeta_0\| \leq \|A_\epsilon(x_\epsilon - \zeta_0)\| + \|(A_\epsilon - A)\zeta_0\| .$$

But it is known that $\|A_\epsilon\|$ is bounded, since A_ϵ converges pointwise. Hence, we obtain that $\|A_\epsilon x_\epsilon - A\zeta_0\| \to 0$ as $\epsilon \to 0$. #

Theorem 2.16. *In a finite dimensional normed linear space X, any operators A_ϵ which converge pointwise to A, approximate A regularly.*

The proof follows from the fact that every bounded set of X is compact.

Theorem 2.17. *Let E_1, E_2 be given Banach spaces, M a compact set in E_1 where $0 \notin \bar{M}$, and let the linear operators A and A_ϵ map from E_1 into E_2 where A_ϵ converge pointwise to A as $\epsilon \to 0$. If there exists a sequence $\{\psi_\epsilon\} \subset M$ such that*

$$\lim_{\epsilon\to 0} \|A_\epsilon \psi_\epsilon\| = 0 , \tag{2.46}$$

then the operator A does not have a left inverse.

Proof. Without loss of generality, we may assume that ψ_ϵ is a convergent sequence. Let ψ_ϵ converge to $\zeta_0 \in E_1$. Obviously $\|\zeta_0\| > 0$, but from the given conditions

$$\|A_\epsilon\psi_\epsilon - A\zeta_0\| \leq \|A(\psi_\epsilon - \zeta_0)\| + \|(A_\epsilon - A)\zeta_0\| \to 0 \text{ as } \epsilon \to 0$$

and

$$\|A\zeta_0\| \leq \|A\psi_\epsilon\| + \|A_\epsilon\psi_\epsilon - A\zeta_0\| \to 0 \text{ as } \epsilon \to 0 .$$

Consequently, $A\zeta_0 = 0$ and hence the operator A does not have a left inverse. #

Corollary. *If the operator A has a left inverse and the operators A_ϵ converge pointwise to A then, for sufficiently small ϵ,*

$$\|A_\epsilon x\| \geq m\|x\|, \quad x \in M , \tag{2.47}$$

where $m > 0$ does not depend on ϵ or x and where M is a compact set with $0 \notin \bar{M}$.

Theorem 2.18. *Under the conditions of Theorem 2.17, it follows that if*

1) *A has a left inverse;*
2) *the equations (2.43) and $A_\epsilon x_\epsilon = y_\epsilon$ have solutions $\overset{*}{x}$ and $\overset{*}{x}_\epsilon$ respectively;*

3) $\|y_\epsilon - y\| \to 0$ *as* $\epsilon \to 0$;

4) *for sufficiently small* ϵ, $\overset{*}{x}_\epsilon$ *belongs to a compact set, then*

$$\|\overset{*}{x} - \overset{*}{x}_\epsilon\| \to 0 \text{ as } \epsilon \to 0.$$

If, in addition,

5) *there exists a constant* C_ϵ *such that* $C_\epsilon(\overset{*}{x} - \overset{*}{x}_\epsilon) \in M$, *then*

$$\|\overset{*}{x}_\epsilon - \overset{*}{x}\| < \frac{1}{m}\left(\|(A_\epsilon - A)\overset{*}{x}\| + \|y_\epsilon - y\|\right) \quad , \tag{2.48}$$

where m *does not depend on* ϵ.

Proof. Let us assume that $\{\overset{*}{x}_\epsilon\}$ does not converge to $\overset{*}{x}$. Then there exists a convergent sequence $\{\overset{*}{x}_{\epsilon_i}\}$ such that $\|\overset{*}{x}_{\epsilon_i} - \overset{*}{x}\| \geq d > 0$. From the corollary to Theorem 2.17, we have

$$\|\overset{*}{x}_{\epsilon_i} - \overset{*}{x}\| \leq \frac{1}{m_1} \|A_{\epsilon_i}(\overset{*}{x}_{\epsilon_i} - \overset{*}{x})\|$$

$$\leq \frac{1}{m_1}\left(\|(A_{\epsilon_i} - A)\overset{*}{x}\| + \|y_{\epsilon_i} - y\|\right) \to 0, \text{ as } \epsilon_i \to 0 \ ,$$

and must therefore be less than d. The resulting contradiction proves the first part of the theorem.

If there exists a constant C_ϵ such that $C_\epsilon(\overset{*}{x} - \overset{*}{x}_\epsilon) \in M$ (where $0 \notin \bar{M}$) then

$$\|\overset{*}{x}_\epsilon - \overset{*}{x}\| \leq \frac{1}{m}\|A_\epsilon(\overset{*}{x}_\epsilon - \overset{*}{x})\| \leq \frac{1}{m}\left(\|(A_\epsilon - A)\overset{*}{x}\| + \|y_\epsilon - y\|\right) \ ,$$

which establishes the theorem. #

In order to examine conditions under which $\overset{*}{x}_\epsilon$ belongs to a compact set, Ivanov [68] introduced the concept of a *quasi-solution.*

Definition 3. *Let* M *be a compact set of a given space* E_1. *We say that an element* $\overset{o}{x} \in \bar{M}$ *is a quasi-solution of* (2.43) *on* M *if the functional* $I(x) = \|Ax-y\|$ *attains its minimum on* $\bar{M}$ *at* $\overset{o}{x}$.

As a consequence of the compactness of M, there exists a quasi-solution for every $y \in E_2$. When the exact solution of (2.43) belongs to $\bar{M}$, it coincides with the quasi-solution on M. Consequently, the quasi-solution can be considered as a generalization of the exact solution.

Theorem 2.19. (V.K. Ivanov). *Let* A_ϵ *converge pointwise to* A, $\overset{o}{x}_\epsilon$ *be a quasi-solution of* $A_\epsilon x_\epsilon = y_\epsilon$ *on a compact set* M, *and* $\|y_\epsilon - y\| \to 0$ *as* $\epsilon \to 0$. *If* M *contains* $\overset{*}{x}$ *(the unique solution of* (2.43)), *then* $\|\overset{o}{x}_\epsilon - \overset{*}{x}\| \to 0$ *as* $\epsilon \to 0$.

It should be noted that Ivanov [69] examined the properties of a quasi-solution when (2.43) does not have a unique solution.

Let us assume that we are able to construct a minimizing sequence $x_n^{(\epsilon)} \in M$, a compact set[9]:

$$I_\epsilon(x_n^{(\epsilon)}) = \|A\, x_n^{(\epsilon)} - y_\epsilon\| \to \inf_{x\in M} \|A_\epsilon x - y_\epsilon\| = I_\epsilon(\overset{o}{x}_\epsilon)\ . \tag{2.49}$$

<u>Theorem 2.20</u>. *Under the conditions of Theorem* 2.19, *the elements* $x_n^{(\epsilon)}$ *approximate* $\overset{*}{x}$ *arbitrarily closely for sufficiently small* ϵ, *or sufficiently large* n.

<u>Proof</u>. We assume that $\|x_n^{(\epsilon)} - \overset{*}{x}\| \geq d > 0$ for sufficiently small ϵ or sufficiently large n. Since the sequence $x_n^{(\epsilon)} - \overset{*}{x}$ is compact and its closure does not contain 0, it follows from the corollary to Theorem 2.17 that

$$\begin{aligned} \|x_n^{(\epsilon)} - \overset{*}{x}\| &\leq \frac{1}{m} \|A_\epsilon(x_n^{(\epsilon)} - \overset{*}{x})\| \\ &\leq \frac{1}{m} \{I_\epsilon(x_n^{(\epsilon)}) - I_\epsilon(\overset{o}{x}_\epsilon) + I_\epsilon(\overset{o}{x}_\epsilon) \\ &\qquad + \|y_\epsilon - y\| + \|(A - A_\epsilon)\overset{*}{x}\|\}\ . \end{aligned}$$

But

$$I_\epsilon(\overset{o}{x}_\epsilon) = \|A_\epsilon \overset{o}{x}_\epsilon - y_\epsilon\| \leq \|A_\epsilon \overset{*}{x} - y_\epsilon\| \leq \|(A_\epsilon - A)\overset{*}{x}\| + \|y - y_\epsilon\|\ . \tag{2.50}$$

Hence

$$\|x_n^{(\epsilon)} - \overset{*}{x}\| \leq \frac{1}{m}\ \{I_\epsilon(x_n^{(\epsilon)}) - I_\epsilon(\overset{o}{x}_\epsilon) + 2\|y - y_\epsilon\| + 2\|(A - A_\epsilon)\overset{*}{x}\|\}\ , \tag{2.51}$$

which must be less than d. The resulting contradiction proves the theorem. #

<u>Remark</u>. If, as $\epsilon \to 0$ and $n \to \infty$, $(x_n^{(\epsilon)} - \overset{*}{x}) / \|x_n^{(\epsilon)} - \overset{*}{x}\|$ belongs to a compact set, then the bound (2.51), where m does not depend on ϵ or n, is valid for $\|x_n^{(\epsilon)} - \overset{*}{x}\|$.

A compact set M can be defined with the help of various sufficient conditions. For example, in the space of continuous functions C, if the solution of $Ax = y$ is continuously differentiable or if it has a derivative which is square summable, then M can be defined by either the condition

$$\|x\|_C^2 + \|x'\|_C^2 \leq C_1\ , \tag{2.52}$$

or

$$\|x\|_{L_2}^2 + \|x'\|_{L_2}^2 \leq C_1\ , \tag{2.53}$$

---

(9) Some algorithms for the construction of a minimizing sequence when M is a closed convex compact set, will be considered in §6.2.

where C_1 is a given constant, and x' denotes the derivative of x.

A very general and suitable definition of a compact set M is (see [91]):

$$M = B(\tilde{S}(0,1)), \tag{2.54}$$

where B is a completely continuous operator.(10)

It is not difficult to see that the problem of constructing a quasi-solution on a convex-compact set is a problem in convex programming (see [52]). In fact, for a linear operator A, the function $I(x) = ||Ax-y||$ is convex:

$$I\left(\frac{x_1 + x_2}{2}\right) \leq \frac{I(x_1) + I(x_2)}{2} \quad ,$$

and the admissible set M is convex. Generally speaking, the specific functional $I(x)$ and the bounded functionals (2.52) or (2.53) are not continuously differentiable, which contrasts with the usual problem in convex programming. Some algorithms for the construction of the minimizing sequence $x_n^{(\epsilon)}$ will be given in §6.

We note that in [130] the method of regularization, which is analogous to the method of §2, has been developed. It can be applied in a Banach space (or even in a metric space) for suitable assumptions about the way in which A_ϵ approximates A. The essential advantage of this method (A.N. Tikhonov) over the method of quasi-solutions (V.K. Ivanov) is that the first method reduces the problem of the approximate solution of (2.43) to the minimization of a certain auxiliary functional independent of additional restrictions. However, in this case, the properties of the method are more difficult to establish.

§2.3 Non-Linear Problems

We examine the equation

$$Ax = y \quad , \tag{2.55}$$

where A is a non-linear continuous operator with domain $D(A) \subseteq E_1$ and range $R(A) \subseteq E_2$, where E_1 and E_2 are Banach spaces and $y \in E_2$.

Let M be a compact set in $D(A)$. We assume that a quasi-solution of (2.55) is any element $\overset{o}{x} \in \bar{M}$ at which the functional

(10) An operator B is called *completely continuous* if it transforms bounded sets into compact sets.

$$I(x) = \|Ax - y\| \tag{2.56}$$

attains its infimum on M.

In general, we replace (2.55) by the equation

$$A_\epsilon x = y_\epsilon \ , \tag{2.57}$$

where A_ϵ is a continuous operator which satisfies

$$\|y_\epsilon - y\| \leq \epsilon, \ \|A_\epsilon x - Ax\| \leq \epsilon(x) \to 0, \ \epsilon \to 0 \ , \tag{2.58}$$

and $\epsilon(x)$ is a continuous functional on $D(A)$.

Lemma 1. *The relationship*

$$\tilde{\epsilon} = \sup_{x \in M} \{\epsilon(x)\} \to 0 \ , \ \epsilon \to 0 \ , \tag{2.59}$$

is valid.

Proof. Let $x_k \in M$ and $\epsilon(x_k) \to \tilde{\epsilon}$ as $k \to \infty$. Since M is compact, we may assume $x_k \to \tilde{x}$ say, and we have $\tilde{\epsilon} = \epsilon(\tilde{x})$. Thus $\tilde{\epsilon} \to 0$ as $\epsilon \to 0$. #

We denote by N the range of the operator A defined on M.

Lemma 2. (A.N. Tikhonov [128]). *If an operator A defines a one-to-one mapping of M onto N, and A is continuous, then there exists an inverse operator A^{-1}, which is continuous on N.*

Theorem 2.21. *Let* (2.55) *have a solution $\overset{*}{x} \in \bar{M}$ and the operator A be a one-to-one mapping of $\bar{M}$ onto $\bar{N}$. Then the quasi-solutions $\overset{o}{x}_\epsilon$ of* (2.57) *converge in norm to $\overset{*}{x}$ as $\epsilon \to 0$.*

Proof. On the basis of Lemma 1, we have

$$\begin{aligned} \|A\overset{*}{x} - A\overset{o}{x}_\epsilon\| &\leq \|A\overset{*}{x} - y\| + \|y - y_\epsilon\| + \|y_\epsilon - A_\epsilon \overset{o}{x}_\epsilon\| + \|A_\epsilon \overset{o}{x}_\epsilon - A\overset{o}{x}_\epsilon\| \\ &\leq \epsilon + \|y_\epsilon - A_\epsilon \overset{*}{x}\| + \epsilon(\overset{o}{x}_\epsilon) \\ &\leq 2(\epsilon + \tilde{\epsilon}) \to 0, \ \epsilon \to 0. \end{aligned} \tag{2.60}$$

But, from Lemma 2, infinitesimal variations in $\bar{M}$ correspond to infinitesimal variations in $\bar{N}$. Hence the theorem is proved. #

We construct an $\tilde{\epsilon}$-net on M, and denote it by $M_{\tilde{\epsilon}}$. We assume that $\bar{x}_{\tilde{\epsilon}}$ is the element of $M_{\tilde{\epsilon}}$ at which $\|A_\epsilon x - y_\epsilon\|$ attains its smallest value on $M_{\tilde{\epsilon}}$.

Theorem 2.22. *Under the assumptions of the previous theorem, the elements $\bar{x}_{\tilde{\epsilon}}$ converge to $\overset{*}{x}$ as $\epsilon \to 0$.*

Proof. By analogy with (2.60), we obtain

$$\|A\overset{*}{x} - A\bar{x}_{\tilde{\epsilon}}\| \leq \epsilon + \|y_\epsilon - A_\epsilon \bar{x}_{\tilde{\epsilon}}\| + \epsilon(\bar{x}_{\tilde{\epsilon}})$$
$$\leq \epsilon + \|y_\epsilon - A_\epsilon \bar{x}^*_{\tilde{\epsilon}}\| + \tilde{\epsilon} ,$$

where $\bar{x}^*_{\tilde{\epsilon}} \in M_{\tilde{\epsilon}}$ and $\|\overset{*}{x} - \bar{x}^*_{\tilde{\epsilon}}\| \leq \tilde{\epsilon}$. Hence

(2.61) $$\|A\overset{*}{x} - A\bar{x}_{\tilde{\epsilon}}\| \leq 2(\epsilon + \tilde{\epsilon}) + \|A\overset{*}{x} - A\bar{x}^*_{\tilde{\epsilon}}\| \to 0 \text{ as } \tilde{\epsilon} \to 0.$$

But in this case we also have $\|\overset{*}{x} - \bar{x}_{\tilde{\epsilon}}\| \to 0$ as $\epsilon \to 0$, which proves the theorem. #

The determination of $\bar{x}_{\tilde{\epsilon}}$ can be done by an algorithm which sorts through the elements of $M_{\tilde{\epsilon}}$. This approach can be useful in certain cases, particularly when $\tilde{\epsilon}$ is not too small. Therefore, we have to construct an ϵ-net with as small a number of elements as possible. Methods for its construction have been investigated in detail in [23]. In [111] an ϵ-net is used to search for all solutions of (2.55) in the case when A is a completely continuous operator.

The direct determination of a quasi-solution $\overset{\circ}{x}_\epsilon$ is obviously a general problem in non-linear programming in a Banach space. One of the algorithms for the approximate solution of similar types of problems is described in §6.

Just as for linear equations, the method of regularization can be used instead of the method of quasi-solutions to solve some forms of non-linear equations (see [130]).

For the approximate solution of non-linear operator equations, various iterative methods including simple iteration are frequently applied (see [75, 115]). We now formulate conditions which guarantee the convergence of simple iteration in the case when $E_1 = E_2$.

Theorem 2.23. (see, for example, [83]). *Equation* (2.55) *can always be represented as*

(2.62) $$u = Du + v ,$$

where the operator D satisfies $D0 = 0$, u is unknown and v is known.
If the following conditions are satisfied,

1) $\|Du_1 - Du_2\| \leq C(\rho)\ \|u_1 - u_2\|$, $\|u_i\| \leq \rho$, $i = 1,2,$;

2) $\gamma = C(\rho) < 1$;

3) $\|v\| \leq (1 - \gamma)\rho$;

then (2.62) *has a unique solution $u^* \in E_1$, where $\|u^*\| < \rho$. This solution can be* *defined by the method of simple iteration:*

(2.64) $$u^{(k)} = Du^{(k-1)} + v,\ k = 1,2,\dots;\quad u^{(0)} = v,$$

where

(2.65) $$\|u^* - u^{(k)}\| \leq \frac{\gamma^{k+1}}{1-\gamma}\ \|v\|.$$

In general, the solution of non-linear problems can be reduced to the determination of the minimum of a non-linear functional $I(x)$ (which is not necessarily of the form (2.56)), defined on a given set M. For this, it is necessary to distinguish between two essentially different representations for the approximate solutions of such problems.

Definition 1. *Any element $x_\delta \in M$ which satisfies*

(2.66) $$I(x_\delta) \leq \inf_{x\in M} \{I(x)\} + \delta\ ,$$

we shall call a β-approximation with accuracy δ.

Definition 2. *Let us assume that there exists $\overset{o}{x} \in M$ such that $I(\overset{o}{x}) = \inf_{x\in M} \{I(x)\}$. Any element $x_\delta \in M$ which satisfies*

(2.67) $$\|\overset{o}{x} - x_\delta\| \leq \delta\ ,$$

we shall call an α-approximation with accuracy δ.

The estimation of the total error which, in general, contains error caused by inaccuracies in the input data, error due to the method or algorithm, and round-off error, has considerable interest for the practical approximation of non-linear problems.

Following [4], we shall estimate the total error for the approximate solution of (2.62) by the method of simple iteration, under the conditions of Theorem 2.23.

In practice, use is usually made of an approximating equation of the form

(2.68) $$u = D_\epsilon u + v_\epsilon\ ,$$

which is also considered in E_1. The non-linear operator D_ϵ is related to D by

(2.69) $$\|Du - D_\epsilon u\| \leq \eta_1(\epsilon, u)$$

where, for every fixed $u \in E_1$, $\eta_1(\epsilon, u) \to 0$ as $\epsilon \to 0$ and $D_\epsilon 0 = D0 = 0$.

Below, we shall need to assume that some or all of the following conditions are satisfied:

(2.70)
1) $\|D_\epsilon u_1 - D_\epsilon u_2\| \leq C_\epsilon(\rho_\epsilon)\|u_1 - u_2\|$ for $\|u_i\| \leq \rho_\epsilon$, $i = 1,2,$;
2) $\|v - v_\epsilon\| \leq \eta_2(\epsilon)$;
3) $|C(\rho) - C_\epsilon(\rho_\epsilon)| < \eta_3(\epsilon)$;

where $\eta_2(\epsilon) \to 0$, $\eta_3(\epsilon) \to 0$ as $\epsilon \to 0$.

We shall examine the following questions:

a). Does the solvability of the approximate equation follow from the solvability of the exact equation, and conversely?

b). How large is the error in the solution when (2.62) is replaced by (2.68)?

Theorem 2.24. *Let us assume that the conditions* (2.63) - 1), 2), (2.69) *and* (2.70) *are satisfied.*

a) If

$$\delta = (1 - \gamma)\rho - \|v\| > 0 , \tag{2.71}$$

then there exists an $\bar{\epsilon}_1$ *such that*

$$\gamma_\epsilon = C_\epsilon(\rho_\epsilon) \leq \gamma + \eta_3(\epsilon) < 1 , \tag{2.72}$$

$$\|v_\epsilon\| \leq (1 - \gamma_\epsilon)\rho_\epsilon , \tag{2.73}$$

for any $\epsilon \leq \bar{\epsilon}_1$ *and* (2.68) *has a unique solution* u_ϵ^* *in the sphere* $\|u\| \leq \rho_\epsilon = \rho$.

b) If (2.63)-3) *is also satisfied then for* $\gamma_\epsilon \leq \gamma$ *and for any* ϵ *such that*

$$\eta_2(\epsilon) \leq (\gamma - \gamma_\epsilon)\rho , \tag{2.74}$$

(2.68) *has a unique solution in the sphere* $\|u\| \leq \rho_\epsilon = \rho$. *For* $\gamma_\epsilon > \gamma$ *and for*

$$\rho_\epsilon = \frac{(1 - \gamma)\rho + \eta_2(\epsilon)}{1 - \gamma_\epsilon} , \tag{2.75}$$

there exists an $\bar{\epsilon}_2$ *such that for every* $\epsilon \leq \bar{\epsilon}_2$, (2.72) *and* (2.73) *are satisfied and* (2.68) *has a unique solution in the sphere* $\|u\| \leq \rho_\epsilon$.

In either case a) or b), the solution can be found by the method of simple iteration $u_\epsilon^{(k)} = D_\epsilon u_\epsilon^{(k-1)} + v_\epsilon$, *where* $u_\epsilon^{(0)} = v_\epsilon$. *As* $\epsilon \to 0$ *and* $k \to \infty$ *the sequence* $u_\epsilon^{(k)}$ *converges to a solution* u^* *of* (2.62), *where the error is given by*

$$(2.76)\qquad \|u^* - u_\epsilon^{(k)}\| \le \frac{\eta_1(\epsilon, u^*) + \eta_2(\epsilon)}{1 - \gamma - \eta_3(\epsilon)} + \frac{[\gamma + \eta_3(\epsilon)]^{(k+1)}}{1 - \gamma - \eta_3(\epsilon)}\,[\eta_2(\epsilon) + \|v\|]\,.$$

Proof. From (2.63)-2), (2.70)-3) we obtain

$$(2.77)\qquad \gamma_\epsilon = C_\epsilon(\rho_\epsilon) \le \gamma + \eta_3(\epsilon)\quad .$$

As $\epsilon \to 0$, there exists an ϵ_3 such that for $\epsilon \le \epsilon_3$, (2.72) is satisfied. Let $Q(\epsilon) = \rho\,\eta_3(\epsilon) + \eta_2(\epsilon)$. As $\epsilon \to 0$, $Q(\epsilon) \to 0$ and, consequently, there exists an ϵ_2 such that, for $\epsilon \le \epsilon_2$, $Q(\epsilon) \le \delta$. Hence

$$\|v_\epsilon\| \le \|v\| + \eta_2(\epsilon) = (1 - \gamma)\rho - \delta + \eta_2(\epsilon)$$

$$\le (1 - \gamma_\epsilon)\rho + \rho\eta_3(\epsilon) + \eta_2(\epsilon) - \delta \le (1 - \gamma_\epsilon)\rho\ .$$

Thus, for $\epsilon \le \bar{\epsilon}_1 = \max(\epsilon_2, \epsilon_3)$, the conditions (2.72) and (2.73) are satisfied when $\rho = \rho_\epsilon$. Under the conditions of Theorem 2.23, it follows that the equation (2.68) has a unique solution in the sphere $\|u\| \le \rho_\epsilon = \rho$. Part a) of the theorem is now proved.

Let (2.63)-3) be satisfied. Hence, if $C_\epsilon(\rho_\epsilon) \le C(\rho)$, we obtain that $\gamma_\epsilon \le \gamma < 1$. Then

$$\|v_\epsilon\| \le \|v\| + \eta_2(\epsilon) \le (1 - \gamma)\rho + (\gamma - \gamma_\epsilon)\rho \le (1 - \gamma_\epsilon)\rho\ ,$$

which proves (2.73) with $\rho_\epsilon = \rho$. Setting $\rho_\epsilon = \rho$ in (2.70)-1), we satisfy the conditions of Theorem 2.23 and thereby we prove the first part of (b).

Finally we examine the case when $C_\epsilon(\rho_\epsilon) > C(\rho)$; i.e. $\gamma_\epsilon > \gamma$. The validity of (2.72) can be proved as above. From (2.68) it follows that

$$\|u\| \le \|D_\epsilon u - D_\epsilon 0\| + \|v_\epsilon\| \le C_\epsilon(\rho_\epsilon)\|u\| + \|v_\epsilon\|$$

$$\le \gamma_\epsilon \|u\| + (1 - \gamma)\rho + \eta_2(\epsilon)\ ,$$

whence

$$\|u\| \le \frac{(1 - \gamma)\rho + \eta_2(\epsilon)}{1 - \gamma_\epsilon} = \rho_\epsilon\ .$$

As $\epsilon \to 0$, we have that $\rho_\epsilon \ge \rho$ and $\rho_\epsilon \to \rho$. We now show that, for a given ρ_ϵ, the inequality (2.73) is satisfied. In fact

$$\|v_\epsilon\| \le \|v\| + \eta_2(\epsilon) \le (1 - \gamma)\rho + \eta_2(\epsilon)$$

$$= (1 - \gamma)\,\frac{(1 - \gamma_\epsilon)\rho_\epsilon - \eta_2(\epsilon)}{1 - \gamma} + \eta_2(\epsilon)$$

$$= (1 - \gamma_\epsilon)\rho_\epsilon,$$

and (2.73) is proved. Thus all the conditions of Theorem 2.23 are satisfied, and equation (2.68) has a unique solution in the sphere $\|u\| \le \rho_\epsilon$, where ρ_ϵ is given by (2.75).

We now prove (2.76). From (2.62) and (2.68) we obtain

$$\|u^* - u_\epsilon^*\| \le \|Du^* - D_\epsilon u_\epsilon^*\| + \|v - v_\epsilon\|$$

$$\le \|Du^* - D_\epsilon u^*\| + \|D_\epsilon u^* - D_\epsilon u_\epsilon^*\| + \|v - v_\epsilon\|$$

$$\le \eta_1(\epsilon, u^*) + C_\epsilon(\rho_\epsilon)\|u^* - u_\epsilon^*\| + \eta_2(\epsilon)$$

$$\le \eta_1(\epsilon, u^*) + \gamma_\epsilon\|u^* - u_\epsilon^*\| + \eta_2(\epsilon)$$

$$\le \eta_1(\epsilon, u^*) + \eta_2(\epsilon) + [\gamma + \eta_3(\epsilon)]\|u^* - u_\epsilon^*\| ,$$

whence

$$\|u^* - u_\epsilon^*\| \le \frac{\eta_1(\epsilon, u^*) + \eta_2(\epsilon)}{1 - \gamma - \eta_3(\epsilon)} .$$

By analogy with (2.65), we obtain

$$\|u_\epsilon^* - u_\epsilon^{(k)}\| \le \frac{\gamma_\epsilon^{(k+1)}}{1-\gamma_\epsilon}\,\|v_\epsilon\| \le \frac{[\gamma + \eta_3(\epsilon)]^{(k+1)}}{1 - \gamma - \eta_3(\epsilon)}\,[\eta_2(\epsilon) + \|v\|] ,$$

where $u_\epsilon^{(k)}$ is the kth iterate of equation (2.68) with $u_\epsilon^{(0)} = v_\epsilon$. By adding the last two inequalities we obtain (2.76). #

The following theorem examines the converse: does the solvability of the exact equation follow from the solvability of the approximate equation?

Theorem 2.25. *Let the conditions* (2.63)-1), (2.69), (2.70), *and* (2.72) *be satisfied.*

a) If there exists

(2.78) $$\delta_\epsilon = (1 - \gamma_\epsilon)\rho_\epsilon - \|v_\epsilon\| > 0 ,$$

then for

(2.79) $$\eta_3(\epsilon) < 1 - \gamma_\epsilon ,$$

(2.80) $$\rho_\epsilon\,\eta_3(\epsilon) + \eta_2(\epsilon) \le \delta_\epsilon ,$$

the conditions (2.63)-2),-3) *will be satisfied and equation* (2.62) *will have a unique solution* u^* *in the sphere* $\|u\| \le \rho = \rho_\epsilon$.

b) *If* (2.73) *is satisfied, then for* $\gamma \le \gamma_\epsilon$ *and*

$$\eta_2(\epsilon) \le (\gamma_\epsilon - \gamma)\rho_\epsilon \ , \tag{2.81}$$

equation (2.62) *will have a unique solution in the sphere* $\|u\| \le \rho = \rho_\epsilon$. *For* $\gamma > \gamma_\epsilon$ *under the condition* (2.79), *the equation* (2.62) *will have a unique solution in the sphere* $\|u\| \le \rho$ *where*

$$\rho = \frac{(1 - \gamma_\epsilon)\ \rho_\epsilon + \eta_2(\epsilon)}{1 - \gamma} \ . \tag{2.82}$$

The sequence $u_\epsilon^{(k)}$, *which is defined by* (2.68), *converges to this solution when* $\epsilon \to 0$ *and* $k \to \infty$; *the estimated error is*

$$\|u^* - u_\epsilon^{(k)}\| \le \frac{\eta_1(\epsilon,\ u_\epsilon^*) + \eta_2(\epsilon)}{1 - \gamma_\epsilon - \eta_3(\epsilon)} + \frac{\gamma_\epsilon^{(k+1)}}{1-\gamma_\epsilon}\ \|v_\epsilon\| \ . \tag{2.83}$$

Proof. From (2.70)-3) and (2.79) we find $\gamma < \gamma \ + \eta_3(\epsilon) < 1$. Using this inequality, we obtain

$$\|v\| \le \|v_\epsilon\| + \eta_2(\epsilon) = (1 - \gamma_\epsilon)\rho_\epsilon - \delta_\epsilon + \eta_2(\epsilon)$$
$$\le (1 - \gamma)\rho_\epsilon + \rho_\epsilon\ \eta_3(\epsilon) + \eta_2(\epsilon) - \delta_\epsilon \le (1 - \gamma)\rho_\epsilon \ .$$

If we assume $\rho = \rho_\epsilon$ in (2.63), then the conditions of Theorem 2.23 will be satisfied and equation (2.62) will have a unique solution in the sphere $\|u\| \le \rho = \rho_\epsilon$. Thus the first part of the theorem is proved.

We now examine the case when (2.73) is satisfied. If $\gamma \le \gamma_\epsilon$, we have $\gamma < 1$, and

$$\|v\| \le \|v_\epsilon\| + \eta_2(\epsilon) \le (1 - \gamma_\epsilon)\rho_\epsilon + (\gamma_\epsilon - \gamma)\rho_\epsilon = (1 - \gamma)\rho_\epsilon.$$

Setting $\rho = \rho_\epsilon$ in (2.63)-1), the conditions of Theorem 2.23 are satisfied.

Finally, let $\gamma > \gamma_\epsilon$. The validity of (2.63)-2) has been proved earlier. From (2.73), (2.70)-2) and (2.82) we obtain

$$\|v\| \le \|v_\epsilon\| \ + \eta_2(\epsilon) \le (1 - \gamma_\epsilon)\rho_\epsilon + \eta_2(\epsilon)$$
$$\le (1 - \gamma_\epsilon)\ \frac{(1 - \gamma)\rho - \eta_2(\epsilon)}{1 - \gamma_\epsilon} + \eta_2(\epsilon) = (1 - \gamma)\rho \ .$$

Hence the conditions of Theorem 2.23 are satisfied and equation (2.62) has a unique solution in the sphere $\|u\| \leq \rho$ where ρ is defined by (2.82).

The inequality (2.83) can be established by analogy with (2.76), but in this case it is necessary to consider that $\rho_\epsilon \leq \rho$. #

It frequently occurs that an approximate equation is examined in a finite dimensional subspace E_n of the space E. Naturally, the question then arises as to how the approximate equation in E_n will be related to equations (2.62) and (2.68). In this case it can be shown that theorems similar to Theorems 2.24 and 2.25 will also be valid.

In fact, let P_n denote the projection operator of E into the finite dimensional space E_n;

$$P_n(E) = E_n, \; P_n^2 = P_n \; .$$

Obviously the operator P_n leaves unchanged the elements of E_n. In the space E, we examine the equation

$$u = D_{\epsilon n} u + v_{\epsilon n} \quad , \tag{2.84}$$

where $D_{\epsilon n}$ is the operator which maps from the space E into the subspace E_n with $D_{\epsilon n} 0 = 0$. We assume that the operator $D_{\epsilon n}$ satisfies a Lipschitz condition, i.e.,

$$\|D_{\epsilon n} u_1 - D_{\epsilon n} u_2\| \leq C_{\epsilon n}(\rho_{\epsilon n}) \|u_1 - u_2\| \text{ for } \|u_i\| \leq \rho_{\epsilon n} \; , \; i = 1,2 \; .$$

The conditions (2.69) and (2.70)-2),-3) take respectively the forms

$$\|Du - D_{\epsilon n} u\| \leq \eta_1(\epsilon, n, u) \; ,$$

$$\|v - v_{\epsilon n}\| \leq \eta_2(\epsilon, n) \; ,$$

$$|C(\rho) - C_{\epsilon n}(\rho_{\epsilon n})| \leq \eta_3(\epsilon, n) \; ,$$

where $\eta_1(\epsilon, n, u)$, $\eta_2(\epsilon, n)$, $\eta_3(\epsilon, n) \to 0$ as $\epsilon \to 0$ and $n \to \infty$. Taking these last three conditions into account, Theorems 2.24 and 2.25 can now be applied to the equations (2.62) and (2.84). The sequence $u_{\epsilon n}^{(k)}$, constructed by simple iteration from (2.84), converges to the solution u^* of (2.62) as $\epsilon \to 0$, $n \to \infty$ and $k \to \infty$. The rate of convergence is defined by the inequality

$$\|u^* - u_{\epsilon n}^{(k)}\| \leq \frac{\eta_1(\epsilon, n, u^*) + \eta_2(\epsilon, n)}{1 - \gamma - \eta_3(\epsilon, n)} + \frac{[\gamma + \eta_3(\epsilon, n)]^{(k+1)}}{1 - \gamma - \eta_3(\epsilon, n)} [\eta_2(\epsilon, n) + \|v\|]. \tag{2.85}$$

For the practical realization of an algorithm, it is necessary to solve the new equation

$$u = D_{\epsilon n\tau}\, u + v_{\epsilon n\tau} \quad , \tag{2.86}$$

where $D_{\epsilon n\tau}$ is the pseudo-operator which maps from the space E into the subspace $E_{n\tau}$ over the field of numbers with a finite word length τ, where $D_{\epsilon n\tau}\, 0 = 0$, and $v_{\epsilon n\tau}$ is the correctly rounded value of $v_{\epsilon n}$. The operator $D_{\epsilon n\tau}$ differs from $D_{\epsilon n}$ since the arithmetical operations which are contained in $D_{\epsilon n}$ are replaced by pseudo-operations. The corresponding errors have been examined in [172]. We assume that the operator $D_{\epsilon n\tau}$ satisfies a Lipschitz condition in the sphere $\|u\| \le \rho_{\epsilon n\tau}$, that is

$$\|D_{\epsilon n\tau}\, u_1 - D_{\epsilon n\tau}\, u_2\| \le C_{\epsilon n\tau}\,(\rho_{\epsilon n\tau})\, \|u_1 - u_2\| \text{ for } \|u_i\| \le \rho_{\epsilon n\tau}\, , i = 1,2.$$

In this case, the conditions (2.69) and (2.70)-2),-3) take respectively the forms

$$\begin{aligned} &\|Du - D_{\epsilon n\tau}\, u\| \le \eta_1(\epsilon, n, \tau, u) \ , \\ &\|v - v_{\epsilon n\tau}\| \le \eta_2(\epsilon, n, \tau) \ , \\ &|C(\rho) - C_{\epsilon n\tau}(\rho_{\epsilon n\tau})| \le \eta_3(\epsilon, n, \tau) \ , \end{aligned} \tag{2.87}$$

where $\eta_1(\epsilon, n, \tau, u) \to 0$, $\eta_2(\epsilon, n, \tau) \to 0$, $\eta_3(\epsilon, n, \tau) < \min(1-\gamma, \delta/\rho)$ as $\epsilon \to 0$, $n \to \infty$ and $\tau \to \infty$.

Obviously, under these conditions, Theorems 2.24 and 2.25 apply to the equations (2.62) and (2.86). The sequence $u^{(k)}_{\epsilon n\tau}$, which is obtained by simple iteration of (2.86), converges to the solution of (2.62) as $\epsilon \to 0$, $n \to \infty$, $\tau \to \infty$ and $k \to \infty$. The rate of convergence is defined by the inequality

$$\begin{aligned} \|u^* - u^{(k)}_{\epsilon n\tau}\| \le \ & \frac{\eta_1(\epsilon, n, \tau, u^*) + \eta_2(\epsilon, n, \tau)}{1 - \gamma - \eta_3(\epsilon, n, \tau)} \ + \\ & + \frac{[\gamma + \eta_3(\epsilon, n, \tau)\,]^{(k+1)}}{1 - \gamma - \eta_3(\epsilon, n, \tau)}\ [\eta_2(\epsilon, n, \tau) + \|v\|] \ . \end{aligned} \tag{2.88}$$

Estimates which follow from Theorem 2.25, and which characterize the rate of convergence, will be different from (2.76), (2.85) and (2.88) because u^*_ϵ, $u^*_{\epsilon n}$, $u^*_{\epsilon n\tau}$ will appear instead of u^*, and γ_ϵ, $\gamma_{\epsilon n}$, $\gamma_{\epsilon n\tau}$ will appear everywhere instead of γ, in the first term of the right hand side of (2.76), (2.85) and (2.88) respectively. This is because, whereas we have

$\rho \le \rho_\epsilon \le \rho_{\epsilon n} \le \rho_{\epsilon n\tau}$ in Theorem 2.24, in Theorem 2.25 we have $\rho_{\epsilon n\tau} \le \rho_{\epsilon n} \le \rho_\epsilon \le \rho$.

Along with the theoretical estimate (2.88), we introduce a corresponding practical estimate. From the above results and inequality (2.83), we find that

$$\|u^* - u^{(k)}_{\epsilon n\tau}\| \le \frac{\eta_1(\epsilon, n, \tau, u^*_{\epsilon n\tau}) + \eta_2(\epsilon, n, \tau)}{1 - \gamma_{\epsilon n\tau} - \eta_3(\epsilon, n, \tau)} + \frac{\gamma^{k+1}_{\epsilon n\tau}}{1-\gamma_{\epsilon n\tau}} \|v_{\epsilon n\tau}\| ,$$

from which we can obtain a practical estimate if we change $\eta_1(\epsilon, n, \tau, u^*_{\epsilon n\tau})$ so that $u^{(k)}_{\epsilon n\tau}$ appears in it instead of $u^*_{\epsilon n\tau}$, the solution of (2.86). We can do this in the following way:

$$\begin{aligned}\|Du^*_{\epsilon n\tau} - D_{\epsilon n\tau} u^*_{\epsilon n\tau}\| &\le \|Du^*_{\epsilon n\tau} - Du^{(k)}_{\epsilon n\tau}\| + \|Du^{(k)}_{\epsilon n\tau} - D_{\epsilon n\tau} u^{(k)}_{\epsilon n\tau}\| + \\ &\quad + \|D_{\epsilon n\tau} u^{(k)}_{\epsilon n\tau} - D_{\epsilon n\tau} u^*_{\epsilon n\tau}\| \\ &\le \gamma\|u^*_{\epsilon n\tau} - u^{(k)}_{\epsilon n\tau}\| + \eta_1(\epsilon, n, \tau, u^{(k)}_{\epsilon n\tau}) + \gamma_{\epsilon n\tau}\|u^{(k)}_{\epsilon n\tau} - u^*_{\epsilon n\tau}\| \\ &\le [\eta_3(\epsilon, n, \tau) + 2\gamma_{\epsilon n\tau}] \frac{\gamma^{k+1}_{\epsilon n\tau}}{1-\gamma_{\epsilon n\tau}} \|v_{\epsilon n\tau}\| + \eta_1(\epsilon, n, \tau, u^{(k)}_{\epsilon n\tau}) \\ &= \eta_1(\epsilon, n, \tau, u^*_{\epsilon n\tau}) .\end{aligned}$$

Then

$$(2.89)\quad \|u^* - u^{(k)}_{\epsilon n\tau}\| \le \frac{\gamma^{k+1}_{\epsilon n\tau}}{1-\gamma_{\epsilon n\tau}} \|v_{\epsilon n\tau}\| + $$
$$+ \frac{[\eta_3(\epsilon, n, \tau) + 2\gamma_{\epsilon n\tau}] \frac{\gamma^{k+1}_{\epsilon n\tau}}{1-\gamma_{\epsilon n\tau}} \|v_{\epsilon n\tau}\| + \eta_1(\epsilon, n, \tau, u^{(k)}_{\epsilon n\tau}) + \eta_2(\epsilon, n, \tau)}{1 - \gamma_{\epsilon n\tau} - \eta_3(\epsilon, n, \tau)} .$$

Not infrequently the approximate equation is examined in a finite dimensional space $\tilde{E}_{n\tau}$ which is isomorphic to the subspace $E_{n\tau}$. In this case we shall assume that there exists a linear operator ψ_0 which is a one-to-one mapping of the subspace $E_{n\tau}$ onto the space $\tilde{E}_{n\tau}$. These conditions ensure the existence and linearity of the inverse operator ψ_0^{-1}. Because there exists a one-to-one correspondence between the elements of $E_{n\tau}$ and $\tilde{E}_{n\tau}$, equation (2.86), which is examined in $E_{n\tau}$, can be transformed into an equivalent equation in the finite-dimensional space $\tilde{E}_{n\tau}$. In order to do this, we have to substitute $u = \psi_0^{-1}\tilde{u}$ in (2.86) and apply the operator ψ_0 to it. The transformed equation will be of the form

$$\bar{u} = \psi_0 \, D_{\epsilon n\tau} \, \psi_0^{-1} \, \bar{u} + \psi_0 \, v_{\epsilon n\tau} \quad .$$

Denoting by $\bar{D}_{\epsilon n\tau}$ the operator which maps $\tilde{E}_{n\tau}$ into itself, such that

$$\bar{D}_{\epsilon n\tau} = \psi_0 \, D_{\epsilon n\tau} \, \psi_0^{-1} , \; D_{\epsilon n\tau} = \psi_0^{-1} \, \bar{D}_{\epsilon n\tau} \, \psi_0 \; ,$$

and letting $\bar{v}_{\epsilon n\tau} = \psi_0 \, v_{\epsilon n\tau}$, the previous equation can be written in the form

$$\bar{u} = \bar{D}_{\epsilon n\tau} \, \bar{u} + \bar{v}_{\epsilon n\tau} \; . \tag{2.90}$$

If the conditions of (2.87) are satisfied, then equation (2.90) will have a unique solution $\bar{u}^*_{\epsilon n\tau} \in \tilde{E}_{n\tau}$ and it can be defined by the method of simple iteration: $\bar{u}^{(k)}_{\epsilon n\tau} = \bar{D}_{\epsilon n\tau} \, \bar{u}^{(k-1)}_{\epsilon n\tau} + \bar{v}_{\epsilon n\tau}$. In fact, the solvability of (2.86) follows from the conditions (2.87) and the solvability of (2.62). Obviously, the solution belongs to the space $E_{n\tau}$. Since there exists a one-to-one correspondence between elements of $E_{n\tau}$ and $\tilde{E}_{n\tau}$, the solution $u^*_{\epsilon n\tau} \in E_{n\tau}$ will correspond to the solution $\bar{u}^*_{\epsilon n\tau} \in \tilde{E}_{n\tau}$, and conversely. Thus, the solvability of (2.90) follows from the solvability of (2.62), and conversely.

In conclusion, it should be noted that, in practice, we take the kth approximation of (2.86) instead of the solution u^* of (2.62). In this case, for the estimation of the error of the kth approximation, we can do without conditions of the form (2.70)-3). In fact

$$\|u^* - u^{(k)}_{\epsilon n\tau}\| \le \|u^* - u^{(k)}\| + \|u^{(k)} - u^{(k)}_{\epsilon n\tau}\| \; .$$

According to Theorem 2.23, the iteration process converges to a solution and, for any $\bar{\epsilon} > 0$, k can be selected so that $\|u^* - u^{(k)}\| \le \bar{\epsilon}/2$. Fixing this value of k, we select values of ϵ, n, τ so that $\|u^{(k)} - u^{(k)}_{\epsilon n\tau}\| \le \bar{\epsilon}/2$. In this way, we obtain $\|u^* - u^{(k)}_{\epsilon n\tau}\| \le \bar{\epsilon}$.

§3 THE METHOD OF LEAST SQUARES

§3.1 The Basis of the Method

In the method of least squares (see, for example, [103]) we take the approximate solution of (2.1) in the form

$$x_n = \sum_{k=1}^{n} \alpha_k e_k \quad , \quad e_k \in D(A) \quad , \tag{3.1}$$

and the unknown values(11) α_k we find (using the method of regularization) from the minimization of the quadratic functional,

$$\Delta_n (P_n) = I^{(\alpha)} (x_n) = \|y - \sum_{k=1}^{n} \alpha_k y_k\|^2 + \alpha \|\sum_{k=1}^{n} \alpha_k e_k\|^2 , \tag{3.2}$$

$$y_k = Ae_k \, , \; P_n = P_n(\alpha_1, \alpha_2, \ldots, \alpha_n) \; .$$

Theorem 3.1. *There exists at least one point $\overset{*}{P}_n$ at which the minimum value of $\Delta_n(P_n)$ is attained.*

Proof. At first we assume that $\alpha > 0$. Without loss of generality, we assume that in the system $e_1, e_2, \ldots, e_n$, the first $m(1 \leq m < n)$ elements are linearly independent, and the others can be expressed in terms of them in the form

$$e_j = \sum_{k=1}^{m} \alpha_k^{(j)} e_k, \; j = m+1, m+2, \ldots, n \; . \tag{3.3}$$

Substituting (3.3) into (3.2) we obtain

$$\Delta_n(P_n) = \|y - \sum_{k=1}^{m} \beta_k y_k\|^2 + \alpha \|\sum_{k=1}^{m} \beta_k e_k\|^2 = \Delta_m(Q_m) \, , \tag{3.4}$$

$$\beta_k = \alpha_k + \sum_{j=m+1}^{n} \alpha_j \alpha_k^{(j)} \, , \; Q_m = Q_m(\beta_1, \beta_2, \ldots, \beta_m).$$

Given that $\alpha > 0$ and $e_1, e_2, \ldots, e_m$ are linearly independent it is easy to prove that the minimum value of $\Delta_m(Q_m)$ is attained at a point $\overset{*}{Q}_m (\overset{*}{\beta}_1, \overset{*}{\beta}_2, \ldots, \overset{*}{\beta}_m)$. Setting $\overset{*}{\alpha}_j = 0$ (where $j = m+1, \ldots, n$) and $\overset{*}{\alpha}_k = \overset{*}{\beta}_k$ (where $k = 1,2,\ldots,m$) we define the point $\overset{*}{P}_n (\overset{*}{\alpha}_1, \overset{*}{\alpha}_2, \ldots, \overset{*}{\alpha}_n)$ such that

$$\Delta_n(\overset{*}{P}_n) = \inf_{P_n} \Delta_n(P_n) \; .$$

---

(11) If necessary we add elements ie_k to e_k so that we can assume α_k is real throughout.

If $\alpha = 0$, by selecting a maximal linearly independent system $y_1, y_2, \dots, y_{m'}$, $(1 \le m' \le m \le n)$, and representing the other y_j in the form (3.3) where the e_k are replaced by y_k and m is replaced by m', we obtain (3.4), with the help of which we complete the proof by the well known method (see [96, p.71]). #

As a corollary, we obtain that the algebraic system

(3.5)
$$\sum_{k=1}^{n} \alpha_k \,[\mathrm{Re}(y_k, y_j) + \alpha\, \mathrm{Re}(e_k, e_j)] = \mathrm{Re}(y, y_j)$$

$$j = 1,2,\dots,n ,$$

(which is equivalent to the conditions $\partial\Delta_n(P_n)/\partial\alpha_j = 0$, $j = 1,2,\dots,n$), always has a solution.

We now assume $e_1, e_2, \dots, e_n$ are the first n elements of a sequence $\{e_k\}$ which is complete in $D(A)$.

<u>Theorem 3.2</u>. *The following relationship is valid:*

(3.6)
$$\lim_{n\to\infty} \Delta_n(\overset{*}{P}_n) = \inf_{x\in D(A)} \{I^{(\alpha)}(x)\} .$$

<u>Proof</u>. It is obvious that $\Delta_{n+1}(\overset{*}{P}_{n+1}) \le \Delta_n(\overset{*}{P}_n)$ and so $\lim_{n\to\infty} \Delta_n(\overset{*}{P}_n) = \overset{*}{I}$ must exist. Let us assume that $\overset{*}{I} - \inf_{x\in D(A)} \{I^{(\alpha)}(x)\} = d > 0$. By the definition of the greatest lower bound, there exists an element $\tilde{x}$ such that $I^{(\alpha)}(\tilde{x}) \le \inf_{x\in D(A)} \{I^{(\alpha)}(x)\} + \epsilon$. Because of the completeness of the system $\{e_k\}$, when $\alpha = 0$ there exist n and α_k such that

$$|I^{(\alpha)}(\tilde{x}) - I^{(\alpha)}\left(\sum_{k=1}^{n} \alpha_k e_k\right)| < \epsilon .$$

Thus

$$\overset{*}{I} - \inf_{x\in D(A)} \{I^{(\alpha)}(x)\} \le I^{(\alpha)}\left(\sum_{k=1}^{n} \alpha_k e_k\right) - I^{(\alpha)}(\tilde{x}) + I^{(\alpha)}(\tilde{x}) - \inf_{x\in D(A)} \{I^{(\alpha)}(x)\}$$

$$\le 2\epsilon .$$

It is sufficient to take $\epsilon < d/2$ in order to obtain a contradiction. It therefore follows that $\overset{*}{I} = \inf_{x\in D(A)} \{I^{(\alpha)}(x)\}$, when $\alpha = 0$. For $\alpha > 0$, we obtain from Theorem 2.1 that there exists $\overset{*}{x}{}^{(\alpha)}$ such that $I^{(\alpha)}(\overset{*}{x}{}^{(\alpha)}) = \inf_{x\in D(A)} \{I^{(\alpha)}(x)\}$. Approximating $\overset{*}{x}{}^{(\alpha)}$ arbitrarily closely by a linear combination of $\{e_k\}$, the proof is established because of the boundedness of the operator A. #

Now we examine the behaviour of

$$\overset{*}{x}_n = \sum_{k=1}^{n} \overset{*}{\alpha}_k e_k,\ \overset{*}{P}_n = \overset{*}{P}_n(\overset{*}{\alpha}_1, \overset{*}{\alpha}_2, \dots, \overset{*}{\alpha}_n),$$

(3.7)

$$\Delta_n(\overset{*}{P}_n) = \inf_{P_n}\{\Delta_n(P_n)\} = \inf_{\alpha_k}\{I^{(\alpha)}(\sum_{k=1}^{n} \alpha_k e_k)\}.$$

Theorem 3.3. *For* $\alpha > 0$

(3.8) $$\|\overset{*}{x}_n - \overset{*}{x}^{(\alpha)}\|^2 \le \frac{I^{(\alpha)}(\overset{*}{x}_n) - I^{(\alpha)}(\overset{*}{x}^{(\alpha)})}{\alpha} \to 0,\ n \to \infty.$$

Proof. From equation (2.8) and Theorem 3.2, we have

$$I^{(\alpha)}(\overset{*}{x}_n) - I^{(\alpha)}(\overset{*}{x}^{(\alpha)}) = \alpha\|\overset{*}{x}_n - \overset{*}{x}^{(\alpha)}\|^2 + \|A(\overset{*}{x}_n - \overset{*}{x}^{(\alpha)})\|^2 \to 0$$

as $n \to \infty$. Hence (3.8) follows directly. #

Let us represent any element $x \in D(A)$ as

(3.9) $$x = x^{(1)} + x^{(2)},\ x^{(1)} \in L(A),\ x^{(2)} \in D(A) \ominus L(A).$$

We denote by $T_n(x) = \sum_{k=1}^{n} t_k(x)\, e_k$ the element of best approximation to x among all linear combinations of the form $\sum_{k=1}^{n} \alpha_k e_k$, and by $I_n(x) = \|T_n(x) - x\|$ the corresponding value of the smallest deviation.

Theorem 3.4. *If the operator A is normally solvable then, for* $\alpha = 0$,

$$\inf\|\overset{*}{x}_n - \overset{o}{x}\| = \|\sum_{k=1}^{n} \overset{*}{\alpha}_k e_k^{(2)} - \overset{*}{x}\| \le \|A^{-1}\|\,\|A\overset{*}{x}_n - A\overset{*}{x}\| \le \|A^{-1}\|\,\|A\|I_n(\overset{*}{x}),$$

(3.10)

where $$I(\overset{o}{x}) = \inf_{x \in D(A)}\{I(x)\},$$

and A^{-1} is the left inverse of A : $D(A^{-1}) = R(A)$, $R(A^{-1}) = D(A) \ominus L(A)$.

Proof. It is easy to see that every element $\overset{o}{x}$ which satisfies $I(\overset{o}{x}) = \inf_{x \in D(A)}\{I(x)\}$ can be represented in the form $\overset{o}{x} = \overset{o}{x}^{(1)} + \overset{o}{x}^{(2)}$, where $\overset{o}{x}^{(2)} = \overset{*}{x}$ ($\overset{*}{x}$ is the solution of $A^*Ax = A^*y$ and $\overset{o}{x}^{(1)} \in L(A)$). Hence

$$\|\overset{*}{x}_n - \overset{o}{x}\|^2 = \|\sum_{k=1}^{n} \overset{*}{\alpha}_k e_k^{(1)} - \overset{o}{x}^{(1)}\|^2 + \|\sum_{k=1}^{n} \overset{*}{\alpha}_k e_k^{(2)} - \overset{*}{x}\|^2$$

from which it obviously follows that

$$\inf_{\overset{o}{x}} \|\overset{*}{x}_n - \overset{o}{x}\| = \|\sum_{k=1}^{n} \overset{*}{\alpha}_k e_k^{(2)} - \overset{*}{x}\| .$$

It remains to note that

$$\|\sum_{k=1}^{n} \overset{*}{\alpha}_k e_k^{(2)} - \overset{*}{x}\| = \|A^{-1}A(\sum_{k=1}^{n} \overset{*}{\alpha}_k e_k^{(2)} - \overset{*}{x})\|$$

$$\leq \|A^{-1}\| \|A \overset{*}{x}_n - A \overset{*}{x}\| \leq \|A^{-1}\| \|AT_n(\overset{*}{x}) - A\overset{*}{x}\|$$

$$\leq \|A^{-1}\| \|A\| I_n(\overset{*}{x}). \quad \#$$

An important role in many relationships is played by the quantity

$$m_n = \min_{\alpha_k} \|\sum_{k=1}^{n} \alpha_k y_k\| , \; y_k = Ae_k , \tag{3.11}$$

$$\|\sum_{k=1}^{n} \alpha_k e_k\| = 1 .$$

If we denote by $\bar{m}_n$ the reciprocal of the norm of the inverse matrix of (3.5), then from [8], $m_n = \bar{m}_n^{1/2}$.

Theorem 3.5. *For $\alpha = 0$, the estimate*

$$\|\overset{*}{x}_n - \overset{*}{x}\| \leq (1 + 2\|A\|/m_n) I_n(\overset{*}{x}) \tag{3.12}$$

is valid.

Proof. From the definition of m_n, we obtain

$$m_n\|\overset{*}{x}_n - T_n(\overset{*}{x})\| \leq \|A(\overset{*}{x}_n - T_n(\overset{*}{x}))\| \leq \|A\| \|T_n(\overset{*}{x}) - \overset{*}{x}\| + \|A(\overset{*}{x}_n - \overset{*}{x})\|$$

$$\leq \|A\| \|T_n(\overset{*}{x}) - \overset{*}{x}\| + \|A(T_n(\overset{*}{x}) - \overset{*}{x})\| \leq 2\|A\| I_n(\overset{*}{x}),$$

and hence

$$\|\overset{*}{x}_n - \overset{*}{x}\| \leq \|\overset{*}{x}_n - T_n(\overset{*}{x})\| + \|T_n(\overset{*}{x}) - \overset{*}{x}\| \leq (1 + 2\|A\|/m_n) I_n(\overset{*}{x}) . \quad \#$$

It should be noted that, if we have found m_n and obtained an estimate of the form $I_n(\overset{*}{x}) \leq \epsilon_n\|\overset{*}{x}\|$, then, for sufficiently small ϵ_n ($2\epsilon_n\|A\|/m_n < 1$), it is not difficult to obtain an estimate of the error from (3.12).

In practice, the method of least squares is applied for the minimization of the functional $I^{\alpha,\epsilon}(x)$ for the specific relationships (2.2) and (2.3). In this case, denoting the minimizing element of the form (3.1) by $\overset{*}{x}_n^{\alpha,\epsilon}$, and repeating the arguments of Theorems 3.1-3.5, we obtain, for $\alpha > 0$,

(3.13)
$$\|\overset{*}{x}_n^{\alpha,\epsilon} - \overset{*}{x}^{\alpha,\epsilon}\|^2 \le \frac{I^{\alpha,\epsilon}(\overset{*}{x}_n^{\alpha,\epsilon}) - I^{\alpha,\epsilon}(\overset{*}{x}^{\alpha,\epsilon})}{\alpha},$$
$$\|A_\epsilon(\overset{*}{x}_n^{\alpha,\epsilon} - \overset{*}{x}^{\alpha,\epsilon})\|^2 \le I^{\alpha,\epsilon}(\overset{*}{x}_n^{\alpha,\epsilon}) - I^{\alpha,\epsilon}(\overset{*}{x}^{\alpha,\epsilon}) ;$$

and, for $\alpha = 0$,

(3.14)
$$\inf\|\overset{*}{x}_n^{0,\epsilon} - \overset{\circ}{x}\| \le \|A_\epsilon^{-1}\|\,\|A_\epsilon\|\, I_n(\overset{*}{x}^{0,\epsilon}) ,$$
$$I^{0,\epsilon}(\overset{\circ}{x}) = \inf_{x\in D(A)} \{I^{0,\epsilon}(x)\}$$
$$\|\overset{*}{x}_n^{0,\epsilon} - \overset{*}{x}^{0,\epsilon}\| \le (1 + 2\|A_\epsilon\|/m_n^{(\epsilon)})I_n(\overset{*}{x}^{0,\epsilon}) ,$$

where

(3.15)
$$m_n^{(\epsilon)} = \inf_{\alpha_k} \frac{\|A_\epsilon(\sum_{k=1}^{n} \alpha_k e_k)\|}{\|\sum_{k=1}^{n} \alpha_k e_k\|} .$$

On the basis of these estimates and the results of §2, we can obtain expressions relating $\overset{*}{x}_n^{\alpha,\epsilon}$ and $\overset{*}{x}$. In other words, we can determine α- and β- approximations with a given accuracy δ (see Definitions 1 and 2 of §2), for the method of least squares.

Theorem 3.6. *Let the equation* (2.1) *have the solution* $\overset{*}{x} = A^*\overset{*}{w}$. *Then the following estimates are valid for* $\alpha > 0$*:*

(3.16)
$$\|\overset{*}{x}_n^{\alpha,\epsilon} - \overset{*}{x}\| \le \left\{\frac{I^{\alpha,\epsilon}(\overset{*}{x}_n^{\alpha,\epsilon}) - I^{\alpha,\epsilon}(\overset{*}{x}^{\alpha,\epsilon})}{\alpha}\right\}^{1/2} + C_1^{(\epsilon)}\frac{\epsilon}{\alpha} + C_2\alpha^{1/2} ,$$
$$\|A(\overset{*}{x}_n^{\alpha,\epsilon} - \overset{*}{x})\| \le \{I^{\alpha,\epsilon}(\overset{*}{x}_n^{\alpha,\epsilon}) - I^{\alpha,\epsilon}(\overset{*}{x}^{\alpha,\epsilon})\}^{1/2} + C_1^{(\epsilon)}\epsilon\alpha^{-1/2} + 2C_2\alpha ,$$

where

(3.17)
$$C_1^{(\epsilon)} = \|y_\epsilon\| + \epsilon + \|A_\epsilon\| + (\epsilon + 2\|A_\epsilon\|)\left[\|\overset{*}{x}_n^{\alpha,\epsilon}\| + \left\{\frac{I^{\alpha,\epsilon}(\overset{*}{x}_n^{\alpha,\epsilon}) - I^{\alpha,\epsilon}(\overset{*}{x}^{\alpha,\epsilon})}{\alpha}\right\}^{1/2}\right],$$
$$C_2 \le \|\overset{*}{w}\|/2 ;$$

and, for $\alpha = 0$:

(3.18)
$$\inf\|\overset{*}{x}_n^{0,\epsilon} - \overset{\circ}{x}\| \le \|A_\epsilon^{-1}\| \frac{I^{(\epsilon)}(\overset{*}{x}_n^{0,\epsilon}) + \epsilon(1 + \|A_\epsilon^{-1}\|\,\|y_\epsilon\|)}{1 - \epsilon\|A_\epsilon^{-1}\|} ,$$
$$I(\overset{\circ}{x}) = \inf_{x\in D(A)} \{I(x)\} ,$$

$$\|\overset{*}{x}_n^{0,\epsilon} - \overset{*}{x}\| \le (1 + 2\|A_\epsilon\|/m_n^{(\epsilon)})I_n(\overset{*}{x}) + 2\epsilon(1 + \|\overset{*}{x}\|)/m_n^{(\epsilon)} \quad ,$$

$$\|A(\overset{*}{x}_n^{0,\epsilon} - \overset{*}{x})\| \le I^{(\epsilon)}(\overset{*}{x}_n^{0,\epsilon}) + \epsilon(1 + \|\overset{*}{x}_n^{0,\epsilon}\|) \quad .$$

Proof. The estimates (3.16) are consequences of the inequalities (3.13), (2.12), (2.15), and (2.25). In addition,it is necessary to note that the estimate (3.17) for $C_1^{(\epsilon)}$ is obtained by replacing $\|y\|$, $\|A\|$ and $\|\overset{*}{x}^{\alpha,\epsilon}\|$ in the square brackets of (2.12) by $\|y_\epsilon\|+\epsilon$, $\|A_\epsilon'\|+\epsilon$ and $\|\overset{*}{x}_n^{\alpha,\epsilon}\| + \{[I^{\alpha,\epsilon}(\overset{*}{x}_n^{\alpha,\epsilon})-I^{\alpha,\epsilon}(\overset{*}{x}^{\alpha,\epsilon})]/\alpha\}^{\frac{1}{2}}$ respectively. The validity of the last replacement follows from (3.13). It should also be noted that the term $\epsilon C_1^{(\epsilon)}/\alpha^{\frac{1}{2}}$ in the second estimate (3.16) is obtained from the results

$$\begin{aligned}
\|A(\overset{*}{x}^{(\alpha)} - \overset{*}{x}^{\alpha,\epsilon})\|^2 &= (A^*A(\overset{*}{x}^{(\alpha)} - \overset{*}{x}^{\alpha,\epsilon}),\ (\overset{*}{x}^{(\alpha)} - \overset{*}{x}^{\alpha,\epsilon}))\\
&= -\alpha\|\overset{*}{x}^{(\alpha)} - \overset{*}{x}^{\alpha,\epsilon}\|^2 + (A^*y - A_\epsilon^*y_\epsilon + (A_\epsilon^*A_\epsilon - A^*A)\overset{*}{x}^{\alpha,\epsilon},\ (\overset{*}{x}^{(\alpha)} - \overset{*}{x}^{\alpha,\epsilon}))\\
&\le (A^*y - A_\epsilon^*y_\epsilon + (A_\epsilon^*A_\epsilon - A^*A)\overset{*}{x}^{\alpha,\epsilon},\ (\overset{*}{x}^{(\alpha)} - \overset{*}{x}^{\alpha,\epsilon}))\\
&\le \frac{1}{\alpha}\{[\|y\| + \|A_\epsilon\| + (\|A_\epsilon\| + \|A\|)\overset{*}{x}^{\alpha,\epsilon}\|]\}^2 \quad ,
\end{aligned}$$

which follow readily from the conditions

$$\alpha\overset{*}{x}^{\alpha,\epsilon} + A_\epsilon^*A_\epsilon\,\overset{*}{x}^{\alpha,\epsilon} = A_\epsilon^*\,y_\epsilon \ , \qquad \alpha\overset{*}{x}^{(\alpha)} + A^*A\overset{*}{x}^{(\alpha)} = A^*y \ .$$

The first estimate (3.18) is a consequence of (2.30). For the proof of the second estimate, we note that

$$\begin{aligned}
m_n^{(\epsilon)}\|\overset{*}{x}_n^{0,\epsilon} - T_n(\overset{*}{x})\| &\le \|A_\epsilon(\overset{*}{x}_n^{0,\epsilon} - T_n(\overset{*}{x}))\|\\
&\le \|A_\epsilon(\overset{*}{x}_n^{0,\epsilon} - \overset{*}{x}^{(\epsilon)})\| + \|A_\epsilon(\overset{*}{x}^{(\epsilon)} - \overset{*}{x})\| + \|A_\epsilon\|I_n(\overset{*}{x})\\
&\le 2\|A_\epsilon\|\,I_n(\overset{*}{x}) + 2\|A_\epsilon(\overset{*}{x}^{(\epsilon)} - \overset{*}{x})\|\\
&\le 2\|A_\epsilon\|\,I_n(\overset{*}{x}) + 2\epsilon\ (1 + \|\overset{*}{x}\|) \ .
\end{aligned}$$

Hence

$$\|\overset{*}{x}_n^{0,\epsilon} - T_n(\overset{*}{x})\| \le \frac{2\|A_\epsilon\|}{m_n^{(\epsilon)}}\ I_n(\overset{*}{x}) + \frac{2\epsilon(1 + \|\overset{*}{x}\|)}{m_n^{(\epsilon)}}$$

and

$$\begin{aligned}
\|\overset{*}{x}_n^{0,\epsilon} - \overset{*}{x}\| &\le \|\overset{*}{x}_n^{0,\epsilon} - T_n(\overset{*}{x})\| + I_n(\overset{*}{x})\\
&\le \left(1 + \frac{2\|A_\epsilon\|}{m_n^{(\epsilon)}}\right)I_n(\overset{*}{x}) + \frac{2\epsilon(1 + \|\overset{*}{x}\|)}{m_n^{(\epsilon)}} \quad .
\end{aligned}$$

The last estimate (3.18) follows readily from the inequality

$$\|A\overset{*}{x}{}_n^{0,\epsilon} - y\| \le \|A_\epsilon \overset{*}{x}{}_n^{0,\epsilon} - y_\epsilon\| + \|(A_\epsilon - A)\overset{*}{x}{}_n^{0,\epsilon}\| + \|y - y_\epsilon\| .$$

This completes the proof of the theorem. #

The estimates (3.16) will be useful if the quantities $I_{n,\alpha,\epsilon} = I^{\alpha,\epsilon}(\overset{*}{x}{}_n^{\alpha,\epsilon}) - I^{\alpha,\epsilon}(\overset{*}{x}{}^{\alpha,\epsilon})$ and $\|\overset{*}{w}\|$, can be estimated. From (2.3), using (2.4) and (3.2) (in the last system we replace A,y by A_ϵ,y_ϵ respectively) we obtain

$$I^{\alpha,\epsilon}(\overset{*}{x}{}^{\alpha,\epsilon}) = (y_\epsilon - A_\epsilon\overset{*}{x}{}^{\alpha,\epsilon}, y_\epsilon) \tag{3.19}$$

$$I^{\alpha,\epsilon}(\overset{*}{x}{}_n^{\alpha,\epsilon}) = (y_\epsilon - A_\epsilon\overset{*}{x}{}_n^{\alpha,\epsilon}, y_\epsilon) ,$$

whence

$$I_{n,\alpha,\epsilon} = (A_\epsilon(\overset{*}{x}{}^{\alpha,\epsilon} - \overset{*}{x}{}_n^{\alpha,\epsilon}), y_\epsilon) . \tag{3.20}$$

In practice, $I_{n,\alpha,\epsilon}$ may be evaluated approximately from the relation

$$I_{n,\alpha,\epsilon} \approx (A_\epsilon(\overset{*}{x}{}_{n+p}^{\alpha,\epsilon} - \overset{*}{x}{}_n^{\alpha,\epsilon}), y_\epsilon),\ p = 1,2,\dots , \tag{3.21}$$

although more precisely it is possible to assert only that

$$I^{\alpha,\epsilon}(\overset{*}{x}{}_n^{\alpha,\epsilon}) - I^{\alpha,\epsilon}(\overset{*}{x}{}_{n+p}^{\alpha,\epsilon}) \le I_{n,\alpha,\epsilon} \le I^{\alpha,\epsilon}(\overset{*}{x}{}_n^{\alpha,\epsilon}) . \tag{3.22}$$

Following the method for the approximate estimation of $\overset{*}{w}$ which is mentioned in [157, 153, 138], we apply the method of least squares to the solution of the equation $A^*\overset{*}{w} = \overset{*}{x}$, where $A\overset{*}{x} \equiv y$.(12) The approximate solution of this equation will be sought in the form

$$w_n = \sum_{k=1}^{n} \beta_k e_k, \qquad e_k \in D(A^*) , \tag{3.23}$$

by minimizing the quadratic functional

$$\phi(Q_n) = \|\overset{*}{x} - \sum_{k=1}^{n} \beta_k z_k\|^2 , \ z_k = A^* e_k,\ Q_n = Q_n(\beta_1,\beta_2,\dots,\beta_n) . \tag{3.24}$$

The unknowns β_k are solutions of the system

$$\sum_{k=1}^{n} \beta_k \operatorname{Re}(z_k, z_j) = \operatorname{Re}(\overset{*}{x}, z_j) \equiv \operatorname{Re}(y, e_j) ,\ j = 1,2,\dots,n , \tag{3.25}$$

and so w_n can be found in explicit form, but the quantity

(12) The approximate estimation of $\overset{*}{w}$ can also be obtained by solving the equation $AA^*w = y$ by the same general scheme used for equation (2.1).

(3.26) $$\delta_n = \|\overset{*}{x} - A^* w_n\|$$

remains unknown. We define $\{w_n^{(\epsilon)}\}$ to be the sequence obtained analogously to $\{w_n\}$, with y replaced by y_ϵ and z_k replaced by $z_k^{(\epsilon)} = A_\epsilon^* e_k$ in (3.25). In order to obtain an estimate of C_2, we may substitute $w_n^{(\epsilon)}$ in (3.17) instead of $\overset{*}{w}$. For sharp estimation it is sufficient to have the estimate in the form $\delta_n = \|\overset{*}{x} - A^* w_n\| \leq \epsilon_n \|\overset{*}{x}\|$, where ϵ_n is a sufficiently small known constant. This type of estimate must be examined individually for each class of operator equations. It should also be noted that $A_\epsilon^* w_n^{(\epsilon)}$ by itself can be accepted as an approximate solution of (2.1).

When applying (3.16) for fixed ϵ, it is not necessary to take the parameter α to be arbitrarily small. Some practical recommendations on the choice of α are contained in [131, 132].

If the operator A is normally solvable, it is possible to put $\alpha = 0$ and apply the estimates (3.18). From the second estimate (3.18), it is easy to obtain an effective estimate if $I_n(\overset{*}{x}) \leq \epsilon_n \|\overset{*}{x}\|$, where ϵ_n is a sufficiently small known constant, which is again only possible for particular equations (see §12).

We now examine the case when A is not necessarily restricted to be a closed operator. We apply the method of least squares as above, i.e. we find the approximate solution in the form of (3.1) under the condition that it minimizes the function (3.2). Instead of Theorem 3.2 we now have the following:

Theorem 3.7. *The relationship* (3.6) *is valid:*

a) *for* $\alpha = 0$, *if the sequence* $\{Ae_k\}$ *is complete in* $R(A)$;
b) *for* $\alpha > 0$, *if* $e_k = A^* w_k$ *and* $\{w_k\}$ *is a basis of the matrix representation for the operator* AA^* (*see* [3, *p.*148]).

Proof. The case a) is proved in almost the same way as Theorem 3.2. For $\alpha > 0$, by Theorem 2.10, there exists $\overset{*}{x}{}^{(\alpha)}$ such that $I^{(\alpha)}(\overset{*}{x}{}^{(\alpha)}) = \inf_{x \in D(A)} \{I^{(\alpha)}(x)\}$, where $\overset{*}{x}{}^{(\alpha)} = A^* \overset{*}{w}{}^{(\alpha)}$ and $\overset{*}{w}{}^{(\alpha)} \in D(AA^*)$. Therefore, taking into account that $e_k = A^* w_k$, we obtain

$$I^{(\alpha)}\Big(\sum_{k=1}^{n} \alpha_k e_k\Big) - I^{(\alpha)}(\overset{*}{x}{}^{(\alpha)}) = \alpha\|\overset{*}{x}{}^{(\alpha)} - \sum_{k=1}^{n} \alpha_k e_k\|^2 + \|A\overset{*}{x}{}^{(\alpha)} - \sum_{k=1}^{n} \alpha_k Ae_k\| \tag{3.27}$$

$$= \alpha\Big(AA^*\big(\overset{*}{w}{}^{(\alpha)} - \sum_{k=1}^{n} \alpha_k w_k\big), \big(\overset{*}{w}{}^{(\alpha)} - \sum_{k=1}^{n} \alpha_k w_k\big)\Big) + \|AA^*\big(\overset{*}{w}{}^{(\alpha)} - \sum_{k=1}^{n} \alpha_k w_k\big)\|^2$$

$$\leq \alpha\|AA^*\big(\overset{*}{w}{}^{(\alpha)} - \sum_{k=1}^{n} \alpha_k w_k\big)\| \times \|\overset{*}{w}{}^{(\alpha)} - \sum_{k=1}^{n} \alpha_k w_k\| + \|AA^*\big(\overset{*}{w}{}^{(\alpha)} - \sum_{k=1}^{n} \alpha_k w_k\big)\|^2 .$$

Under the given conditions, there exists a subsequence $\psi_j = \sum_{k=1}^{n_j} \alpha_k w_k$ such that $\|\overset{*}{w}^{(\alpha)} - \psi_j\|$ and $\|AA^*(\overset{*}{w}^{(\alpha)} - \psi_j)\|$ converge to 0 simultaneously as $j \to \infty$. Hence, from (3.27), we obtain the required result. #

We show that under some very general assumptions about the determination of α- and β- approximate solutions, it is possible to do without the method of regularization.

We impose on the operators A and A_ϵ weaker restrictions than (2.2) and (2.34), namely

$$(3.28) \qquad \|(A - A_\epsilon)x\| \leq \epsilon(x)\ \|x\|,\ x \in D(A),\ \epsilon(x) \to 0 \text{ as } \epsilon \to 0\ ,$$

for arbitrary $x \in D(A)$. As before we assume that $\|y - y_\epsilon\| < \epsilon$.

We denote by $\overset{*}{x}{}_n^{0,\epsilon} = \sum_{k=1}^{n} \overset{*}{\alpha}{}_k^{0,\epsilon} e_k$ the element which minimizes $\|y_\epsilon - \sum_{k=1}^{n} \alpha_k y_k^{(\epsilon)}\|$, $y_k^{(\epsilon)} = A_\epsilon e_k$.

<u>Theorem 3.8</u>. *If the system $\{y_k\}_1^\infty$ is complete in $R(A)$ and y_k (where $k = 1,2,\ldots,n$) are linearly independent, then for every $\delta > 0$ there exist $n(\delta)$ and $\epsilon(\delta)$ such that*

$$\|A\overset{*}{x}{}_n^{0,\epsilon} - y\| \leq \inf_{x\in D(A)} \|Ax - y\| + \delta$$

i.e., $\overset{*}{x}{}_n^{0,\epsilon}$ is a β- approximate solution of the equation $Ax = y$ with accuracy δ.

<u>Proof</u>. Let x_η be a β- approximate solution of the equation with accuracy η, $\overset{*}{x}_n = \sum_{k=1}^{n} \overset{*}{\alpha}_k e_k$ and $\overset{*}{x}_{\eta n} = \sum_{k=1}^{n} \overset{*}{\alpha}_{\eta k} e_k$ be the elements which realize the least values respectively of $\|y - \sum_{k=1}^{n} \alpha_k y_k\|$ and $\|Ax_\eta - \sum_{k=1}^{n} \alpha_k y_k\|$, where $y_k = Ae_k$. Let us write

$$(3.30) \qquad \epsilon_n = \max\,[\epsilon,\ \epsilon(e_1),\ldots,\epsilon(e_n)]\ ,$$

$$a_n = \max_{1\leq k\leq n} \|e_k\|,\ b_n = \max_{1\leq k\leq n} \|y_k\|\ .$$

Then we obtain

$$\|A\overset{*}{x}{}_n^{0,\epsilon} - y\| \leq \epsilon + \|A\overset{*}{x}{}_n^{0,\epsilon} - y_\epsilon\|$$

$$\leq \epsilon + \|A_\epsilon \overset{*}{x}{}_n^{0,\epsilon} - y_\epsilon\| + \|\sum_{k=1}^{n} \overset{*}{\alpha}{}_k^{0,\epsilon}(A_\epsilon - A)e_k\|$$

$$\leq \epsilon + \epsilon_n a_n \sum_{k=1}^{n} |\overset{*}{\alpha}{}_k^{0,\epsilon}| + \|A_\epsilon \overset{*}{x}_{\eta n} - y_\epsilon\|$$

$$\leq 2\epsilon + \epsilon_n a_n \sum_{k=1}^{n} |\overset{*}{\alpha}{}_k^{0,\epsilon} - \overset{*}{\alpha}_k| + \epsilon_n a_n \sum_{k=1}^{n} |\overset{*}{\alpha}_k| + \|Ax_\eta - y\| +$$

$$+ \|A x_\eta - A \overset{*}{x}_{\eta n}\| + \|\sum_{k=1}^{n} \overset{*}{\alpha}_{\eta k} (A - A_\epsilon) e_k\|$$

$$\leq \epsilon_n[2 + a_n \sum_{k=1}^{n} |\overset{*}{\alpha}{}_k^{0,\epsilon} - \overset{*}{\alpha}_k| + a_n(\sum_{k=1}^{n} |\overset{*}{\alpha}_k| + \sum_{k=1}^{n} |\overset{*}{\alpha}_{\eta k}|)] +$$

$$+ \|A x_\eta - A \overset{*}{x}_{\eta n}\| + \eta + \inf_{x \in D(A)} \|Ax - y\| .$$

Because $\{y_k\}_1^\infty$ is complete in $R(A)$, we may choose an $n(\delta)$ and $\eta(\delta)$ such that $\|A x_\eta - A \overset{*}{x}_{\eta n}\| \leq \delta/3$. To estimate $\sum_{k=1}^{n} |\overset{*}{\alpha}{}_k^{0,\epsilon} - \overset{*}{\alpha}_k|$, we compare the two systems of equations:

$$\sum_{k=1}^{n} \overset{*}{\alpha}_k \operatorname{Re}(y_k, y_j) = \operatorname{Re}(y, y_j), \; j = 1,2,\ldots,n ,$$

and

$$\sum_{k=1}^{n} \overset{*}{\alpha}{}_k^{0,\epsilon} \operatorname{Re}(y_k^{(\epsilon)}, y_j^{(\epsilon)}) = \operatorname{Re}(y_\epsilon, y_j^{(\epsilon)}), \; j = 1,2,\ldots,n .$$

It is easy to show that

$$\text{(3.31)} \qquad |\operatorname{Re}(y_k, y_j) - \operatorname{Re}(y_k^{(\epsilon)}, y_j^{(\epsilon)})| \leq \epsilon_n a_n (2b_n + \epsilon_n a_n) \to 0, \; \epsilon \to 0,$$

$$|\operatorname{Re}(y, y_j) - \operatorname{Re}(y_\epsilon, y_j^{(\epsilon)})| \leq \epsilon_n(b_n + \|y\|a_n + \epsilon_n a_n) \to 0, \; \epsilon \to 0.$$

Due to the linear independence of the y_k (where $k = 1,2,\ldots,n$) the first system has a unique solution. Consequently, for sufficiently small ϵ, the second system also has a unique solution, so that

$$\sum_{k=1}^{n} |\overset{*}{\alpha}_k - \overset{*}{\alpha}{}_k^{0,\epsilon}| \to 0 \text{ as } \epsilon \to 0 .$$

To complete the proof it remains to choose $\epsilon(\delta)$ such that the condition

$$2\epsilon_n + \epsilon_n a_n \sum_{k=1}^{n} |\overset{*}{\alpha}_k - \overset{*}{\alpha}_k^{0,\epsilon}| + \epsilon_n a_n \left(\sum_{k=1}^{n} |\overset{*}{\alpha}_k| + \sum_{k=1}^{n} |\overset{*}{\alpha}_{\eta_k}|\right) \leq \delta/3 ,$$

is satisfied. #

Remark 1. In the case when $y \in R(A)$ and $\inf_{x \in D(A)} \|Ax - y\| = 0$, the simple estimate

(3.32)
$$\|A \overset{*}{x}_n^{0,\epsilon} - y\| \leq \|(A - A_\epsilon)\overset{*}{x}_n^{0,\epsilon}\| + \|A_\epsilon \overset{*}{x}_n^{0,\epsilon} - y\|$$
$$\leq \epsilon + \epsilon_n a_n \sum_{k=1}^{n} |\overset{*}{\alpha}_k^{0,\epsilon}| + \|y_\epsilon - A_\epsilon \overset{*}{x}_n^{0,\epsilon}\| ,$$

can give a practically effective method for estimation of the error of the β- approximate solution.

Remark 2. If $e_1, e_2, \ldots, e_n$ are linearly independent and the estimate (2.2) is valid, then a sufficient condition for the linear independence of y_k (where $k = 1,2,\ldots, n$) is that

(3.33)
$$m_n^{(\epsilon)} = \inf_{\alpha_k} \frac{\|A_\epsilon (\sum_{k=1}^{n} \alpha_k e_k)\|}{\|\sum_{k=1}^{n} \alpha_k e_k\|} > \epsilon \geq 0 .$$

In fact,

$$\|\sum_{k=1}^{n} \alpha_k y_k\| \geq \|A_\epsilon \sum_{k=1}^{n} \alpha_k e_k\| - \|(A - A_\epsilon) \sum_{k=1}^{n} \alpha_k e_k\|$$

$$\geq (m_n^{(\epsilon)} - \epsilon) \|\sum_{k=1}^{n} \alpha_k e_k\| > 0 ,$$

and the system $\{y_k\}_1^n$ cannot be linearly dependent.

Following the method which is given in [157], it is possible to prove a theorem which is analogous to Theorem 3.8 with respect to the α- approximate solution.

Let $\overset{*}{x}$ be such that $\|A \overset{*}{x} - y\| = \inf_{x \in D(A)} \|A x - y\|$

and

(3.34)
$$\|(A^* - A_\epsilon^*)z\| \leq \epsilon^*(z) \|z\|, \|y - y_\epsilon\| \leq \epsilon ,$$

where $\epsilon^*(z) \to 0$ as $\epsilon \to 0$ for every fixed $z \in D(A^*)$.

Theorem 3.9. *If the system $\{z_k\}_1^\infty$, where $z_k = A^* e_k$, is complete in $D(A)$ and z_k (where $k = 1,2,\ldots,n$) are linearly independent, then for every $\delta > 0$ there exist $n(\delta)$ and $\epsilon(\delta)$ such that*

$$(3.35)\qquad \|\overset{*}{x} - \sum_{k=1}^{n} \overset{*}{\beta}{}_k^{(\epsilon)}\, z_k^{(\epsilon)}\| \le \delta,\ z_k^{(\epsilon)} = A^* e_k ,$$

where $\overset{*}{\beta}{}_k^{(\epsilon)}$ *is the solution of the linear algebraic system*

$$\sum_{k=1}^{n} \beta_k \operatorname{Re}(z_k^{(\epsilon)}, z_j^{(\epsilon)}) = \operatorname{Re}(y_\epsilon, e_j),\ j = 1,2,\dots,n .$$

Proof. Let us write

$$(3.37)\qquad \epsilon_n^* = \max\,[\epsilon, \epsilon^*(e_1), \epsilon^*(e_2),\dots,\epsilon^*(e_n)] ,$$

$$c_n = \max_{1\le k\le n} \|z_k\|,\ z_k = A^* e_k .$$

Then

$$\|\overset{*}{x} - \sum_{k=1}^{n} \overset{*}{\beta}{}_k^{(\epsilon)} z_k^{(\epsilon)}\| \le \|\overset{*}{x} - \sum_{k=1}^{n} \overset{*}{\beta}_k z_k\| + \|\sum_{k=1}^{n} \overset{*}{\beta}_k z_k - \sum_{k=1}^{n} \overset{*}{\beta}{}_k^{(\epsilon)} z_k^{(\epsilon)}\|$$

$$\le \|\overset{*}{x} - \sum_{k=1}^{n} \overset{*}{\beta}_k z_k\| + \|\sum_{k=1}^{n} \overset{*}{\beta}_k (z_k - z_k^{(\epsilon)})\| +$$

$$+ \|\sum_{k=1}^{n} (\overset{*}{\beta}_k - \overset{*}{\beta}{}_k^{(\epsilon)})(z_k^{(\epsilon)} - z_k)\| + \|\sum_{k=1}^{n} (\overset{*}{\beta}_k - \overset{*}{\beta}{}_k^{(\epsilon)}) z_k\|$$

$$\le \|\overset{*}{x} - \sum_{k=1}^{n} \overset{*}{\beta}_k z_k\| + \epsilon_n^* a_n \sum_{k=1}^{n} |\overset{*}{\beta}_k| + \epsilon_n^* a_n \sum_{k=1}^{n} |\overset{*}{\beta}_k - \overset{*}{\beta}{}_k^{(\epsilon)}|$$

$$+ c_n \sum_{k=1}^{n} |\overset{*}{\beta}_k - \overset{*}{\beta}{}_k^{(\epsilon)}| .$$

Due to the completeness of $\{z_k\}_1^\infty$ in $D(A)$, it follows that there exist an $n(\delta)$ and $\overset{*}{\beta}_k$ such that $\|\overset{*}{x} - \sum_{k=1}^{n} \overset{*}{\beta}_k z_k\| \le \delta/2$. In particular, as $\overset{*}{\beta}_k$ we may take the solution of the system (3.25).

We now prove that $\sum_{k=1}^{n} |\overset{*}{\beta}_k - \overset{*}{\beta}{}_k^{(\epsilon)}| \to 0$ as $\epsilon \to 0$.

Since

$$(3.38)\qquad |(z_k, z_j) - (z_k^{(\epsilon)}, z_j^{(\epsilon)})| \le \epsilon_n^* a_n(2 c_n + \epsilon_n^* a_n) \to 0,\ \epsilon \to 0,$$

$$|(y_\epsilon, e_j) - (y, e_j)| \le \epsilon_n^* a_n \to 0,\ \epsilon \to 0.$$

We obtain the required conclusion by using reasoning analogous to that of the previous theorem when estimating $\sum_{k=1}^{n} |\overset{*}{\alpha}_k - \overset{*}{\alpha}{}_k^{0,\epsilon}|$.

To complete the proof we must now choose $\epsilon(\delta)$ such that

$$\epsilon_n^* a_n \left(\sum_{k=1}^{n} |\overset{*}{\beta}_k| + \sum_{k=1}^{n} |\overset{*}{\beta}_k - \overset{*}{\beta}_k^{(\epsilon)}| \right) + c_n \sum_{k=1}^{n} |\overset{*}{\beta}_k - \overset{*}{\beta}_k^{(\epsilon)}| \leq \delta/2 . \quad \#$$

§3.2 An Orthogonalization Algorithm

We assume that $\{e_k\}$ is an orthonormal system. Then

(3.39) $$\Delta_n(P_n) = I^{(\alpha)}(x_n) = \|y - \sum_{k=1}^{n} \alpha_k y_k\|^2 + \alpha \sum_{k=1}^{n} \alpha_k^2 , \qquad y_k = Ae_k .$$

We first examine the case when $\alpha = 0$.

In order to solve the problem of the minimization of the function $\Delta_n(P_n)$, we construct a system of orthogonal vectors $\{\tilde{y}_k\}_1^m$, where $m \leq n$, which is equivalent to the known system $\{y_k\}_1^n$. We use the well-known orthogonalization method due to Schmidt†:

if $y_1 \neq 0$, then we let

$$\tilde{y}_1 = y_1,$$

$$\tilde{y}_2 = -\alpha_1^{(2)}\tilde{y}_1 + y_2 = \beta_1^{(2)} y_1 + y_2; \qquad \alpha_1^{(2)} = \frac{(y_2, \tilde{y}_1)}{(\tilde{y}_1, \tilde{y}_1)} , \qquad \beta_1^{(2)} = -\alpha_1^{(2)} ;$$

if $\tilde{y}_2 \neq 0$, then

$$\tilde{y}_3 = -\alpha_1^{(3)}\tilde{y}_1 - \alpha_2^{(3)}\tilde{y}_2 + y_3 = \beta_1^{(3)} y_1 + \beta_2^{(3)} y_2 + y_3; \qquad \alpha_1^{(3)} = \frac{(y_3, \tilde{y}_1)}{(\tilde{y}_1, \tilde{y}_1)} ,$$

$$\alpha_2^{(3)} = \frac{(y_3, \tilde{y}_2)}{(\tilde{y}_2, \tilde{y}_2)} , \quad \beta_1^{(3)} = -\alpha_1^{(3)} - \alpha_2^{(3)}\beta_1^{(2)} , \quad \beta_2^{(3)} = -\alpha_2^{(3)} \text{ etc} ;$$

if $\tilde{y}_s \neq 0$, then

$$\tilde{y}_{s+1} = -\alpha_1^{(s+1)}\tilde{y}_1 - \alpha_2^{(s+1)}\tilde{y}_2 - \dots - \alpha_s^{(s+1)}\tilde{y}_s + y_{s+1}$$

$$= \beta_1^{(s+1)} y_1 + \beta_2^{(s+1)} y_2 + \dots + \beta_s^{(s+1)} y_s + y_{s+1} ;$$

$$\alpha_1^{(s+1)} = \frac{(y_{s+1}, \tilde{y}_1)}{(\tilde{y}_1, \tilde{y}_1)} , \quad \alpha_2^{(s+1)} = \frac{(y_{s+1}, \tilde{y}_2)}{(\tilde{y}_2, \tilde{y}_2)} , \quad \dots , \quad \alpha_s^{(s+1)} = \frac{(y_{s+1}, \tilde{y}_s)}{(\tilde{y}_s, \tilde{y}_s)} ,$$

$$\beta_1^{(s+1)} = -\alpha_1^{(s+1)} - \alpha_2^{(s+1)}\beta_1^{(2)} - \alpha_3^{(s+1)}\beta_1^{(3)} - \dots - \alpha_s^{(s+1)}\beta_1^{(s)} ,$$

$$\beta_2^{(s+1)} = -\alpha_2^{(s+1)} - \alpha_3^{(s+1)}\beta_2^{(3)} - \alpha_4^{(s+1)}\beta_2^{(4)} - \dots - \alpha_s^{(s+1)}\beta_2^{(s)} , \quad \dots ,$$

$$\dots , \ \beta_s^{(s+1)} = -\alpha_s^{(s+1)} \text{ etc.}$$

If $y_1 = 0$, then y_2 plays the role of y_1; if $y_2 = 0$, then y_3 plays

† Often referred to as the (classical) Gram-Schmidt method. (Eds.)

the role of y_2 etc. It is not difficult to show that, beginning with a certain n, all the $\{y_k\}_1^n$ cannot be equal to zero if A is not the zero operator, and if the system $\{Ae_k\}$ is everywhere dense in $R(A)$. In fact, assuming the contrary, we obtain that a certain element $Ax \neq 0$ must be approximated arbitrarily closely by a linear combination of zero elements, which is not possible.

If $\tilde{y}_s = 0$ and $\tilde{y}_{s+1} \neq 0$, then the role of $\tilde{y}_s$ is played by $\tilde{y}_{s+1}$; if $\tilde{y}_s = \tilde{y}_{s+1} = 0$ then the role of $\tilde{y}_s$ is played by $\tilde{y}_{s+2}$ etc. It is easy to show that $\tilde{y}_s = 0$ is a necessary and sufficient condition for the linear dependence of $y_1, y_2, \ldots, y_{s-1}, y_s$. Indeed, from the description of Schmidt's method, it follows that $\tilde{y}_s = 0$, implies the linear dependence of $y_1, y_2, \ldots, y_{s-1}, y_s$. Conversely, if y_s depends linearly on $y_1, y_2, \ldots, y_{s-1}$, then $\tilde{y}_s = 0$, since it may easily be shown that $\alpha_1^{(s-1)} \tilde{y}_1 + \alpha_2^{(s-1)} \tilde{y}_2 + \ldots + \alpha_{s-1}^{(s-1)} \tilde{y}_{s-1}$ is the best approximation to y_s among all linear combinations of $\tilde{y}_1, \tilde{y}_2, \ldots, \tilde{y}_{s-1}$.

Hence, taking into account that some of the y_k and $\tilde{y}_s$ may be equal to zero and that we can apply the method of Schmidt by sorting through the elements in any order, we obtain the following identity:

$$y - \sum_{s=1}^{n} \alpha_{k_s} y_{k_s} \equiv y - \sum_{s=1}^{m} d_{k_s} \tilde{y}_{k_s} \quad , \quad m \le n \quad , \tag{3.40}$$

whence it is easy to express α_{k_s} in terms of d_{k_s}, and conversely. For simplicity, we identify α_{k_s} and d_{k_s} with α_k and d_k respectively and assume that $m = n$. Then we find:

$$y_s = \sum_{k=1}^{s} \alpha_k^{(s)} \tilde{y}_k \,, \quad \alpha_s^{(s)} = 1, \quad \alpha_k^{(s)} = \frac{(y_s, \tilde{y}_k)}{(\tilde{y}_k, \tilde{y}_k)} \quad , \quad k = 1,2,\ldots,s-1 \,,$$

$$\tilde{y}_s = \sum_{k=1}^{s} \beta_k^{(s)} y_k \,, \quad \beta_s^{(s)} = 1, \quad \beta_k^{(s)} = - \sum_{j=k}^{s-1} \alpha_j^{(s)} \beta_k^{(j)} \,, \quad k = 1,2,\ldots,s-1 \,,$$

$$\overset{*}{d}_s = \sum_{j=s}^{n} \overset{*}{\alpha}_j \alpha_s^{(j)} \,, \quad \overset{*}{d}_s = \frac{(y, \tilde{y}_s)}{(\tilde{y}_s, \tilde{y}_s)} \,, \quad s = 1,2,\ldots,n \,,$$

$$\overset{*}{\alpha}_s = \sum_{j=s}^{n} \overset{*}{d}_j \beta_s^{(j)} \,, \quad s = 1,2,\ldots,n \,,$$

$$\Delta_n(\overset{*}{P}_n) = \inf_{P_n} \Delta_n(P_n) = \|y\|^2 - \sum_{s=1}^{n} |\overset{*}{d}_s|^2 \, \|\tilde{y}_s\|^2 \,. \tag{3.41}$$

As is obvious from the orthogonalization algorithm, the behaviour of $(\tilde{y}_s, \tilde{y}_s)$, for $s = 1,2,\ldots$, plays a very important role.

Theorem 3.10. *If, for the operator A, the relationship*

(3.42) $$\|Ax\| \geq m\|x\|, \; x \in D(A) ,$$

is valid, then

(3.43) $$(\tilde{y}_s, \tilde{y}_s) \geq m^2, \; s = 1,2,\ldots \; .$$

Proof. We obtain

$$\|\tilde{y}_s\|^2 = \|y_s\|^2 - \sum_{k=1}^{s-1} |\alpha_k^{(s)}|^2 \|\tilde{y}_k\|^2 = \inf_{\alpha_k} \|y_s - \sum_{k=1}^{s-1} \alpha_k y_k\|^2$$

$$= \inf_{\alpha_k} \|A(e_s - \sum_{k=1}^{s-1} \alpha_k e_k)\|^2 \geq m^2 \inf_{\alpha_k} \|e_s - \sum_{k=1}^{s-1} \alpha_k e_k\|^2 \geq m^2 ,$$

which establishes the theorem. #

More exact and effective estimates are given by the following theorem.

Theorem 3.11. *With m_n as defined by* (3.11), *then*

(3.44) $$\|\tilde{y}_n\|^2 \geq m_n^2 \left(1 + \sum_{k=1}^{n-1} |\beta_k^{(n)}|^2\right), \; n = 1,2,\ldots \; .$$

Proof. Since $\{e_k\}$ is an orthonormal system, we obtain

$$\|\tilde{y}_n\|^2 = \|y_n + \sum_{k=1}^{n-1} \beta_k^{(n)} y_k\|^2 \geq m_n^2 \left(1 + \sum_{k=1}^{n-1} |\beta_k^{(n)}|^2\right) . \quad \#$$

We make two remarks:

1. It is easy to see that $(\tilde{y}_s, y_s) = (\tilde{y}_s, \tilde{y}_s)$, whence

(3.45) $$\|\tilde{y}_s\| \leq \|y_s\| .$$

2. In [93], the *conditioning* of the operator A is defined by

(3.46) $$F(A) = \frac{1}{a} , \; a = \inf_{u \in E} \inf_{(u,x)=1} \|Ax\| ,$$

where E is a certain set of elements from H_1^*.[13] We apply the method of Schmidt for all possible orderings of the indices of the y_k and find the one for which $\|\tilde{y}_n\|$ is minimal. From the results of [93], it follows that if

(13) In [93], conditioning is introduced in the more general case when H_1 is a Banach space. In this case, we mean by (u, x) the value of the functional u at x, where $u \in E \subset H_1^*$, H_1^* being the conjugate space of H_1.

E coincides with the subspace spanned by $e_1, e_2, \ldots, e_n$, then $F(A)$ is the reciprocal of the minimum norm:

$$F(A) = \frac{1}{\|\tilde{y}_n\|} \quad . \tag{3.47}$$

We now consider the case when $\alpha > 0$. We introduce in $D(A)$ the scalar product

$$(x_1, x_2)_1 = (Ax_1, Ax_2) + \alpha(x_1, x_2) \ . \tag{3.48}$$

In this case, the system (3.5) takes the form

$$\sum_{k=1}^{n} \alpha_k \operatorname{Re}(e_k, e_j)_1 = \operatorname{Re}(y, y_j), \ j = 1,2,\ldots,n \ . \tag{3.49}$$

We construct, with respect to the scalar product (3.48), the system $\{\tilde{e}_k\}_1^n$, which is equivalent to $\{e_k\}_1^n$. This is possible using Schmidt's algorithm as described above, except that now we have

$$\alpha_k^{(s)} = \frac{(e_s, \tilde{e}_k)_1}{(\tilde{e}_k, \tilde{e}_k)_1} = \frac{(y_s, \tilde{y}_k) + \alpha(e_s, \tilde{e}_k)}{(\tilde{y}_k, \tilde{y}_k) + \alpha(\tilde{e}_k, \tilde{e}_k)} \quad . \tag{3.50}$$

We can now replace Theorems 3.10 and 3.11 by the following theorem.

Theorem 3.12. *Under the restrictions of Theorems* 3.10 *and* 3.11

$$(\tilde{e}_n, \tilde{e}_n)_1 \geq \alpha + m_n^2 \left(1 + \sum_{k=1}^{n-1} |\beta_k^{(n)}|^2\right) \geq \alpha + m^2, \ n = 1,2,\ldots \ . \tag{3.51}$$

Defining the elements $\tilde{e}_s$, it is possible to find the unknowns $\overset{*}{\alpha}_k$, which minimize $\Delta_n(P_n) = I^{(\alpha)}(x_n)$, in terms of formulae which are analogous to (3.41).

Theorem 3.13. *The unknowns* $\overset{*}{\alpha}_k$ *satisfy*

$$\overset{*}{\alpha}_k = \sum_{s=k}^{n} \overset{*}{d}_s \, \beta_k^{(s)} \ , \quad k = 1,2,\ldots,n \quad , \tag{3.52}$$

where $\overset{*}{d}_s$ *is the solution of the triangular system of equations*

$$d_s(\tilde{e}_s, \tilde{e}_s)_1 + \sum_{k=1}^{s-1} d_k \operatorname{Re}(\tilde{e}_k, e_s)_1 = \operatorname{Re}(y, Ae_s), \ s = 1,2,\ldots,n \ . \tag{3.53}$$

Proof. Substituting (3.52) into (3.49) we obtain,

$$\sum_{k=1}^{n} \sum_{s=k}^{n} \overset{*}{d}_s \, \beta_k^{(s)} \, \mathrm{Re}(e_k, e_j)_1 \equiv \sum_{k=1}^{n} \overset{*}{d}_k \, \mathrm{Re}\left(\sum_{s=1}^{k} \beta_s^{(k)} (e_s, e_j)_1 \right)$$

$$\equiv \sum_{k=1}^{n} \overset{*}{d}_k \, \mathrm{Re}(\tilde{e}_k, e_j)_1 \equiv \sum_{k=1}^{j} \overset{*}{d}_k \, \mathrm{Re}(\tilde{e}_k, e_j)_1$$

$$\equiv \mathrm{Re}(y, Ae_j), \; j = 1,2,\ldots,n \, .$$

Thus the theorem is proved. #

Treated in this way, the case when $\alpha > 0$ does not cause basic difficulties. If $m = 0$, then in the case $\alpha > 0$ we nevertheless have a specific guarantee of the applicability of the orthogonalization algorithm because $\|\tilde{e}_s\|$ is always greater than or equal to α.

As is obvious in the theoretical plan, questions about the construction of the orthogonal system $\{\tilde{y}_k\}_1^m$, which is equivalent to $\{y_k\}_1^m$, are solved very simply. Entirely different conditions hold for the practical realization of Schmidt's method on a computer, when it should be taken into account that the original elements y_k are only given with a known accuracy (i.e. $y_k^{(\epsilon)}$ are known, where $\|y_k^{(\epsilon)} - y_k\| \leq \epsilon$). The calculations are always performed with round-off errors,and approximate formulae are applied for the determination of the required scalar products. For simplicity, the latter errors can always be interpreted as errors in the input data, i.e. $y_k^{(\epsilon)}$ are assumed to be such that required scalar products can be estimated by the exact formulae. Round-off errors will not be examined here†.

The first question with which we have to deal involves a criterion for deciding whether $y_k = 0$. The natural criterion will obviously be $\|y_k^{(\epsilon)}\| < \epsilon$. However, if we leave $y_k^{(\epsilon)}$ as it is when $\|y_k^{(\epsilon)}\| > \epsilon$, we can estimate $\|y_k^{(\epsilon)}\|$ with a relative error which is arbitrarily close to 100%. In order to avoid this, we should take as the criterion for $y_k = 0$ the relationship

(3.54) $$\|y_k^{(\epsilon)}\| < \epsilon/\delta_1, \qquad \delta_1 < 1 \, .$$

An analogous situation occurs for the elements $\tilde{y}_k$. As the criterion for $(\tilde{y}_k, \tilde{y}_k) \neq 0$ we take the relationship[14]

(3.55) $$(\tilde{y}_k^{(\epsilon)}, \tilde{y}_k^{(\epsilon)}) \geq \Delta \, [(\tilde{y}_k^{(\epsilon)}, \tilde{y}_k^{(\epsilon)})]/\delta_2, \qquad \delta_2 < 1 \, .$$

---

(14) We denote the absolute and relative errors of any quantity f by $\Delta[f]$ and $\delta[f]$ respectively.

† An error analysis for classical and modified Gram-Schmidt orthogonalization can be found in Å. Björck, *BIT*, 7(1967)1-21. (Eds.)

We now estimate $\Delta_k = \Delta[(\tilde{y}_k^{(\epsilon)}, \tilde{y}_k^{(\epsilon)})]$ using the equation

$$(\tilde{y}_k, \tilde{y}_k) = (y_k, y_k) - \sum_{s=1}^{k-1} \frac{|(y_k, \tilde{y}_s)|^2}{(\tilde{y}_s, \tilde{y}_s)} . \tag{3.56}$$

We obtain

$$\Delta_k \leq |(y_k, y_k) - (y_k^{(\epsilon)}, y_k^{(\epsilon)})| + \sum_{s=1}^{k-1} \left| \frac{|(y_k, \tilde{y}_s)|^2}{(\tilde{y}_s, \tilde{y}_s)} - \frac{|(y_k^{(\epsilon)}, \tilde{y}_s^{(\epsilon)})|^2}{(\tilde{y}_s^{(\epsilon)}, \tilde{y}_s^{(\epsilon)})} \right|$$

$$\leq \epsilon(\|A\| + \|A_\epsilon\|) + \sum_{s=1}^{k-1} \left| \frac{|(y_k, \tilde{y}_s)|^2 - |(y_k^{(\epsilon)}, \tilde{y}_s^{(\epsilon)})|^2}{(\tilde{y}_s^{(\epsilon)}, \tilde{y}_s^{(\epsilon)})} \right| +$$

$$+ \sum_{s=1}^{k-1} \frac{|(y_k, \tilde{y}_s)|^2}{(\tilde{y}_s, \tilde{y}_s)} \cdot \frac{\Delta_s}{(\tilde{y}_s^{(\epsilon)}, \tilde{y}_s^{(\epsilon)})}$$

$$\leq \epsilon(\|A\| + \|A_\epsilon\|) + \sum_{s=1}^{k-1} \frac{(\epsilon\|A\| + \|A_\epsilon\| \, \|\tilde{y}_s - \tilde{y}_s^{(\epsilon)}\|)(\|A\|^2 + \|A_\epsilon\|^2)}{(\tilde{y}_s^{(\epsilon)}, \tilde{y}_s^{(\epsilon)})} +$$

$$+ \sum_{s=1}^{k-1} \frac{\|A\|^2 \Delta_s}{(\tilde{y}_s^{(\epsilon)}, \tilde{y}_s^{(\epsilon)})} . \tag{3.57}$$

In a similar way, we find that

$$\|\tilde{y}_s - \tilde{y}_s^{(\epsilon)}\| \leq \|y_s - y_s^{(\epsilon)}\| + \sum_{k=1}^{s-1} \left| \frac{(y_s, \tilde{y}_k)}{(\tilde{y}_k, \tilde{y}_k)} \tilde{y}_k - \frac{(y_s^{(\epsilon)}, \tilde{y}_k^{(\epsilon)})}{(\tilde{y}_k^{(\epsilon)}, \tilde{y}_k^{(\epsilon)})} \tilde{y}_k^{(\epsilon)} \right| \tag{3.58}$$

$$\leq \epsilon + \sum_{k=1}^{s-1} \frac{\epsilon\|A\|^2 + (\|A_\epsilon\|^2 + \|A_\epsilon\| \, \|A\|) \, \|\tilde{y}_k - \tilde{y}_k^{(\epsilon)}\|}{(\tilde{y}_k^{(\epsilon)}, \tilde{y}_k^{(\epsilon)})} +$$

$$+ \sum_{k=1}^{s-1} \frac{\|A\| \, \Delta_k}{(\tilde{y}_k^{(\epsilon)}, \tilde{y}_k^{(\epsilon)})} .$$

Taking into account the following simple relationships:

$$\Delta_1 = |(y_1, y_1) - (y_1^{(\epsilon)}, y_1^{(\epsilon)})| \leq \epsilon(\|A\| + \|A_\epsilon\|) , \tag{3.59}$$

$$\|y_1 - y_1^{(\epsilon)}\| \leq \epsilon, \qquad \|A\| \leq \|A_\epsilon\| + \epsilon ,$$

we conclude that (3.57)-(3.59) give the recursion rule for the effective estimation of the desired values of Δ_k.

We denote by $n' = n'(\epsilon, \delta_1, \delta_2)$ the maximum k for which the estimate $\Delta_k \leq (\tilde{y}_k^{(\epsilon)}, \tilde{y}_k^{(\epsilon)})\delta_2$ holds, and we then estimate

$$\Delta_0 = \Delta\left[(y_\epsilon, y_\epsilon) - \sum_{k=1}^{n'} |(y_\epsilon, \tilde{y}_k^{(\epsilon)})|^2/(\tilde{y}_k^{(\epsilon)}, \tilde{y}_k^{(\epsilon)})\right] .$$

By an argument similar to that used for (3.57), we obtain

$$(3.60) \qquad \Delta_0 \leq \epsilon(\|y\| + \|y_\epsilon\|) + \sum_{k=1}^{n'} \left| \frac{|(y, \tilde{y}_k)|^2 - |(y_\epsilon, \tilde{y}_k^{(\epsilon)})|^2}{(\tilde{y}_k^{(\epsilon)}, \tilde{y}_k^{(\epsilon)})} \right| +$$

$$+ \sum_{k=1}^{n'} \frac{|(y, \tilde{y}_k)|^2}{(\tilde{y}_k, \tilde{y}_k)} \cdot \frac{\Delta_k}{(\tilde{y}_k^{(\epsilon)}, \tilde{y}_k^{(\epsilon)})}$$

$$\leq \epsilon(\epsilon + 2\|y_\epsilon\|) +$$

$$+ \sum_{k=1}^{n'} \frac{[\epsilon(\|A_\epsilon\|+\epsilon)+\|A_\epsilon\| \, \|\tilde{y}_k-\tilde{y}_k^{(\epsilon)}\|][(\|y_\epsilon\|+\epsilon)(\|A_\epsilon\|+\epsilon)+\|y_\epsilon\| \, \|A_\epsilon\|]+(\|y_\epsilon\|+\epsilon)^2\Delta_k}{(\tilde{y}_k^{(\epsilon)}, \tilde{y}_k^{(\epsilon)})} .$$

We set

$$(3.61) \qquad \Delta_0 \leq \left[(y_\epsilon, y_\epsilon) - \sum_{k=1}^{n'} \frac{|(y_\epsilon, \tilde{y}_k^{(\epsilon)})|^2}{(\tilde{y}_k^{(\epsilon)}, \tilde{y}_k^{(\epsilon)})}\right]\delta_3, \qquad \delta_3 < 1 ,$$

where δ_3 denotes the rounding error corresponding to the number of significant digits that can be obtained in the value of

$$(y_\epsilon, y_\epsilon) - \sum_{k=1}^{n'} \frac{|(y_\epsilon, \tilde{y}_k^{(\epsilon)})|^2}{(\tilde{y}_k^{(\epsilon)}, \tilde{y}_k^{(\epsilon)})} .$$

Thus, with the help of (3.60), additional restrictions can be placed on n' which ensure the validity of (3.61) for a given value of δ_3.

§3.3 Gaussian Elimination with Pivoting

Consider the following system of n simultaneous linear algebraic equations with n unknowns,

$$(3.62) \qquad \sum_{k=1}^{n} a_{jk}\,\alpha_k = b_j, \; j = 1,2,\ldots,n .$$

Using equivalent transformations, we wish to reduce the system to the form $\alpha_k = b_k^*$ where b_k^* will be the required solution.

We put $a_{jk}^{(1)} = a_{jk}$, $b_j^{(1)} = b_j$ and choose from the elements $a_{jk}^{(1)}$, one with largest modulus (the pivot). Let this be $a_{j_1k_1}^{(1)}$. We call the j_1^{th} equation the *pivotal equation.* Dividing both sides of the pivotal equation by $a_{j_1k_1}^{(1)}$, we obtain

$$\alpha_{k_1} + \sum_{\substack{k=1 \\ k \neq k_1}}^{n} a_{j_1k}^{(2)} \alpha_k = b_{j_1}^{(2)} \ , \ a_{j_1k}^{(2)} = \frac{a_{j_1k}^{(1)}}{a_{j_1k_1}^{(1)}} \ , \ b_{j_1}^{(2)} = \frac{b_{j_1}^{(1)}}{a_{j_1k_1}^{(1)}} \ .$$

Except for the pivotal equation, we remove the unknown α_{k_1} from each of the equations of the system (3.62) by subtracting from these equations the above normalized pivotal equation multiplied by $a_{jk_1}^{(1)}$, $j \neq j_1$. The transformed system will be in the form

$$\sum_{\substack{k=1 \\ k \neq k_1}}^{n} a_{jk}^{(2)} \alpha_k = b_j^{(2)}, \ j = 1,2,\ldots,n; \ \ j \neq j_1 \ , \tag{3.63}$$

$$a_{jk}^{(2)} = a_{jk}^{(1)} - a_{jk_1}^{(1)} \times \frac{a_{j_1k}^{(1)}}{a_{j_1k_1}^{(1)}} \ , \ b_j^{(2)} = b_j^{(1)} - a_{jk_1}^{(1)} \times \frac{b_{j_1}^{(1)}}{a_{j_1k_1}^{(1)}} \ .$$

Defining $a_{j_2k_2}^{(2)}$ as an element with largest modulus of the $a_{jk}^{(2)}$, and transforming the system (3.63) in the same way as for (3.62), we obtain

$$\alpha_{k_2} + \sum_{\substack{k=1 \\ k \neq k_1, k_2}}^{n} a_{j_2k}^{(3)} \alpha_k = b_{j_2}^{(3)} \ , \ a_{j_2k}^{(3)} = \frac{a_{j_2k}^{(2)}}{a_{j_2k_2}^{(2)}} \ , \ b_{j_2}^{(3)} = \frac{b_{j_2}^{(2)}}{a_{j_2k_2}^{(2)}}$$

and

$$\sum_{\substack{k=1 \\ k \neq k_1, k_2}}^{n} a_{jk}^{(3)} \alpha_k = b_j^{(3)} \ , \ j = 1,2,\ldots,n; \ \ j \neq j_1, \ j_2 \ ,$$

$$a_{jk}^{(3)} = a_{jk}^{(2)} - a_{jk_2}^{(2)} \times \frac{a_{j_2k}^{(2)}}{a_{j_2k_2}^{(2)}} \ , \ b_j^{(3)} = b_j^{(2)} - a_{jk_2}^{(2)} \times \frac{b_{j_2}^{(2)}}{a_{j_2k_2}^{(2)}} \ .$$

Continuing in this way we arrive at the system

$$
\begin{aligned}
&\alpha_{k_1} + \sum_{\substack{k=1\\ k\neq k_1}}^{n} a_{j_1 k}^{(2)} \alpha_k = b_{j_1}^{(2)}\\
&\alpha_{k_2} + \sum_{\substack{k=1\\ k\neq k_1,k_2}}^{n} a_{j_2 k}^{(3)} \alpha_k = b_{j_2}^{(3)}\\
(3.64)\qquad &-\,-\,-\,-\,-\,-\,-\,-\,-\,-\,-\,-\,-\,-\,-\\
&\alpha_{k_m} + \sum_{\substack{k=1\\ k\neq k_1,k_2,\dots,k_m}}^{n} a_{j_m k}^{(m+1)} \alpha_k = b_{j_m}^{(m+1)}, \quad m \le n,
\end{aligned}
$$

which is equivalent to the original system (3.62). This procedure for determining (3.64) is called *forward elimination*; the procedure for determining the solution α_k from (3.64) is known as *back substitution*. Together, these two procedures define *Gaussian elimination with pivoting*.

We now apply this process in the case when (3.62) corresponds to (3.5), i.e., when it is the same as (3.49).

Theorem 3.14. *For the system* (3.49), *the pivots at each step of the forward elimination process of Gaussian elimination with pivoting, are on the diagonal.*

Proof. We assume that at the first step, the pivot is equal to $\operatorname{Re}(e_k, e_j)_1$, where $k \neq j$. Then we obtain

$$
\begin{aligned}
(e_k, e_k)_1 &= \|Ae_k\|^2 + \alpha\|e_k\|^2 < |\operatorname{Re}(e_k, e_j)_1|\\
&= |\operatorname{Re}(Ae_k, Ae_j) + \alpha \operatorname{Re}(e_k, e_j)|\\
&\le \|Ae_k\|\,\|Ae_j\| + \alpha\|e_k\|\,\|e_j\|,\\
(e_j, e_j)_1 &= \|Ae_j\|^2 + \alpha\|e_j\|^2 < |\operatorname{Re}(e_k, e_j)_1| \le \|Ae_k\|\,\|Ae_j\| + \alpha\|e_k\|\,\|e_j\|,
\end{aligned}
$$

whence

$$
\|Ae_k\|^2 + \|Ae_j\|^2 + \alpha\|e_k\|^2 + \alpha\|e_j\|^2 < 2\|Ae_k\|\,\|Ae_j\| + 2\alpha\|e_k\|\,\|e_j\| ,
$$

or

$$
(\|Ae_k\| - \|Ae_j\|)^2 + \alpha(\|e_k\| - \|e_j\|)^2 < 0 ,
$$

which is not possible.

If we now apply formulae (3.63), since $j_1 = k_1$ we obtain

$$a_{jk}^{(2)} = \mathrm{Re}(e_k, e_j)_1 - \frac{\mathrm{Re}(e_{j_1}, e_j)_1 \times \mathrm{Re}(e_k, e_{j_1})_1}{(e_{j_1}, e_{j_1})_1}$$

since (e_{j_1}, e_{j_1}) is real. Define

$$\tilde{e}_j^{(2)} = e_j - \frac{\mathrm{Re}(e_{j_1}, e_j)_1}{(e_{j_1}, e_{j_1})_1} e_{j_1} ,$$

then $\mathrm{Re}(\tilde{e}_j^{(2)}, e_{j_1})_1 = 0$. We find that

$$a_{jk}^{(2)} = \mathrm{Re}(\tilde{e}_k^{(2)}, \tilde{e}_j^{(2)})_1 .$$

Thus the matrix of the transformed system has the same structure as the original matrix, and so the pivots of the transformed system must also be on the diagonal. The proof of the theorem is now readily established by induction. #

Theorem 3.15. *If* $\{e_k\}_1^n$ *is an orthonormal system, then for the pivots* δ_k *of the system* (3.49) *the estimate*

$$\delta_k \geq \alpha + m_n^2 \geq \alpha + m^2 \tag{3.66}$$

is valid, where m_n *and* m *are given by the formulae* (3.11) *and* (3.42) *respectively.*

Proof. From the previous theorem and the equations

$$\begin{aligned} \tilde{e}_j^{(3)} &= \tilde{e}_j^{(2)} - \frac{\mathrm{Re}(\tilde{e}_{j_2}^{(2)}, \tilde{e}_j^{(2)})_1}{(\tilde{e}_{j_2}^{(2)}, \tilde{e}_{j_2}^{(2)})_1} \tilde{e}_{j_2}^{(2)} \\ &= e_j - \frac{\mathrm{Re}(e_{j_1}, e_j)_1}{(e_{j_1}, e_{j_1})_1} e_{j_1} - \frac{\mathrm{Re}(\tilde{e}_{j_2}^{(2)}, \tilde{e}_j^{(2)})_1}{(\tilde{e}_{j_2}^{(2)}, \tilde{e}_{j_2}^{(2)})_1} \tilde{e}_{j_2}^{(2)} , \end{aligned} \tag{3.67}$$

$$\mathrm{Re}(\tilde{e}_j^{(3)}, e_{j_1}) = \mathrm{Re}(\tilde{e}_j^{(3)}, \tilde{e}_{j_2}^{(2)}) = 0 ;$$

it follows by induction that the pivots are

$$\delta_k = (\tilde{e}_{jk}^{(k)}, \tilde{e}_{jk}^{(k)})_1 . \tag{3.68}$$

In order to complete the proof, we apply (3.51) and note that

$$(3.69)\qquad m_n = \inf_{\alpha_k} \frac{\|A \sum_{k=1}^{n} \alpha_k e_k\|}{\|\sum_{k=1}^{n} \alpha_k e_k\|} \le m_s = \inf_{\alpha_k} \frac{\|A \sum_{k=1}^{s} \alpha_k e_k\|}{\|\sum_{k=1}^{n} \alpha_k e_k\|}, \quad 1 \le s \le n . \quad \#$$

We now examine estimates for the inherent error† in the solution of the general linear algebraic system (3.62), and the specific system (3.5) by Gaussian elimination with pivoting.(15)

We shall assume that the absolute errors of the coefficients $a_{jk}^{(1)}$ and of the right hand sides $b_j^{(1)}$ do not exceed Δ and that

$$(3.70)\qquad \delta_s = |a_{j_s k_s}^{(s)}| \ge 4^{s-1}\, \Delta/\delta, \; s = 1,2,\ldots,n \; ; \quad \delta < 1 .$$

Then, with an accuracy not exceeding small quantities of order higher than Δ or δ, we find that

$$(3.71)\qquad \Delta[a_{jk}^{(s)}] \le 4^{s-1}\, \Delta, \quad |b_j^{(s)}| \le 2^{s-1}\, \sigma ,$$

$$\Delta[b_j^{(s+1)}] \le 2^s \Delta + s\, 2^s \sigma\delta, \; j \neq j_1, j_2, \ldots, j_s, \qquad s = 1,2,\ldots,n ,$$

$$\sigma = \max_{1 \le j \le n} |b_j^{(1)}| ,$$

and

$$(3.72)\qquad \Delta[a_{j_s k}^{(s+1)}] \le 2\delta , \quad |b_{j_s}^{(s+1)}| \le \frac{\sigma}{2^{s-1}} \cdot \frac{\delta}{\Delta} ,$$

$$\Delta[b_{j_s}^{(s+1)}] \le \frac{\delta}{2^{s-1}} + \frac{s\sigma}{2^{s-1}} \cdot \frac{\delta^2}{\Delta} , \quad s = 1,2,\ldots,n .$$

Removing α_{k_i} from (3.64), starting with α_{k_n} and going to α_{k_1}, we obtain

$$(3.73)\qquad \alpha_{k_i} = \sum_{k=i}^{n} \gamma_k^{(i)}\, b_{j_k}^{(k+1)} , \quad \gamma_i^{(i)} = 1, \; i = 1,2,\ldots,n ,$$

where

$$(3.74)\qquad \gamma_k^{(i)} = - \sum_{s=i+1}^{k} a_{j_i k_s}^{(i+1)}\, \gamma_k^{(s)} , \; k = i+1, i+2, \ldots, n .$$

By an elementary error analysis, we find by induction that

$$(3.75)\qquad \Delta[\gamma_k^{(i)}] \le (k-i+1) 2^{k-i-1} \delta, \; |\gamma_k^{(i)}| \le 2^{k-i-1} , \; k = i+1,\ldots,n; \quad i = 1,2,\ldots,n .$$

---

(15) Assuming that the round-off error at each step is the inherent error for the following steps, it is not difficult to obtain an estimate of the total effect of the initial round-off error.

† Forward error analysis. (Eds.)

Using (3.71)-(3.75), we have finally

$$\Delta[\alpha_{k_i}] \leq \sum_{k=i}^{n} \{\Delta[\gamma_k^{(i)}]|b_{j_k}^{(k+1)}| + \Delta[b_{j_k}^{(k+1)}]\ |\gamma_k^{(i)}|\} \tag{3.76}$$

$$\leq \frac{n-i+2}{2^i}\,\delta + \frac{n(n-i+2)}{2^i}\,\sigma\,\frac{\delta^2}{\Delta}\ ,$$

$$|\alpha_{k_i}| \leq \frac{n-i+2}{2^i}\,\sigma\,\frac{\delta}{\Delta}\,,\quad i = 1,2,\ldots,n\ .$$

We formulate this result as the following theorem.

Theorem 3.16. *If the elements of the system* (3.62) *have an absolute error which does not exceed* Δ, *and the pivots* $a_{j_s k_s}^{(s)}$ *satisfy the relationship* (3.70), *then, with an accuracy not exceeding small quantities of higher order than* Δ, δ *and* δ^2/Δ, *the estimates* (3.76) *are valid for the solution of the system* (3.62).

Simpler estimates can be obtained from the following arguments. Let A and A^Δ, b and b^Δ, α and α^Δ be respectively the matrices, right-hand sides and solutions of the exact and approximate systems of the form (3.62). The identity

$$A^\Delta(\alpha - \alpha^\Delta) \equiv b - b^\Delta + (A^\Delta - A)\alpha \equiv \tilde{b} \tag{3.77}$$

is obviously valid. Defining the norm for any n-dimensional vector $a = (a_1,a_2,\ldots,a_n)$ as $\|a\|_n = \max\limits_{1\leq j\leq n} |a_j|$, we obtain

$$\tilde{\sigma} = \|b - b^\Delta + (A^\Delta - A)\alpha\|_n \leq \Delta(1 + \sum_{j=1}^{n} |\alpha_j|)\ . \tag{3.78}$$

We denote the modulus of the pivots of the approximate system by δ_s^Δ, and apply the estimates (3.71)-(3.75) which were needed to obtain $|\alpha_{k_i}|$ (although we do not assume the relationship (3.70)). We find

$$|\tilde{b}_j^{(s)}| \leq 2^{s-1}\tilde{\sigma}\ ,\ j \neq j_1,j_2,\ldots,j_{s-1}\ ;\ |\tilde{b}_{j_s}^{(s+1)}| \leq \frac{2^{s-1}\tilde{\sigma}}{\delta_s^\Delta} \tag{3.79}$$

and

$$|\alpha_{k_i} - \alpha_{k_i}^\Delta| \leq \sum_{k=i}^{n} |\gamma_k^{(i)}|\ |\tilde{b}_{j_k}^{(k+1)}| \tag{3.80}$$

$$\leq \sum_{k=i}^{n} 2^{k-i-1}\,2^{k-1}\,\frac{\tilde{\sigma}}{\delta_k^\Delta} = \frac{\tilde{\sigma}}{2^i}\sum_{k=i}^{n}\frac{4^{k-1}}{\delta_k^\Delta}\ .$$

From (3.80) and (3.78) it easily follows that

$$\sum_{j=1}^{n} |\alpha_j| - \sum_{j=1}^{n} |\alpha_j^\Delta| \le \sum_{i=1}^{n} |\alpha_{k_i} - \alpha_{k_i}^\Delta| \le \tilde{\sigma} \sum_{k=1}^{n} \frac{4^{k-1}}{\delta_k^\Delta}$$

$$\le \Delta\left(1 + \sum_{j=1}^{n} |\alpha_j|\right) \sum_{k=1}^{n} \frac{4^{k-1}}{\delta_k^\Delta} \quad ;$$

from which,under the condition that

$$\Delta R_n < 1, \quad R_n = \sum_{k=1}^{n} \frac{4^{k-1}}{\delta_k^\Delta} \quad , \tag{3.81}$$

we find that

$$\sum_{j=1}^{n} |\alpha_j| \le \left(\sum_{j=1}^{n} |\alpha_j^\Delta| + \Delta R_n \right) \frac{1}{1-\Delta R_n} \tag{3.82}$$

and

$$|\alpha_{k_i} - \alpha_{k_i}^\Delta| \le \frac{\Delta}{2^i(1-\Delta R_n)} \left[1 + \sum_{j=1}^{n} |\alpha_{k_j}^\Delta| \right] \sum_{k=i}^{n} \frac{4^{k-1}}{\delta_k^\Delta} \quad , \tag{3.83}$$

$$|\alpha_{k_i}| \le |\alpha_{k_i}^\Delta| + |\alpha_{k_i} - \alpha_{k_i}^\Delta| \quad .$$

Theorem 3.17. *From the solution of the approximate system of equations* $A^\Delta \alpha^\Delta = b^\Delta$*, estimates for the absolute errors and the moduli of the components of the exact solution, are given by* (3.83).

It is not difficult to obtain *a priori* estimates assuming that (3.70) is valid. Then

$$R_n = \sum_{k=1}^{n} \frac{4^{k-1}}{\delta_k^\Delta} \le \frac{\delta}{\Delta} n \tag{3.84}$$

and for

$$\delta < \frac{1}{n} \quad , \tag{3.85}$$

taking into account that from (3.76) we obtain

$$|\alpha_{k_i}^\Delta| \le \frac{n-i+2}{2^i} \sigma \frac{\delta}{\Delta} \quad , \tag{3.86}$$

we find from (3.83) that

(3.87) $$|\alpha_{k_i} - \alpha^{\Delta}_{k_i}| \le \frac{(n-i+1)n\sigma}{2^i(1-\delta n)} \cdot \frac{\delta^2}{\Delta} + \frac{(n-i+1)}{2^i(1-\delta n)} \cdot \delta ,$$

$$|\alpha_{k_i}| \le \frac{n-i+2}{2^i}\sigma\frac{\delta}{\Delta} + \frac{(n-i+1)n\sigma}{2^i(1-\delta n)} \cdot \frac{\delta^2}{\Delta} + \frac{(n-i+1)}{2^i(1-\delta n)} \delta ,$$

$$\sigma = \max_{1\le j\le n} |b^{\Delta}_j| = \|b^{\Delta}\|_n ,$$

whence

(3.88) $$\|\alpha - \alpha^{\Delta}\|_n \le \frac{n^2}{2(1-\delta n)} \cdot \frac{\delta^2}{\Delta} \|b^{\Delta}\|_n + \frac{n}{2(1-\delta n)} \delta ,$$

$$\|\alpha\|_n \le \left(\frac{n}{2} \cdot \frac{\delta}{\Delta} + \frac{n^2}{2(1-\delta n)} \cdot \frac{\delta^2}{\Delta} \right) \|b^{\Delta}\|_n + \frac{n}{2(1-\delta n)} \delta .$$

We formulate this result in the following theorem.

Theorem 3.18. *If* $\delta^{\Delta}_s \ge 4^{s-1}\Delta/\delta$ $(s = 1,2,\dots,n)$ *and* $\delta < 1/n$, *then the estimates* (3.88) *are valid.*

In the case of a system of the form (3.5), under the conditions of Theorem 3.15, we obtain

(3.89) $$\delta^{\Delta}_n \ge \alpha + (m^{(\epsilon)}_n)^2 , \quad m^{(\epsilon)}_n = \inf_{\alpha_k} \frac{\|A_\epsilon(\sum_{k=1}^{n} \alpha_k e_k)\|}{\|\sum_{k=1}^{n} \alpha_k e_k\|} .$$

Since

(3.90) $$\max \{\max_{k,j} |\mathrm{Re}(y_k, y_j) - \mathrm{Re}(y^{(\epsilon)}_k, y^{(\epsilon)}_j)|, \max_j |\mathrm{Re}(y,y_j) - \mathrm{Re}(y_\epsilon, y^{(\epsilon)}_j)|\}$$

$$\le \max \{2\epsilon\, b^{(\epsilon)}_n + \epsilon^2, (\|y_\epsilon\| + b^{(\epsilon)}_n)\epsilon + \epsilon^2\} = \Delta^{(\epsilon)}_n ,$$

$$b^{(\epsilon)}_n = \max_{1\le j\le n} \|y^{(\epsilon)}_j\| ,$$

then we assume that $\Delta = \Delta^{(\epsilon)}_n$. In this way (3.70) is valid if

(3.91) $$\alpha + (m^{(\epsilon)}_n)^2 \ge 4^{n-1} \Delta^{(\epsilon)}_n/\delta .$$

Let $\overset{*}{x}{}^{\alpha,\epsilon}_n = \sum_{k=1}^{n} \overset{*}{\alpha}{}^{\alpha,\epsilon}_k e_k$ be the element which minimizes $I^{\alpha,\epsilon}\left(\sum_{k=1}^{n} \alpha_k e_k\right)$.

From (3.87) and (3.89), we obtain that

$$(3.92)\qquad \|\overset{*}{x}{}_n^{\alpha,\epsilon} - \overset{*}{x}_n\|^2 = \sum_{i=1}^{n} |\overset{*}{\alpha}{}_{k_i}^{\alpha,\epsilon} - \overset{*}{\alpha}_{k_i}|^2$$

$$\leq \left[\frac{n\,\sigma\,\delta^2}{(1-\delta n)\Delta_n^{(\epsilon)}} + \frac{\delta}{1-\delta n} \right]^2 \sum_{i=1}^{n} \frac{(n-i+1)^2}{4^i}$$

$$\leq \left[\frac{n\,\sigma\,\delta^2}{(1-\delta n)\Delta_n^{(\epsilon)}} + \frac{\delta}{1-\delta n} \right]^2 \frac{n^2}{3} ,$$

whence

$$(3.93)\qquad \|\overset{*}{x}{}_n^{\alpha,\epsilon} - \overset{*}{x}_n\| \leq \frac{\sigma_n^{(\epsilon)}\, n^2}{(1-\delta n)\sqrt{3}} \cdot \frac{\delta^2}{\Delta_n^{(\epsilon)}} + \frac{n}{(1-\delta n)\sqrt{3}}\,\delta ,$$

$$\sigma_n^{(\epsilon)} = \sigma = \max_{1\leq j\leq n} |\mathrm{Re}(y_\epsilon, y_j^{(\epsilon)})| .$$

Theorem 3.19. *If, for a given n,* (3.91) *is valid and $\delta < 1/n$, then inequality* (3.93) *is valid.*

In conclusion, we show that in the general case the estimate $\Delta[a_{jk}^{(s)}] \leq 4^{s-1}\Delta$ cannot be improved.

Let two matrices $\tilde{A}_n$ and $\tilde{A}_n^{\Delta}$ of order n be given:

$$(3.94)\qquad \tilde{A}_n = \begin{pmatrix} 1 & -1+\epsilon_1 & -1+\epsilon_1 & \cdots\cdots & -1+\epsilon_1 & -1+\epsilon_1 \\ -1+\epsilon_1 & 1 & 1-\epsilon(2-\epsilon_1) & \cdots\cdots & 1-\epsilon(2-\epsilon_1) & 1-\epsilon(2-\epsilon_1) \\ -1+\epsilon_1 & 1-\epsilon(2-\epsilon_1) & 1 & \cdots\cdots & 1-\epsilon^2(2-\epsilon_1) & 1-\epsilon^2(2-\epsilon_1) \\ \vdots & \vdots & \vdots & & \vdots & \vdots \\ -1+\epsilon_1 & 1-\epsilon(2-\epsilon_1) & 1-\epsilon^2(2-\epsilon_1) & \cdots\cdots & 1 & 1-\epsilon^{n-2}(2-\epsilon_1) \\ -1+\epsilon_1 & 1-\epsilon(2-\epsilon_1) & 1-\epsilon^2(2-\epsilon_1) & \cdots\cdots & 1-\epsilon^{n-2}(2-\epsilon_1) & 1 \end{pmatrix}$$

where $0 < \epsilon < 1$, $\epsilon_1 = 1-(1-\epsilon)^{\frac{1}{2}}$, and the elements of $\tilde{A}_n^{\Delta}$ are equal to the corresponding elements of $\tilde{A}_n$ plus Δ. For the matrices $\tilde{A}_n$ and $\tilde{A}_n^{\Delta}$, we denote the pivots at the sth-step of the forward elimination of Gaussian elimination with pivoting by $\tilde{\delta}_s$ and $\tilde{\delta}_s^{\Delta}$ respectively.

Theorem 3.20. *For any ϵ such that $0 < \epsilon < 1$, the equation*

$$(3.95)\qquad \tilde{\delta}_s^{\Delta} - \tilde{\delta}_s = (4 - \epsilon - 2\epsilon_1)^{s-1}\Delta + o(\Delta) , \quad \epsilon_1 = 1 - (1-\epsilon)^{\frac{1}{2}} ,$$

is valid.

Proof. For the matrices $\tilde{A}_n$ and $\tilde{A}_n^{\Delta}$, we introduce the notation which was used in the general scheme for Gaussian elimination with pivoting (see (3.62)-(3.64)), the only difference being that we now use the symbols "~" and "~Δ" respectively.

It is obvious that

$$\tilde{\delta}_1 = \tilde{a}^{(1)}_{j_1k_1} = \tilde{a}^{(1)}_{11} = \tilde{a}^{(1)}_{jj} = 1, \quad \tilde{\delta}_1^{\Delta} = \tilde{a}^{\Delta(1)}_{j_1k_1} = \tilde{a}^{\Delta(1)}_{11} = \tilde{a}^{\Delta(1)}_{jj} = 1+\Delta ,$$

$$\tilde{a}^{(1)}_{j1} = -1+\epsilon_1, \; \tilde{a}^{\Delta(1)}_{j1} = -1+\epsilon_1+\Delta, \quad j = 2,3,\dots,n.$$

Hence (using (3.63)) we find that

$$\tilde{\delta}_2 = \tilde{a}^{(2)}_{jj} = \tilde{a}^{(1)}_{jj} - \tilde{a}^{(1)}_{j1} \times \frac{\tilde{a}^{(1)}_{1k}}{\tilde{a}^{(1)}_{11}} = 1 - (-1+\epsilon_1)^2 = 2\epsilon_1 - \epsilon_1{}^2 = \epsilon, \tag{3.96}$$

$$\tilde{\delta}_2{}^{\Delta} = \tilde{a}^{\Delta(2)}_{jj} = 1 + \Delta - \frac{(-1+\epsilon_1+\Delta)^2}{1+\Delta} = \epsilon + (4-\epsilon-2\epsilon_1)\Delta + o(\Delta) ,$$

$$k \neq 1, \; j = 2,3,\dots,n \;;$$

$$\tilde{a}^{(2)}_{j2} = 1 - \epsilon(2-\epsilon_1) - (-1+\epsilon_1)^2 = \epsilon(-1+\epsilon_1) ,$$

$$\tilde{a}^{\Delta(2)}_{j2} = 1 + \Delta - \epsilon(2-\epsilon_1) - \frac{(-1+\epsilon_1+\Delta)^2}{1+\Delta} = \epsilon(-1+\epsilon_1) + (4-\epsilon-2\epsilon_1)\Delta + o(\Delta) ,$$

$$j = 3,4,\dots,n \;;$$

$$\tilde{a}^{(2)}_{j3} = 1 - \epsilon^2(2-\epsilon_1) - (-1+\epsilon_1)^2 = \epsilon[1-\epsilon(2-\epsilon_1)] ,$$

$$\tilde{a}^{\Delta(2)}_{j3} = 1 + \Delta - \epsilon^2(2-\epsilon_1) - \frac{(-1+\epsilon_1+\Delta)^2}{1+\Delta} = \epsilon[1-\epsilon(2-\epsilon_1)] + (4-\epsilon-2\epsilon_1)\Delta + o(\Delta) ,$$

$$j = 4,\dots,n \;;$$

$$\tilde{a}^{(2)}_{n,n-1} = 1-\epsilon^{n-2}(2-\epsilon_1) - (-1+\epsilon_1)^2 = \epsilon[1-\epsilon^{n-3}(2-\epsilon_1)] ,$$

$$\tilde{a}^{\Delta(2)}_{n,n-1} = 1 + \Delta - \epsilon^{n-2}(2-\epsilon_1) - \frac{(-1+\epsilon_1+\Delta)^2}{1+\Delta} = \epsilon[1-\epsilon^{n-3}(2-\epsilon_1)] + (4-\epsilon-2\epsilon_1)\Delta + o(\Delta).$$

The results obtained imply that after the first step of the forward elimination we obtain matrices $\epsilon\tilde{A}_{n-1}$ and $\epsilon\tilde{A}^{\Delta(1)}_{n-1}$, where the elements of $\tilde{A}^{\Delta(1)}_{n-1}$ are equal to the corresponding elements of $\tilde{A}_{n-1}$ plus $\Delta^{(1)} = \{(4-\epsilon-2\epsilon_1)\Delta/\epsilon\}+o(\Delta)$. Similarly, after the second step of forward elimination, we obtain matrices $\epsilon^2\tilde{A}_{n-2}$ and $\epsilon^2\tilde{A}^{\Delta(2)}_{n-2}$, where the elements of $\tilde{A}^{\Delta(2)}_{n-2}$ are equal to the corresponding elements of $\tilde{A}_{n-2}$ plus $\Delta^{(2)} = \{(4-\epsilon-2\epsilon_1)^2/\epsilon^2\}\Delta+o(\Delta)$, etc. Hence it follows that

$$\tilde{\delta}_s = \epsilon^{s-1},\ \tilde{\delta}_s^{\Delta} = \epsilon^{s-1} + (4-\epsilon-2\epsilon_1)^{s-1}\Delta + o(\Delta),$$

$$\tilde{\delta}_s^{\Delta} - \tilde{\delta}_s = (4-\epsilon-2\epsilon_1)^{s-1}\Delta + o(\Delta) ,$$

which establishes the theorem. #

§3.4 A Coordinate-wise Descent Algorithm (Gauss-Seidel)

We shall now examine the problem of the minimization of the function

$$\begin{aligned} \delta_n(Q_n) &\equiv \delta_n(\beta_1,\beta_2,\ldots,\beta_n) \equiv I^{(\alpha)}(x_n) = \\ &= \|y - \sum_{k=1}^{n} \beta_k u_k\|^2 + \alpha \|\sum_{k=1}^{n} (\beta_k e_k/\|y_k\|)\|^2, \end{aligned} \tag{3.97}$$

$$Q_n = Q_n(\beta_1,\beta_2,\ldots,\beta_n),\ u_k = y_k/\|y_k\| ,\quad y_k = Ae_k.\text{[16]}$$

Let $\beta_1^{(0)},\beta_2^{(0)},\ldots,\beta_n^{(0)}$ denote the initial coefficients, and put $\delta_n^{(0)} = \delta_n(\beta_1^{(0)},\beta_2^{(0)},\ldots,\beta_n^{(0)})$. We determine $b_1^{(1)}$ from the condition $\delta_1^{(1)} = \delta_n(\beta_1^{(0)} + b_1^{(1)},\beta_2^{(0)},\ldots,\beta_n^{(0)}). = \min_b [\delta_n(\beta_1^{(0)} + b,\beta_2^{(0)},\ldots,\beta_n^{(0)})]$. Introducing the notation $u_n^{(0)} = y - \sum_{k=1}^{n} \beta_k^{(0)} u_k$ and $x_n^{(0)} = \sum_{k=1}^{n} (\beta_k^{(0)} e_k/\|y_k\|)$, we obtain

$$\delta_n(\beta_1^{(0)} + b,\beta_2^{(0)},\ldots,\beta_n^{(0)}) = \|u_n^{(0)} - bu_1\|^2 + \alpha\|x_n^{(0)} + (be_1)/\|y_1\|\ \|^2. \tag{3.98}$$

Minimizing (3.98) with respect to b, we find that

$$b_1^{(1)} = \frac{\mathrm{Re}(u_n^{(0)},u_1) - \alpha\ \mathrm{Re}(x_n^{(0)},e_1/\|y_1\|)}{1 + \alpha(e_1,e_1)/(y_1,y_1)} . \tag{3.99}$$

---

[16] If, for some k, $(y_k,y_k) \leq \Delta[(y_k,y_k)]/\delta$, where $\Delta[(y_k,y_k)]$ is the absolute error of (y_k,y_k) and $\delta(\delta < 1)$ is the value which determines the accuracy of the subsequent results, then the corresponding e_k should be removed from the original system $e_1,e_2,\ldots\ldots$.

Having determined $b_1^{(1)}$, we find $b_2^{(1)}$ from the condition

$$\delta_2^{(1)} = \delta_n(\beta_1^{(0)} + b_1^{(1)}, \beta_2^{(0)} + b_2^{(1)}, \beta_3^{(0)}, \ldots, \beta_n^{(0)}) = \\ = \min_b \left[\delta_n(\beta_1^{(0)} + b_1^{(1)}, \beta_2^{(0)} + b, \beta_3^{(0)}, \ldots, \beta_n^{(0)}) \right].$$

Similarly we can evaluate $b_3^{(1)}, b_4^{(1)}, \ldots, b_n^{(1)}$. Next we can proceed to evaluate $b_1^{(2)}$ from the condition

$$\delta_1^{(2)} = \delta_n(\beta_1^{(1)} + b_1^{(2)}, \beta_2^{(1)}, \ldots, \beta_n^{(1)}) = \min_b [\delta_n(\beta_1^{(1)} + b, \beta_2^{(1)}, \ldots, \beta_n^{(1)})] ,$$

where

$$\beta_j^{(1)} = \beta_j^{(0)} + b_j^{(1)} \quad ;$$

$b_2^{(2)}$ from the condition

$$\delta_2^{(2)} = \delta_n(\beta_1^{(1)}+b_1^{(2)}, \beta_2^{(1)}+b_2^{(2)}, \beta_3^{(1)}, \ldots, \beta_n^{(1)}) = \min_b \left[\delta_n(\beta_1^{(1)}+b_1^{(2)}, \beta_2^{(1)}+b, \beta_3^{(1)}, \ldots, \beta_n^{(1)}) \right]$$

and so on up to $b_n^{(2)}$ from

$$\delta_n^{(2)} = \delta_n(\beta_1^{(2)}, \ldots, \beta_{n-1}^{(2)}, \beta_n^{(1)}+b_n^{(2)}) = \min_b \left[\delta_n(\beta_1^{(2)}, \ldots, \beta_{n-1}^{(2)}, \beta_n^{(1)}+b) \right],$$

where $\beta_j^{(2)} = \beta_j^{(1)} + b_j^{(2)}$ etc. In general, at the jth step of the iteration process

$$\delta_s^{(j)} = \delta_n(\beta_1^{(j)}, \ldots, \beta_{s-1}^{(j)}, \beta_s^{(j)}, \beta_{s+1}^{(j-1)}, \ldots, \beta_n^{(j-1)}) \\ = \min_b \left[\delta_n(\beta_1^{(j)}, \ldots, \beta_{s-1}^{(j)}, \beta_s^{(j-1)} + b, \beta_{s+1}^{(j-1)}, \ldots, \beta_n^{(j-1)}) \right] ,$$

$$\beta_s^{(j)} = \beta_s^{(j-1)} + b_s^{(j)} = \beta_s^{(0)} + \sum_{r=1}^{j} b_s^{(r)},$$

$$b_s^{(j)} = \frac{\mathrm{Re}\,(u_{s-1}^{(j)}, u_s) - \alpha\,\mathrm{Re}(x_{s-1}^{(j)}, e_s/\|y_s\|)}{1 + \alpha(e_s, e_s)/(y_s, y_s)},$$

$$(3.100)\qquad \begin{aligned} &u_0^{(j)} = u_n^{(j-1)},\ x_0^{(j)} = x_n^{(j-1)},\\ &u_{s-1}^{(j)} = y - \sum_{k=1}^{s-1} \beta_k^{(j)} u_k - \sum_{k=s}^{n} \beta_k^{(j-1)} u_k\ ,\\ &x_{s-1}^{(j)} = \sum_{k=1}^{s-1} \alpha_k^{(j)} e_k + \sum_{k=s}^{n} \alpha_k^{(j-1)} e_k,\\ &\alpha_s^{(j)} = \beta_s^{(j)}/\|y_s\|\ , s = 1,2,..,n;\ j = 1,2,3,\dots\ . \end{aligned}$$

Theorem 3.21. *The coordinate-wise descent algorithm* (3.100) *always converges in the sense that*

$$(3.101)\qquad \lim_{j\to\infty} \delta_s^{(j)} = \inf_{Q_n} \delta_n(Q_n) = I^{(\alpha)}(\overset{*}{x}_n), s = 1,2,..,n,$$

and

$$(3.102)\qquad \lim_{j\to\infty} \alpha_s^{(j)} = \overset{*}{\alpha}_s, s=1,2,..,n\ ,\ \textit{where}\ \overset{*}{x}_n = \sum_{k=1}^{n} \overset{*}{\alpha}_k e_k\ .$$

Proof. For $\alpha = 0$, the theorem is proved in [59]. We shall therefore assume that $\alpha > 0$. Since at each step of the algorithm the residual $\delta_s^{(j)}$ can only be diminished, and since on the other hand it can never be less than zero, then $\lim_{j\to\infty} \delta_s^{(j)}$ exists. Hence we have (on using (3.100)) that

$$(3.103)\qquad \epsilon_s^{(j)} = \delta_{s-1}^{(j)} - \delta_s^{(j)} = (b_s^{(j)})^2 \left\{ 1 + \alpha \frac{(e_s,e_s)}{(y_s,y_s)} \right\}$$

tends to zero as $j \to \infty$, uniformly with respect to s. From (3.100) it follows that

$$(3.104)\qquad \begin{aligned} &\sum_{k=1}^{n} \alpha_k^{(j)} \left[\mathrm{Re}(y_k,y_s) + \alpha\ \mathrm{Re}\ (e_k,e_s)\right]\\ &= \mathrm{Re}\ (y,y_s) + \sum_{k=s+1}^{n} b_k^{(j)} \left[\mathrm{Re}(y_k,y_s) + \alpha\ \mathrm{Re}\ (e_k,e_s)\right]/\|y_k\|. \end{aligned}$$

Let us compare equations (3.104) and (3.5). First, let us assume that $e_1, e_2, \ldots, e_n$ are linearly independent, then the systems of equations (3.104) and (3.5) each have a unique solution and $|\alpha_k^{(j)} - \overset{*}{\alpha}_k| \to 0$ since $\epsilon_s^{(j)} \to 0$. Hence (3.101) and (3.102) readily follow.

Next, suppose (without loss of generality)that only the first m elements $e_1, e_2, \ldots, e_m$ are linearly independent,where $1 \leq m < n$. Then it is not difficult to show (see [59]), that the last $(n - m)$ equations of the system (3.5) are a consequence of the first m equations. Hence assuming in (3.5) that $k > m$, then

$$\alpha_k \equiv \hat{\alpha}_k = \begin{cases} \alpha_k^{(j)}, & k \leq s-1, \\ \alpha_k^{(j-1)}, & k > s-1, \end{cases}$$

we again obtain $|\alpha_k^{(j)} - \hat{\alpha}_k| \to 0$ as $\epsilon_s^{(j)} \to 0$, from which (3.101) readily follows.

For the proof of the stronger assertion (3.102), (in which $\overset{*}{\alpha}_s$ does not depend on j),we prove first of all that the following identity,which is a finite-dimensional analogue of (2.8),is valid :

$$I^{(\alpha)}(x_n) - I^{(\alpha)}(\overset{*}{x}_n) = \|A(x_n - \overset{*}{x}_n)\|^2 + \alpha\|x_n - \overset{*}{x}_n\|^2 , \tag{3.105}$$

$$x_n = \sum_{k=1}^{n} \alpha_k e_k .$$

In fact,

$$I^{(\alpha)}(x_n) - I^{(\alpha)}(\overset{*}{x}_n) = \|A[\overset{*}{x}_n + (x_n - \overset{*}{x}_n)] - y\|^2 + $$
$$+ \alpha\|\overset{*}{x}_n + (x_n - \overset{*}{x}_n)\|^2 - \|A\overset{*}{x}_n - y\|^2 - \alpha\|\overset{*}{x}_n\|^2$$

$$= \|A(x_n - \overset{*}{x}_n)\|^2 + \alpha\|x_n - \overset{*}{x}_n\|^2 + 2\mathrm{Re}\left[(A\overset{*}{x}_n - y, A(x_n - \overset{*}{x}_n)) + \alpha(\overset{*}{x}_n, x_n - \overset{*}{x}_n)\right].$$

But from (3.5), the last term is identically equal to zero for any α_k, so that (3.105) is established. From (3.105), it follows that for $\alpha > 0$, $\|x_n - \overset{*}{x}_n\| \to 0$, if $I^{(\alpha)}(x_n) - I^{(\alpha)}(\overset{*}{x}_n) \to 0$ and there exists only one element $\overset{*}{x}_n$ for which $I^{(\alpha)}(\overset{*}{x}_n) = \inf_{\alpha_k} \{I^{(\alpha)}(x_n)\}$. In particular, this implies that $\{x_s^{(j)}\}$ converges to $\overset{*}{x}_n$ as $j \to \infty$.

Applying (3.105), we rewrite $\epsilon_s^{(j)}$ as

$$\begin{aligned}
\epsilon_s^{(j)} &= I^{(\alpha)}(x_{s-1}^{(j)}) - I^{(\alpha)}(x_s^{(j)}) \\
&= \left(I^{(\alpha)}(x_{s-1}^{(j)}) - I^{(\alpha)}(\overset{*}{x}_n)\right) - \left(I^{(\alpha)}(x_s^{(j)}) - I^{(\alpha)}(\overset{*}{x}_n)\right) \\
&= \|A(x_{s-1}^{(j)}) - \overset{*}{x}_n)\|^2 + \alpha\|x_{s-1}^{(j)} - \overset{*}{x}_n\|^2 - \|A(x_s^{(j)} - \overset{*}{x}_n)\|^2 - \alpha\|x_s^{(j)} - \overset{*}{x}_n\|^2 \\
&= (z_s^{(j)})^2 - (z_{s+1}^{(j)})^2,
\end{aligned} \tag{3.106}$$

where $\quad (z_s^{(j)})^2 = \|A(x_{s-1}^{(j)} - \overset{*}{x}_n)\|^2 + \alpha\|x_{s-1}^{(j)} - \overset{*}{x}_n\|^2.$

Since from (3.103),

$$|b_s^{(j)}| < (\epsilon_s^{(j)})^{\frac{1}{2}} \leq \left[2z_s^{(j)}\,(z_s^{(j)} - z_{s+1}^{(j)})\right]^{\frac{1}{2}}; \tag{3.107}$$

and from (3.104)

$$z_s^{(j)} \leq C_1 \sum_{k=1}^{n} \left|\alpha_k^{(j)} - \overset{*}{\alpha}_k\right| \leq C_2\left[2z_s^{(j)}(z_s^{(j)} - z_{s+1}^{(j)})\right]^{\frac{1}{2}}, \tag{3.108}$$

we find that

$$z_s^{(j)} \leq 2\,C_2^2\,(z_s^{(j)} - z_{s+1}^{(j)})\ ;$$

then from (3.107) it follows that

$$|b_s^{(j)}| \leq 2C_2\,(z_s^{(j)} - z_{s+1}^{(j)}). \tag{3.109}$$

Noting that $z_s^{(j)} \geq z_{s+1}^{(j)} \geq z_s^{(j+1)}$, we have as a consequence of (3.109) that

$$\begin{aligned}
\sum_{j=1}^{\infty} |b_s^{(j)}| &\leq 2C_2(z_s^{(1)} + z_s^{(2)} - z_{s+1}^{(1)} + z_s^{(3)} - z_{s+1}^{(2)} + \ldots) \\
&\leq 2C_2\ z_s^{(1)}
\end{aligned} \tag{3.110}$$

From the convergence of $\sum_{j=1}^{\infty} |b_s^{(j)}|$, the validity of (3.102) now follows easily.

This proves the theorem. #

When applied to the particular case of solving a system of linear algebraic equations (where A is a matrix, $e_1 = (1,0,\dots,0)$, $e_2 = (0,1,0,\dots,0)$, $\dots$, $e_n = (0,0,\dots,0,1)$), Theorem 3.21 implies that for the system of equations (3.5), the Gauss-Seidel algorithm converges when the matrix of the system is positive definite. Note that the approximate solutions will differ depending on the choice of the initial vector.

The rate of convergence of the coordinate-wise descent algorithm will now be discussed under the assumption that $\{e_k\}$ is an orthonormalized system. A linear algebraic system of equations of order n will be regarded as being in the n-dimensional Euclidean space E_n. We denote the matrix of the system (3.5) by $\alpha I + \overline{A}$, where I is the unit matrix in E_n. The matrix of the system of equations which is obtained as a result of substituting $\alpha_s = \beta_s/\|y_s\|$, and dividing the sth equation in (3.5) by $\|y_s\|$, will be denoted by $\alpha\overline{Y} + \overline{B}$, where $\overline{Y}$ is a diagonal matrix with elements $(y_1,y_1)^{-1}, (y_2,y_2)^{-1}, \dots, (y_n,y_n)^{-1}$. We denote the right hand side of the first system by $\overline{y}$, and the right hand side of the second system by $\overline{u}$. If, in the equation $\beta_s^{(j)} = \beta_s^{(j-1)} + b_s^{(j)}$, instead of $b_s^{(j)}$ we substitute its representation as given in (3.100), then after a simple transformation we obtain the equations

$$\beta_s^{(j)}\left[1 + \alpha/(y_s,y_s)\right] + \sum_{k=1}^{s-1} \beta_k^{(j)} \operatorname{Re}(u_k,u_s) = \operatorname{Re}(y,u_s) - \sum_{k=s+1}^{n} \beta_k^{(j-1)} \operatorname{Re}(u_k,u_s), \tag{3.111}$$

for $s = 1,2,\dots,n$.

We represent the matrix $\overline{B}$ as $\overline{B} = I + \overline{B}_1 + \overline{B}_1^T$, where $\overline{B}_1$ is a lower triangular matrix having the same elements as $\overline{B}$ below the principal diagonal but zero elements on and above it. $\overline{B}_1^T$ is the transpose of $\overline{B}_1$. We can rewrite (3.111) as

$$(I + \alpha\overline{Y} + \overline{B}_1)Q^{(j)} = \overline{u} - \overline{B}_1^T Q^{(j-1)},$$

so that

$$Q^{(j)} = -(I + \alpha\overline{Y} + \overline{B}_1)^{-1} \overline{B}_1^T Q^{(j-1)} + (I + \alpha\overline{Y} + \overline{B}_1)^{-1} \overline{u} . \tag{3.112}$$

Assuming that the matrix $\alpha\bar{Y} + \bar{B}$ is non-singular, it is proved in [8] that the eigenvalues of the matrix $(I + \alpha\bar{Y} + \bar{B}_1)^{-1}\bar{B}_1^T$ are less than one in modulus. This is the condition necessary for the convergence of the process of simple iteration. Having calculated an approximation to the eigenvalue of largest modulus of the matrix $(I + \alpha\bar{Y} + \bar{B}_1)^{-1}\bar{B}_1^T$, and having determined its normal Jordan form (see [8]), we can estimate the rate of convergence of the algorithm. That there are considerable difficulties in this approach is obvious; we shall now describe an alternative approach which is also applicable to the case when the matrix $\alpha\bar{Y} + \bar{B}$ is singular.

As a criterion for terminating the algorithm, we shall take the condition

$$\delta_n^{(j)} - \delta_n^{(j+1)} < \eta\delta_n^{(0)}, \quad \eta > 0,\ j \geq 0. \tag{3.113}$$

Since, except when stopping, $\delta_n^{(j)} - \delta_n^{(j+1)} \geq \eta\delta_n^{(0)}$ at each iteration the maximum number of iterations (j^0) must satisfy the relation

$$\begin{aligned}\eta\delta_n^{(0)} > \delta_n^{(j^0)} - \delta_n^{(j^0+1)} &= -\delta_n^{(0)} + \delta_n^{(0)} - \delta_n^{(1)} + \delta_n^{(1)} - \delta_n^{(2)} + \ldots + \\ &\quad + \delta_n^{(j^0-1)} - \delta_n^{(j^0)} + 2\delta_n^{(j^0)} - \delta_n^{(j^0+1)} \\ &\geq -\delta_n^{(0)} + \eta j^0 \delta_n^{(0)} + 2\delta_n^{(j^0)} - \delta_n^{(j^0+1)},\end{aligned}$$

from which

$$j^0 \leq [1 - \delta_n^{(j^0)}/\delta_n^{(0)}]/\eta \leq 1/\eta. \tag{3.114}$$

Let $\bar{m}_n = \inf_{P_n} [\|\bar{A}P_n\|/\|P_n\|]$. It is then easily seen that

$$\|(\alpha I + \bar{A})P_n\| \geq (\alpha + \bar{m}_n)\|P_n\|.$$

<u>Theorem 3.22</u> *If* (3.113) *is satisfied then*

$$\|x_n^{(j)} - \overset{*}{x}_n\|^2 = \sum_{k=1}^{n}\left|\alpha_k^{(j)} - \overset{*}{\alpha}_k\right|^2 \leq \frac{d_n\delta_n^{(0)}\eta}{(\bar{m}_n + \alpha)^2} \leq \frac{d_n\delta_n^{(0)}}{\alpha^2}\eta, \tag{3.115}$$

and

$$I^{(\alpha)}(x_n^{(j)}) - I^{(\alpha)}(\overset{*}{x}_n) \leq \frac{d_n\delta_n^{(0)}}{\bar{m}_n + \alpha}\eta \leq \frac{d_n\delta_n^{(0)}}{\alpha}\eta, \tag{3.116}$$

where

$$d_n = \sum_{s=1}^{n-1} (n-s)\|y_s\|^2 \leq [n(n-1)/2] \max_{1\leq s\leq n} \|y_s\|^2 . \tag{3.117}$$

Proof. We have that

$$\begin{aligned}\delta_n^{(j)} - \delta_n^{(j+1)} &= \delta_n^{(j)} - \delta_1^{(j+1)} + \delta_1^{(j+1)} - \delta_2^{(j+1)} + \dots + \delta_{n-1}^{(j+1)} - \delta_n^{(j+1)} \\ &= \sum_{s=1}^{n} (b_s^{(j+1)})^2 \Big[1 + \alpha/(y_s,y_s)\Big] \leq \eta\, \delta_n^{(0)}.\end{aligned} \tag{3.118}$$

But

$$\alpha\, \alpha_s^{(j-1)} + \sum_{k=1}^{s-1} \alpha_k^{(j)} \operatorname{Re}(y_k,y_s) + \sum_{k=s}^{n} \alpha_k^{(j-1)} \operatorname{Re}(y_k,y_s)$$

$$= \operatorname{Re}(y,y_s) - b_s^{(j)} \|y_s\| [1 + \alpha/(y_s,y_s)],$$

and recalling that $\alpha_k^{(j)} - \alpha_k^{(j-1)} = b_k^{(j)}/\|y_k\|$, we find

$$\alpha\, \alpha_s^{(j)} + \sum_{k=1}^{n} \alpha_k^{(j)} \operatorname{Re}(y_k,y_s) = \sum_{k=s+1}^{n} b_k^{(j)} \operatorname{Re}(y_k,y_s)/\|y_k\| + \operatorname{Re}(y,y_s).$$

Therefore

$$\begin{aligned}\sum_{k=1}^{n} \left|\alpha_k^{(j)} - \overset{*}{\alpha}_k\right|^2 &\leq \frac{1}{(\bar{m}_n+\alpha)^2} \sum_{s=1}^{n-1} \left(\sum_{k=s+1}^{n} \left| b_k^{(j)}\right| \cdot \|y_s\| \right)^2 \\ &\leq \frac{1}{(\bar{m}_n+\alpha)^2} \sum_{k=1}^{n} \left|b_k^{(j)}\right|^2 \sum_{s=1}^{n-1} (n-s)\|y_s\|^2 \\ &\leq \frac{d_n\, \delta_n^{(0)}}{(\bar{m}_n+\alpha)^2}\, \eta \leq \frac{d_n \delta_n^{(0)}}{\alpha^2}\, \eta ,\end{aligned}$$

which establishes (3.115).

For the proof of (3.116), we note from (3.105) that

$$I^{(\alpha)}(x_n^{(j)}) - I^{(\alpha)}(\overset{*}{x}_n) = \|Ax_n^{(j)} - A\overset{*}{x}_n\|^2 + \alpha\|x_n^{(j)} - \overset{*}{x}_n\|^2$$

$$= \sum_{s=1}^{n}\sum_{k=1}^{n} (\alpha_k^{(j)} - \overset{*}{\alpha}_k)(\alpha_s^{(j)} - \overset{*}{\alpha}_s)(y_k, y_s) \quad +$$

$$+ \alpha\sum_{s=1}^{n}\sum_{k=1}^{n} (\alpha_k^{(j)} - \overset{*}{\alpha}_k)(\alpha_s^{(j)} - \overset{*}{\alpha}_s)(e_k, e_s)$$

$$\leq \|\overset{*}{P}_n - P_n^{(j)}\| . \|(\alpha I + \bar{A})(\overset{*}{P}_n - P_n^{(j)})\|$$

$$\leq \frac{1}{(\bar{m}_n + \alpha)} \|(\alpha I + \bar{A})(\overset{*}{P}_n - P_n^{(j)})\|^2$$

$$\leq \frac{d_n \delta_n^{(0)}}{\bar{m}_n + \alpha} \eta \leq \frac{d_n \delta_n^{(0)}}{\alpha} \eta \quad . \qquad \#$$

As before, we shall assume that the absolute errors in the original coefficients $\mathrm{Re}(y_k, y_j)$ and $\mathrm{Re}\,(y, y_j)$ are less than Δ. We base the choice of η on the accuracy of the initial data. We denote by $\alpha I + \bar{A}$ and $\alpha I + \bar{A}^{\Delta}$, $\bar{y}$ and $\bar{y}^{\Delta}$, $\overset{*}{P}_n$ and $\overset{*}{P}_n^{\Delta}$ the matrices, the right hand sides, and the solutions of the exact and approximate systems (3.5) respectively. Furthermore we shall let $P_n^{(j)}$ and $P_n^{\Delta(j)}$ denote the exact and approximate results of the j th iteration for the solution of these systems by the algorithm (3.100). From the identity

$$(3.119) \qquad (\alpha I + \bar{A}^{\Delta})(\overset{*}{P}_n - \overset{*}{P}_n^{\Delta}) = \bar{y} - \bar{y}^{\Delta} + (\bar{A}^{\Delta} - \bar{A})\overset{*}{P}_n \ ,$$

it follows that

$$(3.120) \qquad \|\overset{*}{x}_n - \overset{*}{x}_n^{\Delta}\| = \|\overset{*}{P}_n - \overset{*}{P}_n^{\Delta}\| \leq \Big[\|\bar{y} - \bar{y}^{\Delta}\| + \|\bar{A}^{\Delta} - \bar{A}\| . \|\overset{*}{P}_n\|\Big] / (\bar{m}_n^{\Delta} + \alpha),$$

where
$$\bar{m}_n^{\Delta} = \inf_{P_n} \{\|\bar{A}^{\Delta} P_n\| / \|P_n\|\}.$$

But

$$\|\bar{y} - \bar{y}^{\Delta}\|^2 = \sum_{j=1}^{n} \left| \operatorname{Re}(y, y_j) - \operatorname{Re}(y^{\Delta}, y_j^{\Delta}) \right|^2 \le n\Delta^2, \tag{3.121}$$

$$\|\bar{A}^{\Delta} - \bar{A}\|^2 \le \sum_{j=1}^{n} \sum_{k=1}^{n} \left| \operatorname{Re}(y_k, y_j) - \operatorname{Re}(y_k^{\Delta}, y_j^{\Delta}) \right|^2 \le n^2 \Delta^2,$$

so that

$$\|\overset{*}{x}_n - \overset{*}{x}_n{}^{\Delta}\| \le \Delta\left[n^{\frac{1}{2}} + n\|\overset{*}{x}_n\|\right] / (\bar{m}_n{}^{\Delta} + \alpha). \tag{3.122}$$

From this and (3.115) we obtain

$$\|x_n^{\Delta(j)} - \overset{*}{x}_n\| = \| P_n^{\Delta(j)} - \overset{*}{P}_n\| \le \| P_n^{\Delta(j)} - \overset{*\Delta}{P}_n\| + \| \overset{*\Delta}{P}_n - \overset{*}{P}_n\| \tag{3.123}$$

$$\le \frac{\left(d_n^{\Delta} \ \delta_n^{(0)}\right)^{\frac{1}{2}}}{(\bar{m}_n^{\Delta} + \alpha)} \eta^{\frac{1}{2}} + \frac{n^{\frac{1}{2}} + n\|\overset{*}{x}_n\|}{(\bar{m}_n{}^{\Delta} + \alpha)} \Delta .$$

Assuming that

$$n\Delta / (\bar{m}_n{}^{\Delta} + \alpha) < 1, \tag{3.124}$$

we find that

$$\|\overset{*}{x}_n\| \le \left[\|x_n^{(j)}\| + \frac{\left(d_n^{\Delta} \ \delta_n^{(0)}\right)^{\frac{1}{2}}}{(\bar{m}_n{}^{\Delta} + \alpha)} \eta^{\frac{1}{2}} + \frac{n^{\frac{1}{2}}\Delta}{(\bar{m}_n^{\Delta} + \alpha)} \right] \left(1 - \frac{n\Delta}{(\bar{m}_n^{\Delta} + \alpha)}\right)^{-1} \tag{3.125}$$

and

$$\|x_n^{\Delta(j)} - \overset{*}{x}_n\| \le \left[\frac{\left(d_n^{\Delta} \ \delta_n^{(0)}\right)^{\frac{1}{2}}}{(\bar{m}_n{}^{\Delta} + \alpha)} \eta^{\frac{1}{2}} + \frac{\left(n^{\frac{1}{2}} + n\|x_n^{\Delta(j)}\|\right)}{(\bar{m}_n{}^{\Delta} + \alpha)} \Delta \right] \left(1 - \frac{n\Delta}{\bar{m}_n{}^{\Delta} + \alpha}\right)^{-1} . \tag{3.126}$$

Estimating d_n^{Δ} to be $n^2 (b_n^{\Delta})^2/2$, $b_n^{\Delta} = \max_{1 \le s \le n} \|y_s^{\Delta}\|$,

and assuming that the first term in the square brackets of (3.126) is equal to the second [17], we obtain

$$\delta_n^{(0)} \eta = 2 \left(n^{-\frac{1}{2}} + \|x_n^{\Delta(j)}\|\right)^2 \Delta^2 / (b_n^{\Delta})^2 . \tag{3.127}$$

---

(17) Obviously, other criteria are possible.

In practice, we apply (3.127) as follows : for a given η, we iterate until (3.113) is satisfied, we determine $\|x_n^{\Delta(j)}\|$ and calculate the right hand side of (3.127). If it is less than the left, we reduce η ; if it is greater, then we assume that η has been chosen correctly.

In addition, it is possible to estimate $\|\overset{*}{x}_n\|$ so that

$$\|\overset{*}{x}_n\| = \|\overset{*}{P}_n\| \leq \frac{\|\bar{y}\|}{(\bar{m}_n + \alpha)} \leq \frac{n^{\frac{1}{2}}\Delta + \|\bar{y}^{\Delta}\|}{\bar{m}_n^{\Delta} + \alpha - n\Delta} , \tag{3.128}$$

and to substitute (3.128) in (3.123). This gives an *a priori* estimate for η.

§3.5 The Steepest Descent Algorithm (L.V. Kantorovich)

It is well known that the function $\Delta_n(P)$ increases most rapidly in a direction which coincides with that of the gradient to the level surface $\Delta_n(P)$ = const, given by the vector

$$\text{grad } \Delta_n = \left(\frac{\partial\Delta_n}{\partial\alpha_1}, \frac{\partial\Delta_n}{\partial\alpha_2}, \ldots, \frac{\partial\Delta_n}{\partial\alpha_n} \right),$$

$$\frac{1}{2}\frac{\partial\Delta_n}{\partial\alpha_s} = \text{Re}\left(\sum_{k=1}^{n} \alpha_k y_k - y, y_s \right) + \alpha \,\text{Re}\left(\sum_{k=1}^{n} \alpha_k e_k, e_s \right), \; s = 1,2,\ldots,n. \tag{3.129}$$

Let us consider the following computational scheme for estimating $\inf\limits_P \Delta_n(P)$. We choose an initial point $P_0(\alpha_1^{(0)}, \alpha_2^{(0)},\ldots,\alpha_n^{(0)})$ and calculate

$$\Delta_n^{(0)} = \Delta_n(P_0) = \|y - \sum_{k=1}^{n} \alpha_k^{(0)} y_k\|^2 + \alpha\|\sum_{k=1}^{n} \alpha_k^{(0)} e_k\|^2 \;;$$

$$\left(\frac{\partial\Delta_n}{\partial\alpha_s}\right)_0 = \left.\frac{\partial\Delta_n}{\partial\alpha_s}\right|_{P=P_0} \quad s = 1,2,\ldots,n \;;$$

$$\sigma_n^{(0)} = \|\sum_{k=1}^{n} \left(\frac{\partial\Delta_n}{\partial\alpha_k}\right)_0 y_k\|^2 + \alpha\|\sum_{k=1}^{n} \left(\frac{\partial\Delta_n}{\partial\alpha_k}\right)_0 e_k\|^2 .$$

If $\sigma_n^{(0)} \neq 0$, then we find

$$t_0 = \left\{ \mathrm{Re}\left(y - \sum_{k=1}^{n} \alpha_k^{(0)} y_k, \sum_{k=1}^{n} \left(\frac{\partial \Delta_n}{\partial \alpha_k}\right)_0 y_k\right) - \alpha\, \mathrm{Re}\left(\sum_{k=1}^{n} \alpha_k^{(0)} e_k, \sum_{k=1}^{n} \left(\frac{\partial \Delta_n}{\partial \alpha_k}\right)_0 e_k\right)\right\} \Big/ \sigma_n^{(}$$

$$\Delta_n(P_0 + t_0 \operatorname{grad} \Delta_n|_{P=P_0}) = \min_t \Delta_n(P_0 + t \operatorname{grad} \Delta_n|_{P=P_0}) ;$$

$$\alpha_s^{(1)} = \alpha_s^{(0)} + t_0 \left(\frac{\partial \Delta_n}{\partial \alpha_s}\right)_0 , \quad s = 1,2,\dots,n ;$$

$$\Delta_n^{(1)} = \Delta_n(P_1), \quad P_1(\alpha_1^{(1)}, \alpha_2^{(1)}, \dots, \alpha_n^{(1)}) ;$$

$$\left(\frac{\partial \Delta_n}{\partial \alpha_s}\right)_1 = \frac{\partial \Delta_n}{\partial \alpha_s}\bigg|_{P=P_1} , \quad s = 1,2,\dots,n ;$$

$$\sigma_n^{(1)} = \left\| \sum_{k=1}^{n} \left(\frac{\partial \Delta_n}{\partial \alpha_k}\right)_1 y_k \right\|^2 + \alpha \left\| \sum_{k=1}^{n} \left(\frac{\partial \Delta_n}{\partial \alpha_k}\right)_1 e_k \right\|^2 .$$

If $\sigma_n^{(1)} \neq 0$, then we find

$$t_1 = \left\{ \mathrm{Re}\left(y - \sum_{k=1}^{n} \alpha_k^{(1)} y_k, \sum_{k=1}^{n} \left(\frac{\partial \Delta_n}{\partial \alpha_k}\right)_1 y_k\right) - \alpha\, \mathrm{Re}\left(\sum_{k=1}^{n} \alpha_k^{(1)} e_k, \sum_{k=1}^{n} \left(\frac{\partial \Delta_n}{\partial \alpha_k}\right)_1 e_k\right)\right\} \Big/ \sigma_n^{(}$$

$$\alpha_s^{(2)} = \alpha_s^{(1)} + t_1 \left(\frac{\partial \Delta_n}{\partial \alpha_s}\right)_1 , \quad s = 1,2,\dots,n ;$$

$$\Delta_n^{(2)} = \Delta(P_2) , \quad P_2(\alpha_1^{(2)}, \alpha_2^{(2)}, \dots, \alpha_n^{(2)}) ;$$

$$\left(\frac{\partial \Delta_n}{\partial \alpha_s}\right)_2 = \frac{\partial \Delta_n}{\partial \alpha_s}\bigg|_{P=P_2} , \quad s = 1,2,\dots,n ;$$

$$\sigma_n^{(2)} = \left\| \sum_{k=1}^{n} \left(\frac{\partial \Delta_n}{\partial \alpha_k}\right)_2 y_k \right\|^2 + \alpha \left\| \sum_{k=1}^{n} \left(\frac{\partial \Delta_n}{\partial \alpha_k}\right)_2 e_k \right\|^2 ,$$

etc. In general, if at the $(j+1)^{\text{th}}$ step $(j = 0,1,2,\dots)$

$$\sigma_n^{(j)} = \left\| \sum_{k=1}^{n} \left(\frac{\partial \Delta_n}{\partial \alpha_k}\right)_j y_k \right\|^2 + \alpha \left\| \sum_{k=1}^{n} \left(\frac{\partial \Delta_n}{\partial \alpha_k}\right)_j e_k \right\|^2 \neq 0 , \tag{3.130}$$

then we find

$$t_j = \left\{ \mathrm{Re}\left(y - \sum_{k=1}^{n} \alpha_k^{(j)} y_k, \sum_{k=1}^{n} \left(\frac{\partial \Delta_n}{\partial \alpha_k}\right)_j y_k\right) - \alpha\, \mathrm{Re}\left(\sum_{k=1}^{n} \alpha_k^{(j)} e_k, \sum_{k=1}^{n} \left(\frac{\partial \Delta_n}{\partial \alpha_k}\right)_j e_k\right)\right\} \Big/ \sigma_n^{(j)}$$

$$\alpha_s^{(j+1)} = \alpha_s^{(j)} + t_j \left(\frac{\partial \Delta_n}{\partial \alpha_s}\right)_j , \quad s = 1,2,\dots,n ;$$

$$\Delta_n^{(j+1)} = \Delta_n(P_{j+1}), \quad P_{j+1}(\alpha_1^{(j+1)}, \alpha_2^{(j+1)}, \dots, \alpha_n^{(j+1)}) ; \tag{3.131}$$

$$\left(\frac{\partial \Delta_n}{\partial \alpha_s}\right)_{j+1} = \frac{\partial \Delta_n}{\partial \alpha_s}\bigg|_{P=P_{j+1}} , \quad s = 1,2,\dots,n ;$$

$$x_n^{(j+1)} = \sum_{k=1}^{n} \alpha_k^{(j+1)} e_k .$$

When $\alpha = 0$, we denote by Q_n the projection operator which maps elements of the form $\sum_{k=1}^{n} \alpha_k e_k$ onto the subspace $L_n(A)$ comprising the zeros of the operator A of the same form; and by $\overset{*}{x}_n = \sum_{k=1}^{n} \overset{*}{\alpha}_k e_k$, the unique element which minimizes $I\,(\sum_{k=1}^{n} \alpha_k e_k)$, and which is orthogonal to $L_n(A)$.

Theorem 3.23. *Algorithm* (3.131) *always converges in the sense that*

(3.132) $$\lim_{j\to\infty} \Delta_n(P_j) \equiv \lim_{j\to\infty} I^{(\alpha)}(x_n^{(j)}) = \inf_P \Delta_n(P) = \inf_{\alpha_k}\left\{I^{(\alpha)}(\sum_{k=1}^{n} \alpha_k e_k)\right\}.$$

In addition, when $\alpha = 0$

(3.133) $$\lim_{j\to\infty} \|x_n^{(j)} - \overset{*}{x}_n - Q_n(\sum_{k=1}^{n} \alpha_k^{(0)} e_k)\| = 0 \;;$$

and when $\alpha > 0$,

(3.134) $$\lim_{j\to\infty} \|x_n^{(j)} - \overset{*}{x}_n\| = 0 \;.$$

The proof of this theorem will not be considered here because, for $\alpha = 0$, it is proved in [59]; it is also a consequence of the more general result of [151] (see also Theorem 5.4). For $\alpha > 0$, Theorem 3.23 is a consequence of Theorem 3.24.

When using this algorithm on a computer, we shall take as a termination criterion either

(3.135) $$\Delta_n^{(j)} - \Delta_n^{(j+1)} < \eta\Delta_n^{(0)} \quad \text{or} \quad \sigma_n^{(j)} < \sigma \;,$$

where σ and η are arbitrarily chosen positive constants.

Theorem 3.24. *If* (3.135) *is satisfied then*

(3.136) $$\|x_n^{(j)} - \overset{*}{x}_n\|^2 \le \frac{M_n^2+\alpha}{(m_n^2+\alpha)^2}\eta\Delta_n^{(0)} \quad \textit{or} \quad \|x_n^{(j)} - \overset{*}{x}_n\|^2 \le \frac{\sigma}{4(m_n^2+\alpha)^3} \;;$$

and

(3.137) $$I^{(\alpha)}(x_n^{(j)}) - I^{(\alpha)}(\overset{*}{x}_n) \le \frac{M_n^2+\alpha}{m_n^2+\alpha}\,\eta\Delta_n^{(0)}$$

$$\textit{or} \quad I^{(\alpha)}(x_n^{(j)}) - I^{(\alpha)}(\overset{*}{x}_n) \le \frac{\sigma}{4(m_n^2+\alpha)^2} \;;$$

where

$$(3.138)\quad m_n = \inf_{\alpha_k} \left\{ \|A(\sum_{k=1}^n \alpha_k e_k)\| \Big/ \|\sum_{k=1}^n \alpha_k e_k\| \right\},\ M_n = \sup_{\alpha_k} \left\{ \|A(\sum_{k=1}^n \alpha_k e_k)\| \Big/ \|\sum_{k=1}^n \alpha_k e_k\| \right\}$$

Proof. After simple manipulations, we obtain

$$\Delta_n^{(j)} - \Delta_n^{(j+1)} = t_j^2 \, \sigma_n^{(j)}$$

$$= \left\{ \mathrm{Re}\Big(y - \sum_{k=1}^n \alpha_k^{(j)} y_k, \sum_{k=1}^n \Big(\frac{\partial \Delta_n}{\partial \alpha_k}\Big)_j y_k\Big) - \alpha \, \mathrm{Re}\Big(\sum_{k=1}^n \alpha_k^{(j)} e_k, \sum_{k=1}^n \Big(\frac{\partial \Delta_n}{\partial \alpha_k}\Big)_j e_k\Big) \right\}^2 \Big/ \sigma_n^{(j)}$$

(3.139)

$$= 4 \Big[\sum_{s=1}^n (\gamma_s^{(j)})^2\Big]^2 \Big/ \sigma_n^{(j)} ,$$

$$\gamma_s^{(j)} = \sum_{k=1}^n \alpha_k^{(j)} \mathrm{Re}[(y_k, y_s) + \alpha(e_k, e_s)] - \mathrm{Re}(y, y_s) .$$

From (3.130) and (3.138), we find

$$\sigma_n^{(j)} = 4\|A(\sum_{s=1}^n \gamma_s^{(j)} e_s)\|^2 + 4\alpha \|\sum_{s=1}^n \gamma_s^{(j)} e_s\|^2 ,$$

(3.140)

$$4(m_n^2 + \alpha) \sum_{s=1}^n (\gamma_s^{(j)})^2 \le \sigma_n^{(j)} \le 4(M_n^2 + \alpha) \sum_{s=1}^n (\gamma_s^{(j)})^2 .$$

From (3.139) and (3.140) it follows that

$$(3.141)\quad \frac{\sum_{s=1}^n (\gamma_s^{(j)})^2}{M_n^2 + \alpha} \le \Delta_n^{(j)} - \Delta_n^{(j+1)} \le \frac{\sum_{s=1}^n (\gamma_s^{(j)})^2}{m_n^2 + \alpha} ,$$

so that from (3.140) we obtain

$$(3.142)\quad \sum_{s=1}^n (\gamma_s^{(j)})^2 < (M_n^2 + \alpha)\eta\Delta_n^{(0)} \quad \text{or} \quad \sum_{s=1}^n (\gamma_s^{(j)})^2 < \sigma/[4(m_n^2 + \alpha)] ,$$

depending upon which of (3.135) is satisfied.

As in the proof of Theorem 3.22, we find

$$(3.143)\quad \|x_n^{(j)} - \overset{*}{x}_n\|^2 = \sum_{k=1}^n |\alpha_k^{(j)} - \overset{*}{\alpha}_k|^2 \le \sum_{s=1}^n (\gamma_s^{(j)})^2 \Big/ (\bar{m}_n + \alpha)^2 ,$$

from which

$$(3.144)\quad \|x_n^{(j)} - \overset{*}{x}_n\|^2 \le \frac{M_n^2 + \alpha}{(\bar{m}_n + \alpha)^2} \eta\Delta_n^{(0)} \quad \text{or} \quad \|x_n^{(j)} - \overset{*}{x}_n\|^2 \le \frac{\sigma}{4(\bar{m}_n + \alpha)^3} .$$

In addition,

$$I^{(\alpha)}(x_n^{(j)}) - I^{(\alpha)}(\overset{*}{x}_n) = \|Ax_n^{(j)} - A\overset{*}{x}_n\|^2 + \alpha\|x_n^{(j)} - \overset{*}{x}_n\|^2$$

$$= \sum_{s=1}^{n} |\alpha_s^{(j)} - \overset{*}{\alpha}_s|\gamma_s^{(j)} \leq \|x_n^{(j)} - \overset{*}{x}_n\| \left(\sum_{s=1}^{n} (\gamma_s^{(j)})^2\right)^{1/2}$$

and

$$(3.145)\quad I^{(\alpha)}(x_n^{(j)})-I^{(\alpha)}(\overset{*}{x}_n) < \frac{M_n^2+\alpha}{\bar{m}_n+\alpha}\,\eta\Delta_n^{(0)} \ \text{ or } \ I^{(\alpha)}(x_n^{(j)})-I^{(\alpha)}(\overset{*}{x}_n) < \frac{\sigma}{4(\bar{m}_n+\alpha)(m_n^2+\alpha)} .$$

It remains to note that

$$(3.146)\qquad \bar{m}_n = \inf_{P_n} \frac{\|\bar{A}P_n\|}{\|P_n\|} = \inf_{P_n} \frac{(\bar{A}P_n, P_n)}{\|P_n\|^2} = \inf_{\alpha_k} \frac{\|A(\sum_{k=1}^{n} \alpha_k e_k)\|^2}{\|\sum_{k=1}^{n} \alpha_k e_k\|^2} = m_n^2 ,$$

and to substitute m_n^2 for $\bar{m}_n$ in (3.144) and (3.145). #

Since, except when terminating, we have at each iteration that $\Delta_n^{(j)} - \Delta_n^{(j+1)} \geq \eta\Delta_n^{(0)}$ or $\sigma_n^{(j)} \geq \sigma$, the maximum number of iterations j^0 must satisfy

$$\Delta_n^{(j^0)}-\Delta_n^{(j^0+1)} = -\Delta_n^{(0)}+\Delta_n^{(0)}-\Delta_n^{(1)}+\Delta_n^{(1)}-\Delta_n^{(2)}+\ldots+\Delta_n^{(j^0-1)}-\Delta_n^{(j^0)}+2\Delta_n^{(j^0)}-\Delta_n^{(j^0+1)}$$

$$\geq -\Delta_n^{(0)} + \eta j^0\Delta_n^{(0)} + 2\Delta_n^{(j^0)} - \Delta_n^{(j^0+1)} ,$$

or (from (3.140) and (3.141))

$$\Delta_n^{(j^0)} - \Delta_n^{(j^0+1)} \geq -\Delta_n^{(0)} + \sigma j^0/[4(M_n^2+\alpha)^2] + 2\Delta_n^{(j^0)} - \Delta_n^{(j^0+1)} .$$

Thus we have

$$(3.147)\qquad j^0 \leq [1 - \Delta_n^{(j^0)}/\Delta_n^{(0)}]/\eta \leq 1/\eta, \quad \text{or} \quad j^0 \leq 4\Delta_n^{(0)}(M_n^2+\alpha)^2/\sigma .$$

We shall now make an estimate of the number of iterations as a function of the error in the initial data. We shall make the same assumptions and use the same notation as in §3.4. From (3.122) and (3.136), we obtain

$$\|x_n^{\Delta(j)} - \overset{*}{x}_n\| \leq \|x_n^{\Delta(j)} - \overset{*\Delta}{x}_n\| + \|\overset{*\Delta}{x}_n - \overset{*}{x}_n\|$$

$$\leq \frac{[(M_n^\Delta)^2+\alpha]^{1/2}}{(m_n^\Delta)^2+\alpha}\,(\eta\Delta_n^{(0)})^{1/2} + \frac{n^{1/2}+n\|\overset{*}{x}_n\|}{(m_n^\Delta)^2+\alpha}\,\Delta$$

or

$$\|x_n^{\Delta(j)} - \overset{*}{x}_n\| \leq \frac{\sigma^{1/2}}{2[(m_n^\Delta)^2+\alpha]^{3/2}} + \frac{n^{1/2}+n\|\overset{*}{x}_n\|}{(m_n^\Delta)^2+\alpha}\,\Delta .$$

Assuming

$$n\Delta/[(m_n^{\Delta})^2+\alpha] < 1 \quad , \tag{3.148}$$

we find that

$$\|\overset{*}{x}_n\| \leq \|x_n^{\Delta(j)}\| + \left\{ \frac{[(M_n^{\Delta})^2+\alpha]^{\frac{1}{2}}}{(m_n^{\Delta})^2+\alpha} (\eta\Delta_n^{(0)})^{\frac{1}{2}} + \frac{n^{\frac{1}{2}}\Delta}{(m_n^{\Delta})^2+\alpha} \right\} \left(1 - \frac{n\Delta}{(m_n^{\Delta})^2+\alpha}\right)^{-1} ,$$

$$\|x_n^{\Delta(j)} - \overset{*}{x}_n\| \leq \left\{ \frac{[(M_n^{\Delta})^2+\alpha]^{\frac{1}{2}}}{(m_n^{\Delta})^2+\alpha} (\eta\Delta_n^{(0)})^{\frac{1}{2}} + \frac{n^{\frac{1}{2}}+n\|x_n^{\Delta(j)}\|}{(m_n^{\Delta})^2+\alpha} \Delta \right\} \left(1 - \frac{n\Delta}{(m_n^{\Delta})^2+\alpha}\right)^{-1} ;$$

(3.149)

or

$$\|\overset{*}{x}_n\| \leq \|x_n^{\Delta(j)}\| + \left\{ \frac{\sigma^{\frac{1}{2}}}{2[(m_n^{\Delta})^2+\alpha]^{3/2}} + \frac{n^{\frac{1}{2}}\Delta}{(m_n^{\Delta})^2+\alpha} \right\} \left(1 - \frac{n\Delta}{(m_n^{\Delta})^2+\alpha}\right)^{-1} ,$$

$$\|x_n^{\Delta(j)} - \overset{*}{x}_n\| \leq \left\{ \frac{\sigma^{\frac{1}{2}}}{2[(m_n^{\Delta})^2+\alpha]^{3/2}} + \frac{n^{\frac{1}{2}}+n\|x_n^{\Delta(j)}\|}{(m_n^{\Delta})^2+\alpha} \Delta \right\} \left(1 - \frac{n\Delta}{(m_n^{\Delta})^2+\alpha}\right)^{-1} .$$

Putting the first term equal to the second in the square brackets of the second expression in (3.149), we obtain

$$\eta\Delta_n^{(0)} = [n^{\frac{1}{2}} + n\|x_n^{\Delta(j)}\|]^2\Delta^2/[(M_n^{\Delta})^2+\alpha] \ , \tag{3.150}$$

or

$$\sigma = 4[n^{\frac{1}{2}} + n\|x_n^{\Delta(j)}\|]^2\Delta^2/[(m_n^{\Delta})^2+\alpha] \ .$$

It is also possible to estimate $\|\overset{*}{x}_n\|$ in advance:

$$\begin{aligned} \|\overset{*}{x}_n\| = \|\overset{*}{P}_n\| &\leq \|\bar{y}\|/(\bar{m}_n+\alpha) \\ &\leq [n^{\frac{1}{2}}\Delta + \|\bar{y}^{\Delta}\|]/[(m_n^{\Delta})^2+\alpha-n\Delta] \ . \end{aligned} \tag{3.151}$$

A priori estimates may now be obtained for η and σ.

§3.6 Some Other Algorithms

Using the ideas of [157, 152, 153], we can introduce coordinate-wise and steepest descent algorithms which minimize directly the norm of the error at each step. For example, we can use the coordinate-wise descent algorithm to minimize $\|\overset{*}{x}-x_{n,s}^{(j)}-az_s\|$ where $\overset{*}{x}$ is the exact solution of $Ax = y$ (assumed to be unique), $x_{n,s}^{(j)}$ is an approximate solution, $z_s = A^*e_s$ (A^* being the operator conjugate to A), and $\{z_s\}$ is a complete system in $D(A)$. The required value a^0 is given by

$$(3.152) \qquad a^0 = \frac{(\overset{*}{x} - x_{n,s}^{(j)}, z_s)}{(z_s, z_s)} = \frac{(y - Ax_{n,s}^{(j)}, e_s)}{(z_s, z_s)} .$$

Theorems analogous to those proved above can be given for the convergence of these algorithms.

The above conditions imply that instead of solving (3.5), we solve the system (3.25). Obviously the Gaussian elimination with pivoting algorithm as given in §3.3 can be applied to the solution of the latter system of equations.

The coordinate-wise and steepest descent algorithms are related to a number of local search algorithms, and they are particular examples of a large class of so-called *steep descent* algorithms [134], which differ from each other both by the way in which the choice of direction of descent of the minimizing function is made, and also by the way in which the least value of this function in the given direction is estimated. As an example of a local search algorithm which differs from those described, we can consider an algorithm (see [60]) which is a combination of a steep descent and a random search algorithm. A significant class of steep descent algorithms has been given for the numerical solution of the Cauchy problem for the system of equations

$$(3.153) \qquad \frac{dP_n(t)}{dt} = - \text{grad}\, \Delta_n(P_n) \ , \ P_0 = P(t_0) \ ,$$

which has a solution $P_n(t) \to \overset{0}{P}$ as $t \to \infty$, where $\Delta_n(\overset{0}{P}) = \inf_P \Delta_n(P)$.

As an example of a non-local steep descent algorithm where the direction of descent at each step is chosen from the results of the preceding steps, we may consider the so-called *conjugate gradient* algorithms. However these algorithms require that $\inf \Delta_n(P)$ be known. Since the systems (3.5) and (3.25) always have solutions, the conjugate gradient algorithm (and many others) can be used for their approximate solution. For details, see [136]. It should be noted that little investigation has yet been done on problems connected with the estimation of the effects of round-off errors.

§4 THE RITZ-GALERKIN METHOD

§4.1 Connection with the Method of Least Squares

Let H_1 and H_2 be Hilbert spaces, and $Ax = y$ be an equation with a linear operator A which maps from H_1 into H_2, with domain $D(A) \subseteq H_1$. We assume that $e_1, e_2, \ldots, e_n, \ldots$ and $v_1, v_2, \ldots, v_n, \ldots$ are two complete systems in $D(A)$ and H_2 respectively. We denote by $\tilde{v}_1, \tilde{v}_2, \ldots, \tilde{v}_m$ $(m \leq n)$, an orthogonal linearly independent system of elements which is equivalent to $v_1, v_2, \ldots, v_n$; and by P_n the projection operator in H_2 defined by,

$$P_n v = \sum_{s=1}^{m} \frac{(v, \tilde{v}_s)}{(\tilde{v}_s, \tilde{v}_s)} \tilde{v}_s \quad , \quad v \in H_2. \tag{4.1}$$

If $\tilde{v}_1, \tilde{v}_2, \ldots, \tilde{v}_m$ is a maximal linearly independent system of elements from $v_1, v_2, \ldots, v_n$ then the following relations are valid, in which $\beta_s^{(j)}$ and $\gamma_j^{(s)}$ are constants:

$$\begin{aligned} v_j &= \sum_{s=1}^{j-1} \beta_s^{(j)} \tilde{v}_s + \tilde{v}_j \ , \ j = 1,2,\ldots,m \ , \\ \tilde{v}_s &= \sum_{j=1}^{s-1} \gamma_j^{(s)} v_j + v_s \ , \ s = 1,2,\ldots,m \ , \\ v_j &= \sum_{s=1}^{m} \beta_s^{(j)} \tilde{v}_s \ , \ j = m+1, m+2,\ldots,n \ . \end{aligned} \tag{4.2}$$

We seek an approximate solution of the equation $Ax = y$ in the form $x_n = \sum_{k=1}^{n} \alpha_k e_k$, using the condition

$$\sum_{k=1}^{n} \alpha_k (Ae_k, v_j) = (y, v_j), \ j = 1,2,\ldots,n \ . \tag{4.3}$$

From equations (4.2), we see that (4.3) is equivalent to the system of equations

$$\sum_{k=1}^{n} \alpha_k (Ae_k, \tilde{v}_j) = (y, \tilde{v}_j) \ , \ j = 1,2,\ldots,m \ , \tag{4.4}$$

which can be written as

$$P_n A x_n = P_n y \ . \tag{4.5}$$

If (4.5) has a solution, then it obviously minimises

$$(4.6)\qquad \delta_n(P) = \|y - \sum_{j=1}^{m} \frac{(Ax_n, \tilde{v}_j)}{(\tilde{v}_j, \tilde{v}_j)} \tilde{v}_j\| = \|y - \sum_{k=1}^{n} \alpha_k P_n Ae_k\| ,$$

$$P = P(\alpha_1, \alpha_2, \ldots, \alpha_n) .$$

It may happen that (4.5), or equivalently (4.3), does not possess a solution. However the problem of the calculation of the values $\mathring{\alpha}_k$ which minimize (4.6), is always solvable (see [59] and Theorem 3.1), and the unknowns $\mathring{\alpha}_k$ satisfy the system of equations

$$(4.7)\qquad \sum_{k=1}^{n} \alpha_k (P_n Ae_k, P_n Ae_j) = (y, P_n Ae_j),\ j = 1,2,\ldots,n .$$

If (4.3) is written in matrix form as

$$(4.8)\qquad UP = \bar{y},$$

then it is readily verified that the system of equations (4.7) is equivalent to

$$(4.9)\qquad U^*UP = U^*\bar{y} ,$$

where U^* is the complex conjugate transpose of the matrix U.

Let us formulate the above results in the form of a theorem.

Theorem 4.1. *If the algebraic system* (4.3) *defined by the Ritz-Galerkin method has a solution, then this solution minimizes* (4.6). *On the other hand, the problem of the minimization of* (4.6) *is always solvable, and is equivalent to the problem of solving* (4.7), *which is in turn equivalent to the system of equations* (4.9).

In this way, the Ritz-Galerkin(18) method in a Hilbert space is equivalent to the method of least squares when $P_n Ae_k$ replaces Ae_k $(= y_k)$. However it follows that in a Hilbert space, the Ritz-Galerkin method is worse than the least squares method because the best operator P_n from the point of view of ease of calculation of $P_n y_k$ and the solution of (4.5), is obviously the operator P_n^0 $(P_n^0 y_k = y_k,\ k = 1,2,\ldots,n)$.

If A is a positive operator(19), and the system $\{v_k\}$ coincides with $\{e_k\}$, then the corresponding Ritz-Galerkin method is examined in [103]. Problems concerning the solvability of the system (4.3) no longer arise, since they are equivalent to the problems of the solvability of systems arising from the method of least squares for the solution of $Bx = v$, where $Bv = y$,

(18) In the literature, this method is also called the "projection" method.

(19) An operator A in a Hilbert space H is called *positive* $(A \geq 0)$, if $(Ax, x) \geq 0$ for every $x \in H$.

and B is the square root of the operator A i.e. $B^2 = A$(see [75]).

Important recommendations concerning the choice of the systems $\{e_k\}$ and $\{v_k\}$ are given in [103, 105].

§4.2 Properties of the Ritz-Galerkin Method

Properties of the Ritz-Galerkin method are difficult to establish. Let us give a very general theorem concerning its convergence.

Theorem 4.2. *In the case when both $Ax = y$ and the system* (4.3) *have solutions, then for every infinite sequence $\{n_r\}$, the convergence of the projection of the residual is ensured, i.e. $\|P_{n_r}(y - Ax_{n_r})\| \to 0$ as $n_r \to \infty$.*

Proof. We have that

$$\|P_n(y - Ax_n)\| \le \|P_n y - y\| + \|y - \sum_{k=1}^{n} \overset{0}{\alpha}_k P_n A e_k\|$$

$$(4.10) \qquad \le \|P_n y - y\| + \|y - \sum_{k=1}^{n} \overset{*}{\alpha}_k P_n A e_k\|$$

$$\le 2\|P_n y - y\| + \|y - \sum_{k=1}^{n} \overset{*}{\alpha}_k A e_k\| ,$$

where $\|y - \sum_{k=1}^{n} \overset{*}{\alpha}_k A e_k\| = \inf_{\alpha_k} \|y - \sum_{k=1}^{n} \alpha_k A e_k\|$.

The proof of the theorem now readily follows[†]. #

Obviously, the residual will only converge under the condition that the angle between the residual and its projection either remains less than $\pi/2$ for all n greater than a given n_0, or converges sufficiently slowly to $\pi/2$ as n increases. More precisely, if

$$(4.11) \qquad \|P_{n_r}(y - \sum_{k=1}^{n_r} \overset{0}{\alpha}_k A e_k)\| \ge C_{n_r} \|y - \sum_{k=1}^{n_r} \overset{0}{\alpha}_k A e_k\| \quad , \quad 0 < C_{n_r} < 1 ,$$

then

$$(4.12) \qquad \|y - \sum_{k=1}^{n_r} \overset{0}{\alpha}_k A e_k\| \le \left\{ 2\|P_{n_r} y - y\| + \|y - \sum_{k=1}^{n_r} \overset{0}{\alpha}_k A e_k\| \right\} / C_{n_r} ,$$

i.e. the norm of the residual tends to zero if $\|P_{n_r} y - y\| / C_{n_r}$ tends to zero as $n_r \to \infty$.

A very general condition which ensures the convergence of the residual, has been established in [117].

---

† Conditions on A and $\{e_k\}$ must be such that $\lim_{n\to\infty} \|y - \sum_{k=1}^{n} \overset{*}{\alpha}_k A e_k\| = 0$. (Eds.)

Theorem (4.3). (N.I. Pol'skiy [117]). *Suppose the equation* $Ax = y$ *has a solution and that for all* n *(greater than a given* n_0*) there exists* $C > 0$ *(independent of* n*) such that*

$$(4.13) \qquad \|P_n \sum_{k=1}^{n} \alpha_k A e_k\| \geq C \|\sum_{k=1}^{n} \alpha_k A e_k\|$$

for any set $\{\alpha_k\}$, *where* $\{Ae_k\}$ *are linearly independent. Then for all* $n > n_0$, *equation* (4.5) *(and hence the system* (4.3)*) possesses a solution and* $Ax_n \to Ax$ *with respect to the norm in the space* H_2.(20)

The basic advantages of the Ritz-Galerkin method become apparent when it is applied to the approximation of eigenvalues (see [116]) and to the approximate solution of equations in a Banach space.

As is shown in [75], the widely used method of collocation is a special case of the Ritz-Galerkin method in a Banach space. In the collocation method we write $x_n = \sum_{k=1}^{n} \alpha_k e_k$, and choose the coefficients α_k such that the equation $Ax = y$ is satisfied exactly by x_n at certain prescribed points t_j. Let A be a linear operator which maps a Banach space X into itself. We choose a complete system of elements $\{e_k\}$ in X, and a complete system of functionals $\{\psi_j\}$ in a conjugate space X^*. The coefficients α_k are defined by the conditions

$$(4.14) \qquad \psi_j(Ax_n) = \psi_j(y) \;, \; j = 1,2,\ldots,n \;.$$

It is obvious that the method of collocation can be considered as a Ritz-Galerkin method, if we take each ψ_j to be defined as the value of the function at a given point t_j.

(20) In [117], the corresponding theorem is stated and proved in the more general case when H_1 and H_2 are Banach spaces.

§5 ITERATIVE METHODS

§5.1 Simple Iteration and Analytic Continuation

Assuming that the operator A maps from a Hilbert space H into itself, we can write equation (2.1) in the form

$$x = x - Ax + y \quad \text{or} \quad x = Tx + y,\ T = I - A. \tag{5.1}$$

If

$$\|T\| \le q < 1\ , \tag{5.2}$$

then equation (5.1) has a unique solution x which can be determined by simple iteration:

$$\begin{aligned} x_n &= Tx_{n-1} + y\ , \qquad n = 1,2,\dots\ , \\ &= y + Ty + \dots + T^{n-1}y + T^n x_0,\quad x_0 \in H\ , \\ x &= x_0 + (x_1 - x_0) + (x_2 - x_1) + \dots + (x_n - x_{n-1}) + \dots \\ &= x_0 + (Tx_0 + y - x_0) + T(Tx_0 + y - x_0) + \dots + \\ &\qquad + T^{n-1}(Tx_0 + y - x_0) + \dots \\ &= y + Ty + T^2 y + \dots + T^n y + \dots\ . \end{aligned} \tag{5.3}$$

Hence,

$$\|x - x_n\| \le \frac{q^n}{1-q}\ \|Tx_0 + y - x_0\| = \frac{q^n}{1-q}\ \|y - Ax_0\|\ . \tag{5.4}$$

If condition (5.2) is not satisfied by $T = I - A$, then in certain cases it is possible to transform equation (2.1) into an equivalent equation of the form $x = Tx + y$, where T does now satisfy condition (5.2). We shall show how this may be done in the case when A is a positive operator. The importance of this case follows from the observation that every solution of $Ax = y$ (when A is an arbitrary linear operator) is also a solution of the equation $A^*Ax = A^*y$ (A^* is conjugate to A), and A^*A is a positive operator.

Thus, let A satisfy the condition

$$(Ax,\ x) \ge 0\ , \tag{5.5}$$

for every $x \in H$. We denote the upper and lower bounds of the operator A by M and m respectively, so that

$$M = \sup_{\|x\|=1} (Ax,\ x)\ ,\ m = \inf_{\|x\|=1} (Ax,\ x)\ . \tag{5.6}$$

We examine two cases:

1. $m > 0$. The simplest method (see [26]) is to transform equation (2.1) into the form

$$(5.7) \qquad x = x + \beta(y-Ax), \quad \text{or} \quad x = Tx + \beta y,\ T = I - \beta A .$$

The parameter β is chosen so that $\|T\| = \|I-\beta A\|$ is minimised. Since (see [75, Ch. 9 and 16]) the norm of the operator $I-\beta A$ is equal to the maximum of $|1-\beta m|$ and $|1-\beta M|$, we find that

$$(5.8) \qquad \beta = \frac{2}{M+m}, \quad \text{and} \quad \|T\| = \|I - \frac{2}{M+m}A\| = \frac{M-m}{M+m} < 1 .$$

Solving (5.7) by simple iteration, we find in a manner analogous to that used in the derivation of (5.4),

$$(5.9) \qquad \|x - x_n\| \le \frac{1}{m}\left(\frac{M-m}{M+m}\right)^n \|y - Ax_0\| .$$

This procedure can be generalized (see [10, 26]). We transform equation (2.1) to the form

$$(5.10) \qquad x = x + P_r(A)(y - Ax) ,$$

$$\text{or } x = Tx + P_r(A)y,\ T = I - P_r(A)A ,$$

where P_r is a polynomial of degree r. It is known that (see [75])

$$(5.11) \qquad \|T\| = \|I - P_r(A)A\| \le \max_{t\in[m,M]} |1 - tP_r(t)| .$$

We want to choose the polynomial P_r so that $\|T\|$ is minimized. We shall then have a polynomial Q_{r+1} which is such that $Q_{r+1}(0) = 1$, and deviates least from zero on the interval $[m,M]$. But (see [109]) the Chebyshev polynomial $T_{r+1}(u) = 2^{-r}\cos[(r+1)\arccos u]$, deviates least from zero on the interval $[-1,1]$. Hence $\tilde{T}_{r+1}(t) = T_{r+1}\{[2t - (M+m)]/(M-m)\}$ deviates least from zero on $[m,M]$. Its constant term (i.e. coefficient of t^0) is given by $\tilde{T}_{r+1}(0) = T_{r+1}[(M+m)/(M-m)]$.(21)

It is clear that the required polynomial Q_{r+1} is given by

$$(5.12) \qquad Q_{r+1}(t) = 1 - tP_r(t) = \frac{T_{r+1}\{[2t - (M+m)]/(M-m)\}}{T_{r+1}\{(M+m)/(M-m)\}} ,$$

---

(21) Since $T_{r+1}(u) = 2^{-r}\cos[(r+1)\arccos u]$ is a polynomial, it is defined for all u. In practice, for the evaluation of $T_{n+1}(u)$, it is convenient to use the recurrence relation

$$T_{n+1}(u) = u\,T_n(u) - \frac{1}{4}T_{n-1}(u) , \quad T_0(u) = 2, \quad T_1(u) = u .$$

and the desired minimum norm is such that

$$(5.13)\qquad \|T\| = \|I - AP_r(A)\| \leq \frac{1}{2^r\, T_{r+1}\,\{(M+m)/(M-m)\}} \leq \left(\frac{M-m}{M+m}\right)^{r+1} .$$

The last inequality is a consequence of the Chebyshev inequality, see [109, p.78].

Solving (5.10) by simple iteration we obtain

$$(5.14)\qquad \|x - x_n\| \leq \frac{\{2^r\, T_{r+1}[(M+m)/(M-m)] + 1\}\, \|y - Ax_0\|}{m\{2^r\, T_{r+1}[(M+m)/(M-m)] - 1\}\{2^r\, T_{r+1}[(M+m)/(M-m)]\}^n} .$$

2. $m \geq 0$. Following [91, 129], we introduce the operator $A^{(\alpha)} = A + \alpha I$, $\alpha \geq 0$ instead of the operator A, and the equation

$$(5.15)\qquad A^{(\alpha)}x = y ,$$

instead of equation (2.1). We denote by $\overset{*}{x}$ the solution of equation (2.1) which is orthogonal to the null space of A; the corresponding solution of equation (5.15) is denoted by $\overset{*}{x}^{(\alpha)}$.

Theorem 5.1. *As* $\alpha \to 0$, $\|\overset{*}{x} - \overset{*}{x}^{(\alpha)}\| \to 0$. *In addition*

$$(5.16)\qquad \|\overset{*}{x} - \overset{*}{x}^{(\alpha)}\| \leq [\delta_R + (\delta_R^2 + \alpha R^2)^{\frac{1}{2}}]/2 ,$$

and

$$(5.17)\qquad \|A(\overset{*}{x} - \overset{*}{x}^{(\alpha)})\| = \alpha\|\overset{*}{x}^{(\alpha)}\| \leq \alpha\|\overset{*}{x}\| ,$$

where

$$(5.18)\qquad \delta_R = \inf_{\|w\|\leq R} \|\overset{*}{x} - Uw\| , \quad U^2 = A .$$

Proof. We assume that $v = U\overset{*}{x}$, and consider the equation

$$(5.19)\qquad Ux = v .$$

It is readily seen that the solution of (5.19) is orthogonal to the null space of the operator U, and coincides with $\overset{*}{x}$. Thus,equation (5.15) can be written in the form

$$(5.20)\qquad \alpha x + U^2 x = Uv .$$

Hence (see §2.1) $\overset{*}{x}^{(\alpha)}$ minimizes

$$(5.21)\qquad I^{(\alpha)}(x) = \|Ux - v\|^2 + \alpha\|x\|^2 .$$

From (2.19) and (2.22) (where we write U instead of A) we obtain

$$(5.22)\qquad \| U(\overset{*}{x} - \overset{*}{x}{}^{(\alpha)})\|^2 = \alpha \mathrm{Re}(\overset{*}{x} - \overset{*}{x}{}^{(\alpha)}, \overset{*}{x}) - \alpha\|\overset{*}{x} - \overset{*}{x}{}^{(\alpha)}\|^2 ,$$

and from (5.18),

$$\|U(\overset{*}{x} - \overset{*}{x}{}^{(\alpha)})\|^2 \le \alpha\|U(\overset{*}{x} - \overset{*}{x}{}^{(\alpha)})\|R + \alpha\|\overset{*}{x} - \overset{*}{x}{}^{(\alpha)}\|\delta_R - \alpha\|\overset{*}{x} - \overset{*}{x}{}^{(\alpha)}\|^2 ,$$

or

$$(5.23)\qquad (\|U(\overset{*}{x} - \overset{*}{x}{}^{(\alpha)})\| - \alpha R/2)^2 + \alpha(\|\overset{*}{x} - \overset{*}{x}{}^{(\alpha)}\| - \delta_R/2)^2 \le (\alpha^2 R^2 + \alpha\delta_R^2)/4 .$$

From (5.23) we easily obtain (5.16) and

$$(5.24)\qquad \|U(\overset{*}{x} - \overset{*}{x}{}^{(\alpha)})\| \le [\alpha R + (\alpha^2 R^2 + \alpha\delta_R^2)^{\frac{1}{2}}]/2 .$$

For the proof of (5.17), it is sufficient to consider the inequality (2.20), and to note that since $A\overset{*}{x} = y$ and $A\overset{*}{x}{}^{(\alpha)} = y - \alpha\overset{*}{x}{}^{(\alpha)}$, then

$$A(\overset{*}{x} - \overset{*}{x}{}^{(\alpha)}) = \alpha\overset{*}{x}{}^{(\alpha)} .$$

We now show that $\delta_R \to 0$ as $R \to \infty$. Let $R_1 = \overline{R(U)}$, the closure of the range of the operator U, and $R_2 = H \ominus R_1$, the orthogonal complement of R_1 with respect to H. If we now write $\overset{*}{x}$ as $\overset{*}{x} = \overset{*}{x}{}^{(1)} + \overset{*}{x}{}^{(2)}$, $(\overset{*}{x}{}^{(1)} \in R_1, \overset{*}{x}{}^{(2)} \in R_2)$, we obtain $(\overset{*}{x}{}^{(2)}, Uw) = 0$ for every $w \in H$. In particular, if we choose $w = U\overset{*}{x}{}^{(2)}$, then $(\overset{*}{x}{}^{(2)}, U^2\overset{*}{x}{}^{(2)}) = \|U\overset{*}{x}{}^{(2)}\| = 0$, so that $\overset{*}{x}{}^{(2)}$ is a zero of the operator U. But $\|\overset{*}{x}{}^{(2)}\|^2 = (\overset{*}{x}{}^{(2)}, \overset{*}{x}-\overset{*}{x}{}^{(1)}) = (\overset{*}{x}{}^{(2)}, \overset{*}{x}) - (\overset{*}{x}{}^{(2)}, \overset{*}{x}{}^{(1)}) = 0$. Hence $\overset{*}{x} = \overset{*}{x}{}^{(1)} \in R_1$, which establishes the required result.

To complete the proof, it remains to choose, for arbitrarily small $\epsilon > 0$, an R such that $\delta_R \le \epsilon/2$, and α such that $[\epsilon + (\epsilon^2 + 8\alpha R^2)^{\frac{1}{2}}]/4 \le \epsilon$ for this fixed R. #

For the operator $A^{(\alpha)}$, its upper and lower bounds are given by $M+\alpha$ and $m+\alpha$, respectively. If we transform (5.15) into the form of (5.7), and solve it by simple iteration, we obtain as a consequence of (5.9) that

$$(5.25)\qquad \|\overset{*}{x}{}^{(\alpha)} - x_n^{(\alpha)}\| \le \frac{1}{(m+\alpha)}\left(\frac{M-m}{M+m+2\alpha}\right)^n \|y - \alpha x_0 - Ax_0\| .$$

Combining (5.16) and (5.25), we have

$$(5.26)\qquad \|\overset{*}{x} - x_n^{(\alpha)}\| \le \frac{1}{(m+\alpha)}\left(\frac{M-m}{M+m+2\alpha}\right)^n \|y - \alpha x_0 - Ax_0\| + [\delta_R + (\delta_R^2 + \alpha R^2)^{\frac{1}{2}}]/2 .$$

Let us now assume that a positive operator A_ϵ and an element y_ϵ are given instead of A and y respectively, where $\|A - A_\epsilon\| \le \epsilon$ and $\|y - y_\epsilon\| \le \epsilon$. Let $x_n^{\epsilon,\alpha}$ denote the result of applying n iterations to the equation of the form

(5.7) in which we now have $T = T_\epsilon^{(\alpha)} \equiv I - \beta^{\epsilon,\alpha} A_\epsilon^{(\alpha)}$. Assuming for simplicity that $x_0 = 0$, then

$$\|x_n^{(\alpha)} - x_n^{\epsilon,\alpha}\| = \|T^{(\alpha)}x_{n-1}^{(\alpha)} - T_\epsilon^{(\alpha)}x_{n-1}^{\epsilon,\alpha} + \beta^{(\alpha)}y - \beta^{\epsilon,\alpha}y_\epsilon\|$$

$$\le \|T^{(\alpha)}\|\,\|x_{n-1}^{(\alpha)} - x_{n-1}^{\epsilon,\alpha}\| + \|T^{(\alpha)} - T_\epsilon^{(\alpha)}\|\,\|x_{n-1}^{\epsilon,\alpha}\| + \beta^{(\alpha)}\epsilon + |\beta^{(\alpha)} - \beta^{\epsilon,\alpha}|\,\|y_\epsilon\|$$

$$\le \|T^{(\alpha)}\|\,\|x_{n-1}^{(\alpha)} - x_{n-1}^{\epsilon,\alpha}\| + \frac{\|T^{(\alpha)} - T_\epsilon^{(\alpha)}\|}{1 - \|T_\epsilon^{(\alpha)}\|}\beta^{\epsilon,\alpha}\|y_\epsilon\| + \beta^{(\alpha)}\epsilon + |\beta^{(\alpha)} - \beta^{\epsilon,\alpha}|\,\|y_\epsilon\|$$

$$\le \|T^{(\alpha)}\|^2\|x_{n-2}^{(\alpha)} - x_{n-2}^{\epsilon,\alpha}\| + (1 + \|T^{(\alpha)}\|)\left\{\frac{\|T^{(\alpha)} - T_\epsilon^{(\alpha)}\|}{1 - \|T_\epsilon^{(\alpha)}\|}\beta^{\epsilon,\alpha}\|y_\epsilon\| + \beta^{(\alpha)}\epsilon + |\beta^{(\alpha)} - \beta^{\epsilon,\alpha}|\,\|y$$

(5.27)

$$\le \frac{1}{(1 - \|T^{(\alpha)}\|)(1 - \|T_\epsilon^{(\alpha)}\|)}\left\{\|T^{(\alpha)} - T_\epsilon^{(\alpha)}\|\beta^{\epsilon,\alpha}\|y_\epsilon\| + (1 - \|T_\epsilon^{(\alpha)}\|)(\beta^{(\alpha)}\epsilon + |\beta^{(\alpha)} - \beta^{\epsilon,\alpha}|\,\|y_\epsilon\|\right.$$

Since $\|A\| = M$, $\|A_\epsilon\| = M_\epsilon$, $\beta^{(\alpha)} = 2/(M+m+2\alpha)$, $\beta^{\epsilon,\alpha} = 2/(M_\epsilon+m_\epsilon+2\alpha)$, it follows that

$$\|T^{(\alpha)} - T_\epsilon^{(\alpha)}\| = \|\beta^{\epsilon,\alpha}(\alpha I + A_\epsilon) - \beta^{(\alpha)}(\alpha I + A)\| \le \frac{2\epsilon}{M_\epsilon + m_\epsilon + 2\alpha} + \frac{4\epsilon(\|A\| + \alpha)}{(M_\epsilon + m_\epsilon + 2\alpha)(M+m+2\alpha)}$$

$$\le \frac{6\epsilon}{M_\epsilon + m_\epsilon + 2\alpha},$$

$$\frac{1}{1 - \|T^{(\alpha)}\|} = \frac{M+m+2\alpha}{2m+2\alpha}, \qquad \frac{1}{1 - \|T_\epsilon^{(\alpha)}\|} = \frac{M_\epsilon + m_\epsilon + 2\alpha}{2m_\epsilon + 2\alpha}$$

and

$$|\beta^{(\alpha)} - \beta^{\epsilon,\alpha}| \le \frac{4\epsilon}{(M_\epsilon + m_\epsilon + 2\alpha)(M+m+2\alpha)}.$$

Hence we obtain

$$\|x_n^{(\alpha)} - x_n^{\epsilon,\alpha}\| \le \left(\frac{\epsilon}{(m+\alpha)(m_\epsilon+\alpha)}\right)\left(\frac{3M+m+2m_\epsilon+6\alpha}{M_\epsilon+m_\epsilon+2\alpha}\right)\|y_\epsilon\| + \frac{\epsilon}{m+\alpha}$$

(5.28)
$$\le \frac{\epsilon}{\alpha^2}\left(3 + \frac{4\epsilon}{M_\epsilon + 2\alpha}\right)\|y_\epsilon\| + \frac{\epsilon}{\alpha}.$$

From (5.26) and (5.28) we obtain

$$\|\overset{*}{x} - x_n^{\epsilon,\alpha}\| \le \frac{1}{(m+\alpha)}\left(\frac{M-m}{M+m+2\alpha}\right)^n \|y\| + \frac{\delta_R}{2} + \frac{(\delta_R^2+\alpha R^2)^{\frac{1}{2}}}{2} +$$

$$+ \left(\frac{\epsilon}{(m+\alpha)(m_\epsilon+\alpha)}\right)\left(\frac{3M+m+2m_\epsilon+6\alpha}{M_\epsilon+m_\epsilon+2\alpha}\right)\|y_\epsilon\| + \frac{\epsilon}{m+\alpha}$$

$$\le \left[\frac{1}{\alpha}\left(\frac{M_\epsilon+\epsilon}{M_\epsilon-\epsilon+2\alpha}\right)^n (\|y_\epsilon\|+\epsilon)\right] + \left[\frac{\delta_R + (\delta_R^2+\alpha R^2)^{\frac{1}{2}}}{2}\right] +$$

$$(5.29) \qquad + \left[\frac{\epsilon}{\alpha^2}\left(3 + \frac{4\epsilon}{M_\epsilon+2\alpha}\right)\|y_\epsilon\| + \frac{\epsilon}{\alpha}\right] .$$

On equating the first and third square brackets on the right hand side of this inequality we can define the number of iterations n as a function of ϵ and α.

An estimate similar to (5.29) can be obtained in a different way,if we use the obvious inequality

$$(5.30) \qquad \|\overset{*}{x} - x_n^{\epsilon,\alpha}\| \le \|\overset{*}{x} - \overset{*}{x}^{(\alpha)}\| + \|\overset{*}{x}^{(\alpha)} - \overset{*}{x}^{\epsilon,\alpha}\| + \|\overset{*}{x}^{\epsilon,\alpha} - x_n^{\epsilon,\alpha}\| .$$

We can estimate $\|\overset{*}{x}^{\epsilon,\alpha} - x_n^{\epsilon,\alpha}\|$ similarly to (5.25), and $\|\overset{*}{x}^{(\alpha)} - \overset{*}{x}^{\epsilon,\alpha}\|$ by using (2.14) in which it is assumed that $A^* = A_\epsilon^* = I$. As a result we can prove the following theorem.

<u>Theorem 5.2</u>. *The estimate*

$$(5.31) \qquad \begin{aligned} \|\overset{*}{x} - x_n^{\epsilon,\alpha}\| &\le \frac{1}{2\alpha}\left(\frac{M_\epsilon}{M_\epsilon+2\alpha}\right)^n \|y_\epsilon - \alpha x_0 - A_\epsilon x_0\| + \\ &+ \frac{1}{2}\left[\delta_R + (\delta_R^2 + \alpha R^2)^{\frac{1}{2}}\right] + \frac{\epsilon}{\alpha}(1 + \|\overset{*}{x}\|) , \end{aligned}$$

is valid.

We now examine the method of analytic continuation. Together with equation (2.1), we shall consider the equation

$$(5.32) \qquad (A - \lambda I)x = y ,$$

where λ is a complex parameter. From (5.32) it follows that

$$(5.33) \qquad x = \mu Ax - \mu y , \quad \mu = 1/\lambda .^{(22)}$$

Since the operator A is bounded, we have for

$$(5.34) \qquad |\mu| < \|A\|^{-1} , \ (|\lambda| > \|A\|) ,$$

that the series

---

(22) We recall that those values of λ for which $A-\lambda I$ has an inverse, are called *regular* values of the operator A. All other values of λ form the *spectrum* of A. For points of the spectrum of A, the values $\mu = 1/\lambda$ are called the *characteristic values*.

$$(5.35)\qquad x(\mu) = -\mu y - \mu^2 Ay - \mu^3 A^2 y - \ldots - \mu^n A^{n-1} y - \ldots$$

will converge (with respect to the norm), and its sum $x(\mu)$ will be a solution of (5.33). But x is an analytic function of μ, which is defined by the power series (5.35). The basic problem of analytic continuation is that of determining $x(\mu)$ for any μ at which x is analytic, given only the series (5.35) and perhaps some information about its properties.

This question is discussed in [87, 91], and we shall give a general algorithm later. Initially we restrict attention to a detailed examination of the case when A is a positive operator.

In this case, $x(\mu) = x(1/\lambda)$ is an analytic function for all $\lambda \notin [m,M]$, where m, M are the bounds of the operator A (see (5.6)).

Putting

$$(5.36)\qquad \begin{aligned} &\frac{1}{\mu} = \lambda = \frac{M-m}{2} z + \frac{M+m}{2}\ , \quad z = \frac{1}{2}\left(w + \frac{1}{w}\right)\ , \\ &w = z - (z^2 - 1)^{\frac{1}{2}}\ , \quad z = [2\lambda - (M+m)]/(M-m)\ , \end{aligned}$$

where $(z^2-1)^{\frac{1}{2}}$ is taken as that branch which is positive when $z > 1$, we obtain the function

$$(5.37)\qquad \phi(w) = x\left(\frac{1}{\frac{M-m}{4}\left(w + \frac{1}{w}\right) + \frac{M+m}{2}}\right)\ ,$$

which is analytic in the disc $|w| < 1$. It is clear that $x(\lambda) = \phi[w(\lambda)]$ is an unknown function which is analytic outside of the interval $[m,M]$. Writing $\phi(w)$ in the form of a Taylor series

$$(5.38)\qquad \phi(w) = \sum_{k=0}^{\infty} a_k w^k\ , \quad a_k = \phi^{(k)}(0)/k!\ ,$$

we can use (5.35) to evaluate some of the $\phi^{(k)}(0)$. From (5.36) and (5.37), and observing that $w = 0$ corresponds to $\mu = 0$, we find

$$\phi(0) = x(0) = 0\ ,$$

$$\phi'(0) = x'(0)\frac{4}{(M-m)} = -\frac{4y}{(M-m)}\ ,$$

$$\phi''(0) = x''(0)\frac{16}{(M-m)^2} - x'(0)\frac{16(M+m)}{(M-m)^2}\ ,$$

$$(5.39)\qquad = -2Ay\frac{16}{(M-m)^2} + y\frac{16(M+m)}{(M-m)^2}\ ,$$

$$\phi'''(0) = x'''(0)\,\frac{64}{(M-m)^3} - x''(0)\,\frac{192(M+m)}{(M-m)^3} +$$

$$+\; x'(0)\,\frac{24[4(M+m)^2 - (M-m)^2]}{(M-m)^3} \quad ,$$

$$x'''(0) = -6A^2y \;,\; x''(0) = -2Ay \;,\; x'(0) = -y \;,\; \text{etc.}$$

With the help of the series (5.38), we can evaluate the function x in the neighbourhood of any regular point λ_0. We can write

$$x(\lambda) = \sum_{k=0}^{\infty} \frac{x^{(k)}(\lambda_0)}{k!}\,(\lambda - \lambda_0)^k \;, \tag{5.40}$$

and the values of $x^{(k)}(\lambda_0)$ may be found from the relations

$$x(\lambda_0) = \phi(w_0),\; w_0 = z_0 - (z_0^2-1)^{\frac{1}{2}} \;,\; z_0 = [2\lambda_0 - (M+m)]/(M-m) \;,$$

$$x'(\lambda_0) = \phi'(w_0)\,\frac{dw}{d\lambda}\bigg|_{w=w_0} \;, \tag{5.41}$$

$$x''(\lambda_0) = \phi''(w_0)\left(\frac{dw}{d\lambda}\bigg|_{w=w_0}\right)^2 + \phi'(w_0)\,\frac{d^2w}{d\lambda^2}\bigg|_{w=w_0} \;,\; \text{etc.}$$

As is obvious, the practical application of equations (5.39) and (5.41) can become extremely complicated. If the value of $x(\lambda)$ is not required for real λ to the left of m, or to the right of M, then instead of (5.36) we may use a conformal mapping of the plane with a cut along the real axis from $-\infty$ to m, or M to $+\infty$, on to the disc $|w| < 1$. In the first case, for example, the conformal mapping is given by

$$w = \frac{(\lambda-m)^{\frac{1}{2}} - i}{(\lambda-m)^{\frac{1}{2}} + i} \;,\quad \lambda = m - \left(\frac{1+w}{1-w}\right)^2 \;. \tag{5.42}$$

If it is necessary to determine $x(\lambda)$ only for λ in the upper-half plane, then instead of (5.36), we could use

$$w = \frac{\lambda-i}{\lambda+i} \;,\quad \lambda = i\,\frac{1+w}{1-w} \;, \tag{5.43}$$

which maps the upper-half λ-plane into the disc $|w| < 1$.

It should also be realised that the problem of analytic continuation is not well posed (see [91]). This can be seen by considering an arbitrarily small error in the calculation of the $(k+1)^{\text{th}}$ term of the series (5.40). Thus although $\|\epsilon(\lambda-\lambda_0)^k\| \le \|\epsilon\|$ in the disc $|\lambda-\lambda_0| \le 1$, it can give rise to an arbitrarily large error in the same term for sufficiently large values of $|\lambda-\lambda_0|$.

In connection with this, we recall that the method of analytic continuation (see [91]) can be reduced to the solution of a Fredholm integral equation of the first kind with Cauchy kernel, and to the calculation of Cauchy integrals. Let $\tilde{\mu} = 1/\tilde{\lambda}$ be a regular value of the operator A at which we want to determine the function x. In the complex μ-plane, let Γ be a smooth simple closed curve which contains the points $\tilde{\mu}$ and the circle $|\mu| = \|A\|^{-1}/2$, but none of the singular points of the function x. Consider the following first kind integral equation with Cauchy kernel,

$$Bz \equiv \frac{1}{2\pi i} \int_{\Gamma} \frac{z(s)ds}{s - \mu} = x(\mu) , \quad |\mu| = \|A\|^{-1}/2 . \tag{5.44}$$

Having solved this equation for $z(s)$ on Γ, then from the property of the Cauchy integral we have

$$x(\tilde{\mu}) = \frac{1}{2\pi i} \int_{\Gamma} \frac{z(s)ds}{s - \tilde{\mu}} . \tag{5.45}$$

From analytic function theory, we have that equation (5.44) has a unique solution given by

$$z(s) = B^{-1}x \equiv \frac{1}{2\pi i} \int_{\tilde{\Gamma}} \frac{x(\mu)d\mu}{\mu - s} , \tag{5.46}$$

where $\tilde{\Gamma}$ is any smooth simple closed curve which contains Γ, but does not intersect it.

The relation (5.46) can be interpreted as a completely continuous operator, which transforms a bounded set of functions x ($\|x\| \leq C$)), into a compact set of functions z. For sufficiently large C, this compact set will contain the unknown solution of equation (5.44). Thus, the methods of [91] and of quasi-solutions [69] can be applied for the construction of the required solution.

Observing that the operator B maps from the Hilbert space $L_2(\Gamma)$ of functions which are square integrable on Γ, into the Hilbert space $L_2(\gamma_r)$ of functions which are square integrable on γ_r ($r = \|A\|^{-1}/2$, $\mu = re^{i\theta}$), we can define a conjugate operator B^* by

$$(B^*x,z)_{L_2(\Gamma)} \equiv \int_{\Gamma} (B^*x)\, \bar{z}\,|ds| = (x,Bz)_{L_2(\gamma_r)} \equiv \int_{\gamma_r} x(\mu)\, \overline{(Bz)}\, |d\mu| . \tag{5.47}$$

If we substitute the expression for B from (5.44) into (5.47) we find

$$\int_\Gamma (B^*x)\bar{z}|ds| = \int_{\gamma_r} x(\mu)\,\overline{\left(\frac{1}{2\pi i}\int_\Gamma \frac{z(s)ds}{s-\mu}\right)}\,|d\mu|$$

$$= \int_\Gamma \left(-\frac{1}{2\pi i}\,\frac{\overline{ds}}{|ds|}\int_{\gamma_r}\frac{x(\mu)|d\mu|}{\bar{s}-\bar{\mu}}\right)\overline{z(s)}\,|ds| \quad ,$$

from which we obtain

$$B^*x = -\frac{1}{2\pi i}\,\frac{\overline{ds}}{|ds|}\int_{\gamma_r}\frac{x(\mu)|d\mu|}{\bar{s}-\bar{\mu}} \quad . \tag{5.48}$$

Let us consider now the positive operator B^*B. We have

$$B^*Bz = \left(\frac{1}{2\pi}\right)^2 \frac{\overline{ds}}{|ds|}\int_{\gamma_r}\frac{1}{(\bar{s}-\bar{\mu})}\left(\int_\Gamma \frac{z(\xi)d\xi}{\xi-\mu}\right)|d\mu|$$

$$= \left(\frac{1}{2\pi}\right)^2 \frac{\overline{ds}}{|ds|}\int_\Gamma z(\xi)\left(\int_{\gamma_r}\frac{|d\mu|}{(\xi-\mu)(\bar{s}-\bar{\mu})}\right)d\xi$$

$$= \frac{r}{2\pi\bar{s}}\;\frac{\overline{ds}}{|ds|}\int_\Gamma \frac{z(\xi)d\xi}{\xi - r^2/\bar{s}} \quad . \tag{5.49}$$

We can now estimate bounds for the operator B^*B, so that we may apply the iteration algorithm as described above, to the solution of the equation $B^*Bz = B^*x$. Noting that

$$(B^*Bz,\, z)_{L_2(\Gamma)} = (Bz,\, Bz)_{L_2(\Gamma)} \geq \frac{1}{\|B^{-1}\|^2}\,(z,z)_{L_2(\Gamma)} \quad , \tag{5.50}$$

we can obtain an estimate of the lower bound of $\|B^{-1}\|$. For any two curves Γ_1 and Γ_2, let $L(\Gamma_1)$, $L(\Gamma_2)$ denote their respective lengths, and $\rho(\Gamma_1, \Gamma_2)$ the least distance between them. Then from (5.46), we obtain

$$\|z\|^2_{L_2(\Gamma)} \leq \left(\frac{1}{2\pi}\right)^2 \int_{\tilde{\Gamma}} \|x(\mu)\|^2 |d\mu| \int_\Gamma\int_{\tilde{\Gamma}} \frac{|d\mu|}{|\mu-s|^2}|ds|$$

$$\leq \frac{L(\Gamma)L(\tilde{\Gamma})}{(2\pi)^2\,\rho^2(\Gamma,\tilde{\Gamma})}\int_{\tilde{\Gamma}} \|x(\mu)\|^2\,|d\mu| \quad . \tag{5.51}$$

On the other hand, from equation (5.32),

$$\|x(\mu)\| = \|(A - \frac{1}{\mu}I)^{-1}y\| \leq \|(A - \frac{1}{\mu}I)^{-1}\|\,\|y\| \quad . \tag{5.52}$$

This can be used to obtain a useful estimate when A is a self-adjoint positive operator. In fact, since

(5.53) $$\|\phi(A)\| = \sup_{s\in[m,M]} |\phi(s)| ,$$

we obtain from (5.52),

(5.54) $$\|x(\mu)\| = \|x(\lambda)\| \leq \sup_{s\in[m,M]} |(s-\lambda)^{-1}| \; \|y\| \leq \frac{\|y\|}{\rho(\tilde{\Omega},\ [m,M])} ,$$

$M \notin \tilde{\Gamma}$, where $\tilde{\Omega}$ is the image of $\tilde{\Gamma}$ under the transformation $\lambda = 1/\mu$. To estimate $\|B^{-1}\|$, it remains to note that

$$\|y\| \leq \|A - \frac{1}{\mu} I\|.\|x(\mu)\| \leq \|A\|.\|x(\mu)\|, \quad |\mu| = r = \|A\|^{-1}/2 ,$$

and

$$\|z\|^2_{L_2(\Gamma)} \leq \frac{L(\Gamma)\ L^2(\tilde{\Gamma})\rho^{-2}(\tilde{\Omega},\ [m,M])}{(2\pi)^2\ \rho^2(\Gamma,\tilde{\Gamma})2\pi r} \|A\|^2\ \|x\|^2_{L_2(\gamma_r)} ,$$

which is equivalent to

(5.55) $$\|B^{-1}\|^2 \leq \frac{L(\Gamma)\ L^2(\tilde{\Gamma})}{(2\pi)^2\ \rho^2(\Gamma,\tilde{\Gamma})2\pi r} \ \rho^{-2}(\tilde{\Omega},\ [m,M])\|A\|^2 .$$

From (5.50) and (5.55) it follows that

(5.56) $$(B^*Bz,\ z) \geq \frac{\rho^2(\Gamma,\tilde{\Gamma})\ (2\pi)^2(2\pi r)}{L(\Gamma)\ L^2(\tilde{\Gamma})\ \rho^{-2}(\tilde{\Omega},\ [m,M])\|A\|^2} \ (z,z) ,$$

from which we can obtain an estimate for the lower bound of the operator B^*B. An estimate of the upper bound is given by the following inequality

(5.57) $$(B^*Bz,\ z) = \int_\Gamma \left(\frac{r}{2\pi\bar{s}} \cdot \frac{\overline{ds}}{|ds|} \int_\Gamma \frac{z(\zeta)d\zeta}{\zeta - r^2/\bar{s}} \right) \overline{z(s)}\ |ds|$$

$$\leq \frac{L(\Gamma)\ \rho(\Gamma,\gamma_r)}{2\pi} \ (z,z) .$$

We summarise these results in the following theorem.

<u>Theorem 5.3.</u> *The problem of analytic continuation of the series* (5.35) *to any regular point* μ, *is solved by equation* (5.45), *where* z *is the solution of a uniquely solvable integral equation of the first kind,* $Bz = x$, *in which* B *is given by* (5.44), *and has a conjugate operator* B^* *which is defined by* (5.48). *In the case when* A *is a self-adjoint operator, estimates for bounds of the operator* B^*B *are given by*

(5.58) $$\frac{(2\pi)^3 r \rho^2(\Gamma,\tilde{\Gamma})}{\rho^{-2}(\bar{\Omega}, [m,M]) \; L(\Gamma) \; L^2(\tilde{\Gamma})} \; \frac{1}{\|A\|^2} \leq m(B^*B) \leq M(B^*B) \leq \frac{L(\Gamma) \; \rho(\Gamma,\gamma_r)}{2\pi} .$$

In conclusion we make some observations.

1. The estimate (5.51) and those which follow it are not sharp. It is clear that if we make some smoothness assumption about Γ, we can obtain an estimate with the first power of $\rho(\Gamma,\tilde{\Gamma})$ in the denominator. In place of (5.54), we can obtain an analogous result by taking the required estimate as $\int_\Gamma |x(\mu)|^2 |d\mu|$, instead of $\|x(\mu)\|$. As a result, first powers of the distances $\rho(\Gamma,\tilde{\Gamma})$ and $\rho(\tilde{\Omega}, [m,M])$ will appear in (5.55), (5.56) and (5.58); on the other hand other quantities may become more complicated.

2. For the evaluation of $x(\tilde{\mu})$ from (5.45), and for the solution of the equation $B^*Bz = B^*x$, algorithms for the approximate calculation of Cauchy integrals are required. Some of these algorithms will be examined in Chapter 2 , §8.

3. From Theorem 5.1, it follows that as λ tends to m along the real axis from the left of m, $x(\lambda)$ will approach the solution of the equation $(A-mI)x = y$ which is orthogonal to the null space of the operator $A-mI$, provided such a solution exists. It should be noted that, if λ_* is a point of the discrete spectrum of a self-adjoint operator A, and the equation $(A-\lambda_* I)x = y$ has a solution, then as $\lambda \to \lambda_*$, $x(\lambda)$ will always converge to the solution of this equation which is orthogonal to the null space of the operator $A-\lambda_* I$. This implies that the points λ_* are removable singularities of the analytic function x.

§5.2 Steepest Descent Methods

A general method of steepest descent has been described by Kantorovich and Akilov [75]. Let Φ be a real (non-linear) functional defined on a normed space X. Assuming that the values of the functional Φ are bounded below, we investigate the problem of determining an element $\overset{*}{x} \in X$ (if such exists), which minimizes the functional Φ. The method of *steepest descent* for the solution of this problem proceeds as follows.

Let $x^{(0)}$ be a fixed element from X. We take a *ray* from the point $x^{(0)}$ i.e. the totality of all elements of the form $x^{(0)}+az$, where a is a real non-negative number and $z(\neq 0)$ is an element of X which characterizes the *direction* of the ray. Examination of the functional Φ along the given ray is equivalent to considering the real function

$$\varphi(a;\, x^{(0)},\, z) = \Phi(x^0+az),\ (a \geq 0)\ . \tag{5.59}$$

In addition it is natural to define the derivative of the functional Φ in the z-direction at the point $x^{(0)}$ by

$$\frac{\partial\Phi}{\partial z} = \frac{1}{\|z\|}\left(\frac{\partial\phi}{\partial z}(a;\, x^{(0)},\, z)\right)_{a=0} = \frac{1}{\|z\|}\lim_{h\to 0+}\frac{\Phi(x^{(0)}+hz) - \Phi(x^{(0)})}{h}\ . \tag{5.60}$$

Assuming that the derivative $\frac{\partial\Phi}{\partial z}$ exists for all directions, we further assume that there exists a direction z_* (say) along which this derivative is a minimum. This direction is called the *anti-gradient* (at the point $x^{(0)}$) of the functional Φ. Let $\partial\Phi/\partial z_* < 0$, and let the derivative be continuous along the ray from $x^{(0)}$ in the direction z_*. Under these conditions the function $\phi(a;\, x^{(0)},\, z_*)$ decreases, at least for sufficiently small a. If we denote by a_0 the least (positive) root of the equation

$$\phi'_a(a;\, x^{(0)},\, z_*) = 0\ , \tag{5.61}$$

then we can state that, in the interval $[0, a_0]$, the function ϕ decreases. By the same token, the functional Φ decreases in the direction z_*, at least as far as the point

$$x^{(1)} = x^{(0)} + a_0 z_*\ . \tag{5.62}$$

Repeating these operations with the element $x^{(1)}$ instead of $x^{(0)}$, we obtain the element $x^{(2)}$ etc., the result being a sequence $\{x^{(j)}\}$ which may be a minimizing sequence for the functional Φ, in the sense that

(5.63) $$\lim_{j\to\infty} \Phi(x^{(j)}) = \inf_{x\in X} \Phi(x) \quad .$$

If there exists $\overset{*}{x} \in X$, for which $\inf_{x\in X} \Phi(x) = \Phi(\overset{*}{x})$, then often the minimizing sequences will converge to the required element i.e.

(5.64) $$\lim_{j\to\infty} x^{(j)} = \overset{*}{x} \ .$$

Equation (5.64) can imply either strong convergence (convergence in norm) if

(5.65) $$\|x^{(j)} - \overset{*}{x}\| \to 0 \ , \ (j \to \infty) \ ;$$

or weak convergence if

(5.66) $$f(x^{(j)} - \overset{*}{x}) \to 0, \quad (j \to \infty) \ ,$$

for every functional f from the space X^* which is conjugate to X.

We shall apply the above scheme to the case when $X \equiv H$ (H is a Hilbert space) and

(5.67) $$\Phi(x) \equiv I^{(\alpha)}(x) = \|y - Ax\|^2 + \alpha\|x\|^2, \ \alpha \geq 0 \ ,$$

where A is a linear operator in H. In this case

$$\frac{\partial I^{(\alpha)}}{\partial z} = \lim_{h\to 0+} \frac{\|y - Ax^{(0)} - hAz\|^2 + \alpha\|x^{(0)} + hz\|^2 - \|y - Ax^{(0)}\|^2 - \alpha\|x^{(0)}\|^2}{h\|z\|}$$

(5.68) $$= -2\,\frac{\mathrm{Re}(y - Ax^{(0)}, Az) - \alpha\mathrm{Re}(x^{(0)}, z)}{\|z\|} \ ,$$

and

(5.69) $$z_* = A^*(y - Ax^{(0)}) - \alpha x^{(0)} \quad ,$$

since

$$\frac{\mathrm{Re}(y - Ax^{(0)}, Az) - \alpha\mathrm{Re}(x^{(0)}, z)}{\|z\|} = \frac{\mathrm{Re}(A^*(y - Ax^{(0)}) - \alpha x^{(0)}, z)}{\|z\|}$$

$$\leq \|A^*(y - Ax^{(0)}) - \alpha x^{(0)}\| = \frac{\mathrm{Re}(A^*(y - Ax^{(0)}) - \alpha x^{(0)}, z_*)}{\|z_*\|} \ .$$

Now

(5.70) $$x^{(j+1)} = x^{(j)} + a_j[A^*(y - Ax^{(j)}) - \alpha x^{(j)}], \ j = 0,1,2,\ldots$$

and a_j is determined as above from the functional (5.67), to give

$$(5.71)\qquad a_j = \frac{\|A^*(y-Ax^{(j)}) - \alpha x^{(j)}\|^2}{\|A[A^*(y-Ax^{(j)}) - \alpha x^{(j)}]\|^2 + \alpha\|A^*(y-Ax^{(j)}) - \alpha x^{(j)}\|^2} .$$

In the special case when X is an n-dimensional space of elements of the form $\sum_{k=1}^{n} \alpha_k e_k$, the algorithm (5.70)-(5.71) gives the steepest descent algorithm of §3.5. If $\{e_k\}$ is an orthonormalized system and $\sum_{k=1}^{n} |\alpha_k|^2$ is bounded as $n \to \infty$, then the algorithm (5.70)-(5.71) is a closure of the Sobolev algorithm of §3 (see [124]), i.e. it can be obtained from this algorithm by taking the limit as $n \to \infty$. In fact, if from (5.69) we write z_* as

$$(5.72)\qquad z_* = \sum_{s=1}^{\infty} (A^*(y - A\sum_{k=1}^{\infty} \alpha_k^{(0)} e_k), e_s)e_s - \alpha \sum_{k=1}^{\infty} \alpha_k^{(0)} e_k$$

$$= \sum_{s=1}^{\infty} (y - \sum_{k=1}^{\infty} \alpha_k^{(0)} y_k, y_s)e_s - \alpha \sum_{k=1}^{\infty} \alpha_k^{(0)} e_k ,$$

$$y_k = Ae_k ,$$

and compare (5.72) with (3.129), we see (in the case of a real space) that

$$(5.73)\qquad \|z_* + \frac{1}{2}\sum_{s=1}^{n} \left(\frac{\partial\Delta_n}{\partial\alpha_s}\right)_0 e_s\| \to 0 \text{ as } n \to \infty ,$$

i.e. that a vector which is proportional to the anti-gradient in a coordinate system $\{e_k\}_1^n$ at the point $P_0(\alpha_1^{(0)}, \alpha_2^{(0)},\ldots,\alpha_n^{(0)})$, coincides in the limit with the anti-gradient z_* of an infinite-dimensional Hilbert space H. In a similar way it is not difficult to show that as $n \to \infty$, t_j from (3.131) coincides with a_j defined by (5.71). In general, the algorithm (3.130)-(3.131) is transformed into the algorithm (5.70)-(5.71).

The convergence of the method of steepest descent for $\alpha = 0$ (and of more general methods, see [151]) has been examined in [74, 10, 26, 151]. The results which are obtained can be illustrated by the following theorem.

Theorem 5.4. *The algorithm* (5.70)-(5.71) *always converges in the sense that*

$$(5.74)\qquad \lim_{j\to\infty} I^{(\alpha)}(x^{(j)}) = \inf_{x\in D(A)} I^{(\alpha)}(x) .$$

In addition, if the equation $Ax = y$ *possesses a solution then, for* $\alpha = 0$,

$$(5.75)\qquad \lim_{j\to\infty} \|x^{(j)} - \overset{*}{x} - Qx^{(0)}\| = 0 ;$$

and for $\alpha > 0$,

(5.76) $$\lim_{j\to\infty} \|x^{(j)} - \overset{*}{x}^{(\alpha)}\| = 0 ,$$

where Q is the projection operator of H onto the null space of the operator A, $\overset{*}{x}$ is the unique solution of $Ax = y$ which is orthogonal to this sub-space, and $\overset{*}{x}^{(\alpha)}$ is the element which minimizes $I^{(\alpha)}(x)$ on $D(A)$.

The proof of this theorem will not be considered here because it is a consequence of the more general results of [151], and also of Theorem 5.5. Theorem 5.4 can also be obtained from Theorem 3.23 by considering the limit as $n \to \infty$.

As a criterion for terminating the method of steepest descent on a computer, we take the condition

(5.77) $$I^{(\alpha)}(x^{(j)}) - I^{(\alpha)}(x^{(j+1)}) < \eta I^{(\alpha)}(x^{(0)}) ,$$

where η is some chosen positive constant.

Theorem 5.5. *If (5.77) is satisfied then*

(5.78) $$\|x^{(j)} - \overset{*}{x}^{(\alpha)}\|^2 \le \frac{M^2+\alpha}{(m^2+\alpha)^2} \eta I^{(\alpha)}(x^{(0)})$$

and

(5.79) $$I^{(\alpha)}(x^{(j)}) - I^{(\alpha)}(\overset{*}{x}^{(\alpha)}) \le \frac{M^2+\alpha}{m^2+\alpha} \eta I^{(\alpha)}(x^{(0)}) ,$$

where

(5.80) $$m = \inf_{x\in D(A)} \frac{\|Ax\|}{\|x\|} , \quad M = \sup_{x\in D(A)} \frac{\|Ax\|}{\|x\|} = \|A\| .$$

Proof. After simple manipulations we obtain

(5.81) $$I^{(\alpha)}(x^{(j)}) - I^{(\alpha)}(x^{(j+1)}) = \frac{\|z_j\|^4}{\|Az_j\|^2 + \alpha\|z_j\|^2} = a_j\|z_j\|^2 ,$$

where

(5.82) $$z_j = A^*(y - Ax^{(j)}) - \alpha x^{(j)} = A^*y - (A^*A+\alpha I)x^{(j)} .$$

From (5.82) it follows that

$$x^{(j)} = (A^*A + \alpha I)^{-1} (A^*y - z_j) .$$

But $\overset{*}{x}^{(\alpha)} = (A^*A + \alpha I)^{-1} A^*y$, so that

$$\|x^{(j)} - \overset{*}{x}^{(\alpha)}\|^2 \le \|(A^*A + \alpha I)^{-1}\|^2 \|z_j\|^2$$
$$= \|(A^*A + \alpha I)^{-1}\|^2 (I^{(\alpha)}(x^{(j)}) - I^{(\alpha)}(x^{(j+1)}))/a_j .$$

For the proof of (5.78) it remains to note that

$$\frac{1}{a_j} = \frac{\|Az_j\|^2 + \alpha\|z_j\|^2}{\|z_j\|^2} \leq M^2 + \alpha , \tag{5.83}$$

and

$$\|(A^*A + \alpha I)^{-1}\| \leq 1/(m^2 + \alpha) . \tag{5.84}$$

The last inequality follows from Theorem 2.1, from which we obtain

$$((A^*A+\alpha I)x,\ x) = \|Ax\|^2 + \alpha\|x\|^2 \geq (m^2+\alpha)\|x\|^2$$
$$= (m^2+\alpha)\ \|(A^*A+\alpha I)^{-1}y\|^2, \text{ say },$$

whence

$$\|(A^*A+\ I)^{-1}y\|^2 \leq \frac{1}{(m^2+\alpha)}\ |\ (y,\ (A^*A+\alpha I)^{-1}y)|$$
$$\leq \frac{\|y\|\ \|(A^*A+\alpha I)^{-1}y\|}{(m^2+\alpha)} ,$$

and

$$\|(A^*A+\alpha I)^{-1}y\| \leq \|y\|\ /(m^2+\alpha) .$$

For the proof of (5.79), we note from (2.8) with $\epsilon = 0$, and (5.81)-(5.84), that

$$I^{(\alpha)}(x^{(j)}) - I^{(\alpha)}(\overset{*}{x}^{(\alpha)}) = \|A(x^{(j)} - \overset{*}{x}^{(\alpha)})\|^2 + \alpha\|x^{(j)} - \overset{*}{x}^{(\alpha)}\|^2$$
$$= ((A^*A+\alpha I)(x^{(j)}-\overset{*}{x}^{(\alpha)}),\ (x^{(j)} - \overset{*}{x}^{(\alpha)}))$$
$$\leq \|(A^*A+\alpha I)(x^{(j)}-\overset{*}{x}^{(\alpha)})\|\ \|x^{(j)}-\overset{*}{x}^{(\alpha)}\|$$
$$\leq \|(A^*A+\ I)^{-1}\|\ \|z_j\|^2$$
$$\leq \left(\frac{M^2+\alpha}{m^2+\alpha}\right)\ \eta I^{(\alpha)}(x_0) . \qquad \#$$

It should be noted that inequalities (5.78) and (5.79) are limits of the first inequalities of (3.136) and (3.137) respectively. As in Theorem 3.24, instead of condition (5.77), we can introduce an alternative criterion $a_j < \sigma$, where σ is some given positive constant. Then (5.78) and (5.79) will be replaced by inequalities which are limits of the corresponding second inequalities of (3.136) and (3.137) respectively.

Since, except when stopping, we have at each iteration that $I^{(\alpha)}(x^{(j)}) - I^{(\alpha)}(x^{(j+1)}) \geq \eta I^{(\alpha)}(x^{(0)})$, the maximum number of iterations j^0 must satisfy

$$\begin{aligned} I^{(\alpha)}(x^{j^0}) - I^{(\alpha)}(x^{j^0+1}) &= -I^{(\alpha)}(x^{(0)}) + I^{(\alpha)}(x^{(0)}) - I^{(\alpha)}(x^{(1)}) + I^{(\alpha)}(x^{(1)}) - \\ &- I^{(\alpha)}(x^{(2)}) + \ldots + I^{(\alpha)}(x^{(j^0-1)}) - I^{(\alpha)}(x^{(j^0)}) + \\ &+ 2I^{(\alpha)}(x^{(j^0)}) - I^{(\alpha)}(x^{(j^0+1)}) \\ &\geq -I^{(\alpha)}(x^{(0)}) + j^0 \eta I^{(\alpha)}(x^{(0)}) + 2I^{(\alpha)}(x^{(j^0)}) - I^{(\alpha)}(x^{(j^0+1)}), \end{aligned}$$

whence

$$j^0 \leq [1 - I^{(\alpha)}(x^{(j^0)})/I^{(\alpha)}(x^{(0)})]/\eta \leq 1/\eta \quad . \tag{5.85}$$

L.V. Kantorovich's method can be generalized by assuming that z_* will be the direction of the anti-gradient for Φ_1, and a_0 is defined by the condition that $\Phi_2(x^{(0)}+az_*)$ is minimized. We call the corresponding iterative processes, *methods of steepest descent type.*

V.M. Fridman [138] has observed that we can choose a_j in (5.70) (for $\alpha = 0$) so as to minimize the norm of the error $\|x^{(j+1)} - \mathring{x}\|$, where $\mathring{x}$ is the solution of $Ax = y$, which we assume to be solvable. In fact, on equating the derivative $\frac{d}{da_j}\|x^{(j+1)} - \mathring{x}\|^2$ to zero, we obtain

$$a_j = \frac{\|y - Ax^{(j)}\|^2}{\|A^*(y - Ax^{(j)})\|^2} \quad . \tag{5.86}$$

<u>Theorem 5.6</u>. (V.M. Fridman [138]). *The sequence of elements* $\{x^{(j)}\}$ *which is defined by the recurrence relation*

$$x^{(j+1)} = x^{(j)} + \frac{\|y - Ax^{(j)}\|^2}{\|A^*(y - Ax^{(j)})\|^2} A^*(y - Ax^{(j)}), \; j = 0,1,\ldots, \tag{5.87}$$

converges monotonically and strongly to the solution of the equation $Ax = y$, *and*

$$\mathring{x} = \overset{*}{x} + Qx^{(0)} \quad , \tag{5.88}$$

where Q *and* $\overset{*}{x}$ *have the same meaning as in Theorem 5.4.*

It should be noted that we can also relate both the iterative process with minimum residual (see [84]), and multi-stage iterative processes, to the methods of steepest descent type (see [26]).

We shall now make an estimate for the number of iterations as a function of the error in the input data. Together with the equation $Ax=y$, we examine the approximate equation $A_\epsilon x = y_\epsilon$ in which $\|A_\epsilon - A\| \leq \epsilon$ and $\|y_\epsilon - y\| \leq \epsilon$. Taking into account that the approximating equation has a solution, and applying the iterative process (5.87) to its solution, we obtain on the basis of Theorem 5.6, a sequence $\{x_\epsilon^{(j)}\}$ which converges to the solution $\overset{\circ}{x}_\epsilon$. In general however, the latter does not converge to $\overset{\circ}{x}$ as $\epsilon \to 0$. On the other hand, after r iterations with (5.87) on the exact and the approximating equations, we find $x^{(r)}$ and $x_\epsilon^{(r)}$ with the respective properties that $\|\overset{\circ}{x} - x^{(r)}\| \leq \delta_r \to 0$, $r \to \infty$, and $\|x^{(r)} - x_\epsilon^{(r)}\| \leq \eta_{\epsilon,r} \to 0$ as $\epsilon \to 0$. Hence for any $\epsilon > 0$, there exists an optimum number of iterations, $\overset{\circ}{r}_\epsilon$ say, which minimizes $\|\overset{\circ}{x} - x_\epsilon^{(r)}\|$, and is such that $\|\overset{\circ}{x} - x_\epsilon^{(\overset{\circ}{r}_\epsilon)}\| \to 0$ as $\epsilon \to 0$. We use this last property to determine an approximate value for the number of iterations for a given ϵ. It should be noted that the discussion is also valid for a strong approximation to the operator A, i.e. when $\|(A_\epsilon - A)x\| \leq \epsilon(x) \to 0$ $(\epsilon \to 0)$ for every x; in practice, the conditions are difficult to satisfy because it is difficult to vary ϵ.

We can obtain better results in the case of the method of steepest descent applied to the functional (5.67) when $\alpha > 0$. From (5.78) and (5.84) (where we substitute A_ϵ, y_ϵ for A, y respectively), we obtain

$$\|x_\epsilon^{(j)} - \overset{*}{x}^{(\alpha)}\| \leq \|x_\epsilon^{(j)} - \overset{*}{x}^{\alpha,\epsilon}\| + \|\overset{*}{x}^{\alpha,\epsilon} - \overset{*}{x}^{(\alpha)}\|$$

$$\leq \frac{(M_\epsilon^2+\alpha)^{\frac{1}{2}}(\eta I^{\alpha,\epsilon}(x_\epsilon^{(0)}))^{\frac{1}{2}}}{m_\epsilon^2+\alpha} + \frac{\|(\alpha I+A_\epsilon^* A_\epsilon)(\overset{*}{x}^{\alpha,\epsilon}-\overset{*}{x}^{(\alpha)})\|}{m_\epsilon^2+\alpha} .$$

But

$$\|(\alpha I+A_\epsilon^* A_\epsilon)(\overset{*}{x}^{\alpha,\epsilon}-\overset{*}{x}^{(\alpha)})\| = \|(\alpha I+A_\epsilon^* A_\epsilon)\overset{*}{x}^{\alpha,\epsilon}-(\alpha I+A^* A)\overset{*}{x}^{(\alpha)}+(A^* A-A_\epsilon^* A_\epsilon)\overset{*}{x}^{(\alpha)}\|$$

$$= \|A_\epsilon^* y_\epsilon - A^* y + (A^* A - A_\epsilon^* A_\epsilon)\overset{*}{x}^{(\alpha)}\|$$

$$\leq \epsilon(\|y_\epsilon\| + \|A_\epsilon\| + \epsilon) + \epsilon(2\|A_\epsilon\|+\epsilon)(\|\overset{*}{x}^{(\alpha)} - x_\epsilon^{(j)}\| + \|x_\epsilon^{(j)}\|) .$$

Under the condition that

$$\epsilon(2\|A_\epsilon\| + \epsilon)/(m_\epsilon^2 + \alpha) = q_\epsilon < 1 , \tag{5.89}$$

we obtain

$$\|x_\epsilon^{(j)} - \overset{*}{x}^{(\alpha)}\| \leq \Big\{ [(\|A_\epsilon\|^2+\alpha)(\eta I^{\alpha,\epsilon}(x_\epsilon^{(0)}))]^{\frac{1}{2}} +$$

$$+ \epsilon[(\|y_\epsilon\|+\|A_\epsilon\|+\epsilon)+(2\|A_\epsilon\|+\epsilon)\|x_\epsilon^{(j)}\|]\Big\} \Big/ [(m_\epsilon^2+\alpha)(1-q_\epsilon)] . \tag{5.90}$$

We determine an appropriate number of iterations by equating the first and second terms in the curly brackets of (5.90). This gives

$$\eta I^{\alpha,\epsilon}(x_\epsilon^{(0)}) = \epsilon^2\Big[(\|y_\epsilon\|+\|A_\epsilon\|+\epsilon)+(2\|A_\epsilon\|+\epsilon)\|x_\epsilon^{(j)}\|\Big]^2\Big/\Big[\|A_\epsilon\|^2+\alpha\Big]\,. \tag{5.91}$$

In practice we can apply (5.91) as follows. For a given η, we iterate until (5.77) is satisfied (where $I^{\alpha,\epsilon}$ now appears instead of $I^{(\alpha)}$). With the computed value of $\|x_\epsilon^{(j)}\|$, we calculate the right-hand side of (5.91). If it is less than the left-hand side, then we can reduce η; if it is greater then we consider that η has been chosen properly. An *a priori* estimate for η can be obtained from an estimate of $\|\overset{*}{x}{}^{(\alpha)}\|$.

§6 OTHER METHODS AND ALGORITHMS

§6.1 Yu.V. Vorob'ev's Method of Moments

Let a linear operator A be defined on a Hilbert space H. Choosing an arbitrary element z_0 we construct the sequence of iterates : z_0, $z_1 = Az_0$, $z_2 = Az_1 = A^2 z_0, \ldots, z_n = Az_{n-1} = A^n z_0$. We can define approximate operators A_n by the relation

$$(6.1) \qquad z_k = A^k z_0 = A_n^k z_0 \quad (k = 0,1,2,\ldots,n-1) \; ,$$

$$E_n z_n = E_n A^n z_0 = A_n^n z_0 \; ,$$

where E_n is a projection operator for every element $x \in H$ into the subspace H_n formed by the elements $z_0, z_1, \ldots, z_{n-1}$. It readily follows that

$$(6.2) \qquad A_n = E_n A E_n \; .$$

Let us examine the linear manifold L_{z_0}, which consists of elements of the form $x = P(A)z_0$, where P is an arbitrary polynomial. If to L_{z_0} we add its limit points, we form a subspace H_{z_0}.

Theorem 6.1. (Yu.V. Vorob'ev [24]). *Let A be a bounded self-adjoint positive (or negative) definite operator which is defined on a Hilbert space H, and let z_0 be any element of H. The sequence of solutions $\{x_n\}$ of the equations*

$$(6.3) \qquad E_n A E_n x = E_n y$$

exists, and converges in norm to the solution x of the equation $Ax = y$, under the condition that $y \in H_{z_0}$.

It should be noted that this method is a special case of the Ritz-Galerkin method (see §4). The particular choice of coordinate elements (6.1) implies that $E_n A z_k = A z_k$, $k < n$, i.e. Vorob'ev's method of moments is closely related to the method of least squares in the case when $e_k = z_k = A^k z_0$ (see §3). More precisely, it is not difficult to verify that the solution of equation (6.3) gives in H_n the minimum of the functional

$$(6.4) \qquad \tilde{I}(x) = (Ax, \; x) - 2(y, \, x) \; .$$

The special choice of coordinate elements $e_k = A^k z_0$ allows simple recurrence relations for the solutions x_n to be obtained.

In the particular case when $z_0 = y$, the successive approximations $\{x_n\}$ can be obtained by means of the following algorithm:

$$(6.5)\quad \begin{gathered} x_{n+1} = x_n + h_n g_n,\ g_n = r_n + \ell_{n-1} g_{n-1},\ r_n = r_{n-1} - h_{n-1} A g_{n-1}, \\ h_{n-1} = \frac{\|r_{n-1}\|^2}{(Ag_{n-1}, g_{n-1})},\ \ell_{n-1} = \frac{\|r_n\|^2}{\|r_{n-1}\|^2},\ r_n = y - Ax_n,\ r_0 = g_0 = y. \end{gathered}$$

This algorithm is called *the method of conjugate gradients*, and was proposed (see [159]) for the solution of a system of linear algebraic equations with a positive definite matrix.

Under the conditions of Theorem 6.1, the rate of convergence of the method of moments is given by the inequality

$$(6.6)\qquad \|\overset{*}{x} - x_n\| \le L_n \|y\|/m ,$$

where L_n is the maximum modulus of the Chebyshev polynomial of degree n which deviates least from zero on $[m,M]$ and is equal to 1 at 0; m and M are the lower and upper bounds respectively of the operator A.

§6.2 Generalizations of the Method of Least Squares

Let E_1 and E_2 be normed linear spaces, and consider the solution of the equation $Ax = y$, where A is a linear operator with domain $D(A) \subseteq E_1$ and range $R(A) \subseteq E_2$. We seek an approximate solution of this equation in the form

$$(6.7)\qquad x_n = \sum_{k=1}^{n} \alpha_k e_k , \quad (e_k \in D(A)) ,$$

which minimizes the function

$$(6.8)\quad \begin{gathered} \Delta_n(P_n) = I(x_n) = \|y - \sum_{k=1}^{n} \alpha_k y_k\|^2 , \\ y_k = Ae_k,\ P_n = P_n(\alpha_1, \alpha_2, \dots, \alpha_n) . \end{gathered}$$

If E is a Hilbert space then $I(x_n)$ is a quadratic function of $(\alpha_1, \alpha_2, \dots, \alpha_n)$ and we obtain the method of least squares. In the general case we can only state that the following lemma is valid.

Lemma 1. *The function $\Delta_n(P_n)$ is convex and satisfies a Lipschitz condition.*

Proof. The convexity of $\Delta_n(P_n)$ is equivalent to the inequality

$$\Delta_n\left(\frac{P_n^{(1)} + P_n^{(2)}}{2}\right) = \|(y - \sum_{k=1}^{n} \alpha_k^{(1)} y_k) + (y - \sum_{k=1}^{n} \alpha_k^{(2)} y_k)\|^2/4 \tag{6.9}$$

$$\leq [\Delta_n(P_n^{(1)}) + \Delta_n(P_n^{(2)})]/2 \ ,$$

which follows in an obvious way from the norm properties. The Lipschitz condition is valid since

$$\Delta_n^{\frac{1}{2}}(P_n^{(1)}) - \Delta_n^{\frac{1}{2}}(P_n^{(2)}) \leq \| \sum_{k=1}^{n} (\alpha_k^{(1)} - \alpha_k^{(2)}) y_k \|$$

$$\leq \max_k \|y_k\| \sum_{k=1}^{n} |\alpha_k^{(1)} - \alpha_k^{(2)}| \ . \quad \# \tag{6.10}$$

From the convexity and continuity of $\Delta_n(P_n)$, it follows that $\Delta_n(P)$ has a minimum value which is attained on a connected set of points P_n.

In this more general case, some theorems from §3 which were proved for the function defined in equation (3.2), remain valid for the function defined by (6.8). From the proofs of Theorems3.1 and 3.2, it follows that these theorems are also valid (putting $\alpha = 0$) for the function defined by (6.8). However the system of equations (3.5) will not now occur, since in general the function $\Delta_n(P)$ is not differentiable. Theorems 3.4 and 3.5 will also remain valid, if we assume that the null space $L(A)$ of the operator A is a subspace, $D(A) \ominus L(A)$ is a factor space of $D(A)$ with respect to $L(A)$, and we take the condition of closure of the set $R(A)$ as the definition of the normal solvability of the operator A.

Among the algorithms examined in §3, only the coordinate-wise descent algorithm can be applied to the minimization of the function in (6.8), since it is not based on the concept of orthogonality of vectors and does not require the differentiability of the function being minimized. It is clear that equations like (3.99) no longer exist, but the required value of $b_s^{(j)}$ for which $\delta_n(Q_n) = \|y - \sum_{k=1}^{n} \beta_k u_k\|^2$, $Q_n = Q_n(\beta_1, \beta_2, \ldots, \beta_n)$, $u_k = y_k/\|y_k\|$, attains its minimum value along the sth coordinate at the jth iteration (and such a minimum exists because of the convexity of $\delta_n(Q_n)$), can be found to any degree of accuracy by a very simple algorithm.

In general, it is not possible to approximate $\inf_{P_n} \Delta_n(P_n)$ arbitrarily closely by the coordinate-wise descent algorithm. The obstacle to convergence is the existence of saddle points for which $\Delta_n(P_n)$ does not decrease in any coordinate direction, but at which $\Delta_n(P_n)$ has not attained its minimum.

Theorem 6.2. *If $\Delta_n(P_n)$ is differentiable then the coordinate-wise descent algorithm always converges in the sense that*

$$\lim_{j\to\infty} \Delta_n(P^{(j)}_{n,s}) = \inf_{P_n} \Delta_n(P_n) \quad (s = 1,2,\ldots,n) \,, \tag{6.11}$$

where

$$P^{(j)}_{n,s+1} = P^{(j)}_{n,s} + a^{(j)}_{s+1}\, \tilde{e}_{s+1} \,,$$

$$\begin{aligned} &\Delta_n(P^{(j)}_{n,s+1}) = \inf_a \Delta_n(P^{(j)}_{n,s} + a\tilde{e}_{s+1}) \,, \\ &P^{(j)}_{n,n+1} = P^{(j+1)}_{n,1} \,, \ \tilde{e}_{n+1} = \tilde{e}_1 \,, \end{aligned} \tag{6.12}$$

$$\tilde{e}_1 = (1,0,\ldots,0), \ \tilde{e}_2 = (0,1,0,\ldots,0),\ldots, \ \tilde{e}_n = (0,0,\ldots,0,1) \,.$$

Proof. We select the maximal system of linearly independent elements from the system $\{y_k\}_1^n$. Without loss of generality we can denote this system by $\{y_k\}_1^m$, where $1 \le m \le n$. Substituting into (6.8) the representations of y_k in terms of the first m elements:

$$y_k = \sum_{s=1}^{m} \alpha^{(k)}_s y_s \quad , \ k = m+1, m+2,\ldots,n, \tag{6.13}$$

we obtain

$$\begin{aligned} &\Delta_n(P_n) = \|y - \sum_{k=1}^{n} \alpha_k y_k\|^2 = \|y - \sum_{k=1}^{m} \beta_k y_k\|^2 = \Delta_m(Q_m) \,, \\ &\beta_k = \alpha_k + \sum_{j=m+1}^{n} \alpha_j \alpha^{(j)}_k \,, \ Q_m = Q_m(\beta_1,\beta_2,\ldots,\beta_m) \,. \end{aligned} \tag{6.14}$$

We denote by $Q^{(j)}_{m,s}$ the points which correspond to $P^{(j)}_{n,s}$. From the linear independence of the system $\{y_k\}_1^m$, the sequence of points $Q^{(j)}_{m,s}$ is bounded and consequently we may select from it a convergent subsequence $Q^{(j_r)}_{m,s}$ such that $\lim_{r\to\infty} Q^{(j_r)}_{m,s} = \mathring{Q}_m$. The sequence $\Delta_m(Q^{(j)}_{m,s}) = \Delta_n(P^{(j)}_{n,s})$ does not increase with increasing s and j. Hence

$$\mathring{\Delta}_m = \Delta_m(\mathring{Q}_m) = \lim_{r\to\infty} \Delta_m(Q^{(j_r)}_{m,s}) = \lim_{j\to\infty} \Delta_m(Q^{(j)}_{m,s}) \quad ,$$

where $\Delta_m(Q^{(j)}_{m,s}) \ge \mathring{\Delta}_m$ for all $s (1 \le s \le n)$ and $j = 0,1,2,\ldots$.

We now show that a coordinate direction (in m-dimensional space) along which $\Delta_m(Q_m)$ decreases does not exist at the point $\mathring{Q}_m$. Let us assume the contrary. Let $\Delta_m(\mathring{Q}_m + a\tilde{e}_\nu) < \mathring{\Delta}_m$ where $1 \le \nu \le m$. Then from the continuity of

$\Delta_m(Q_m)$, beginning with a certain r, $\Delta_m(Q_{m,s}^{(j_r)} + a\tilde{e}_\nu) < \mathring{\Delta}_m$ and so *a fortiori* $\Delta_m(Q_{m,s}^{(j_{r+1})} + a\tilde{e}_\nu) \leq \mathring{\Delta}_m$, which is not possible. Consequently all partial derivatives of $\Delta_m(Q_m)$ are equal to zero at the point $\mathring{Q}_m$, and the tangent plane is orthogonal to the axis Δ_m at this point. But, for a convex function the tangent plane is the plane of support, so that all values of $\Delta_m(Q_m)$ are greater than $\mathring{\Delta}_m = \Delta_m(\mathring{Q}_m)$. This proves the theorem. #

The following theorem now follows from the proof of Theorem 6.2.

Theorem 6.3. *Let* $\Delta(P_n)$ *be a differentiable function, and let* $P_{n,s}^{(j)}$ *be constructed by the algorithm of coordinate-wise minimization of* $\Delta(P_n)$. *If the set of* P_n *for which* $\Delta(P_n) \leq \Delta(P_{n,s}^{(j)})$ *(for any fixed* s *and* j*) is bounded, then the coordinate-wise descent algorithm always converges in the sense that any limit point of the sequence* $P_{n,s}^{(j)}$ *is a stationary point for* $\Delta(P_n)$.

In the case of differentiable functions we may also apply the steepest descent algorithm, for which it is not difficult to prove convergence theorems similar to Theorems 6.2 and 6.3.

We shall now examine a *steep descent algorithm* X which is of general applicability and may be used to minimize the function defined in (6.8) when $\Delta_n(P_n)$ is not necessarily differentiable.

Let $\Delta(P)$ denote an arbitrary convex continuous function in an n-dimensional Euclidean space, such that the domain of those P for which $\Delta(P) \leq \Delta(P_0)$ (where P_0 is an initial point) is bounded. The steep descent algorithm X consists of the following three basic parts:-

1. An algorithm for the determination of inf $\Delta(P)$ on a given interval $[P', P'']$

We assume $\Delta(t) = \Delta(P)$ where $P = P' + t(P''-P')$. Calculating $\Delta(t)$ for three values $t = t^{(1)}$ and $t^{(1)}\pm1/2$, we select $t = t^{(2)}$ for which Δ is smallest. If all three values are equal then they coincide with the required value. Otherwise, repeating the operation for the values $t = t^{(2)}$ and $t^{(2)}\pm1/2^2$, we find $t^{(3)}$ for which Δ is least, etc. If $t^{(j)} > 1$ or $t^{(j)} < 0$, then the required value is attained at $t = 1$ or $t = 0$, respectively. This algorithm is called a *successive subdivision algorithm*. It is not difficult to calculate that after k steps of the successive subdivision algorithm (where k is defined by

$$k = \log_2 [\rho(P',P'')/h] , \tag{6.15}$$

$\rho(P',P'')$ being the distance between the points P' and P'') if $\mathring{P}$ denotes the required point, then the quantity $\Delta(\mathring{P}) = \inf_{0 \leq t \leq 1} \Delta(t)$ has been calculated to within an accuracy which does not exceed h.(23)

2. <u>An algorithm for the estimation of inf Δ in a given direction ∇</u>

Let P_0 be an initial point. We form the sequence of points

$$T_r = P_0 + 2^{r-2}\sigma\nabla, \quad r = 2,3,\ldots, \quad (T_1 = T_0 = P_0) \ , \tag{6.16}$$

where σ is a given positive quantity, and calculate

$$\gamma_r = \Delta(T_r) - \Delta(T_{r-1}), \ r = 2,3,\ldots \ . \tag{6.17}$$

If $r_0(r_0 \geq 2)$ is the first value of r for which $\gamma_{r_0} \geq 0$, then the required value of $\mathring{P}$ is attained on the interval $[T_{r_0-2}, T_{r_0}]$, and it may be determined by the successive subdivision algorithm.

For functions of the form (6.8), the problem of the estimation of the least value in a given direction is reduced to the determination of the least value of the function ϕ defined by

$$\phi(a) = \|u - av\| \ , \qquad a \geq 0 \ , \tag{6.18}$$

where $v(\neq 0)$ and u are known elements. Denoting the required value by $\mathring{a}$, we conclude from the simple inequalities $\|u-\mathring{a}v\| \leq \|u\|$ and $\|u-\mathring{a}v\| \geq \mathring{a}\|v\|-\|u\|$, that

$$0 \leq \mathring{a} \leq 2\|u\|/\|v\| \ . \tag{6.19}$$

Consequently, in this particular case, a successive subdivision algorithm can immediately be applied.

3. <u>An algorithm for the determination of a descent direction for Δ</u>

We introduce an n-dimensional cube Π_n with centre at a given point P', planes parallel to the coordinate planes, and with length of edge, 2δ. We denote by $V(P')$ the volume of the domain $D(P')$ being that part of Π_n where the values of $\Delta(P)$ are less than $\Delta(P')-\sigma$. We sample randomly the points $P \in \Pi_n$ with respect to the uniform distribution on Π_n (see [15]). The probability of finding a point P in the domain $D(P')$ at one attempt is obviously equal to

(23) An algorithm for the estimation of the least value of a uni-modal function, (i.e. a function having a unique minimum), which is optimal with respect to the number of calculations of the function, has been described in [5].

(6.20) $$p_1(\delta,\sigma,P') = V(P')/(2\delta)^n \ .$$

Consequently for N experiments, the probability of finding a point P in the domain $D(P')$ is given by

(6.21) $$p_N = 1 - (1 - p_1)^N \ .$$

Definition. *The probability $p_1(\delta,\sigma,P')$ is called the δ,σ-measure of the conditioning of the problem for the determination of a descent direction for Δ at the point P'.*

The smaller the value of p_1, the more difficult is the solution of the problem.

It is clear that for decreasing σ, p_1 can only increase. The dependence of p_1 on δ is considerably more complicated. Assuming that $\sigma = 0$, through the point $(P', \Delta(P'))$ in $(n+1)$ dimensional space, we take the hyperplane $\Delta = \Delta(P')$ which determines a convex region of intersection with the surface $\Delta = \Delta(P)$. The intersection of the projection of this region on to the hyperplane $\Delta = 0$ with the cube Π_n, gives $D(P')$ which is also convex. Since the whole domain $D(P')$ lies along one side of the plane of support of $D(P')$ at the point which is the projection of P' on to $\Delta = 0$, it is obvious that

(6.22) $$p_1(\delta,\sigma,P') \leq 1/2 \ .$$

Also, from the convexity of $D(P')$ it follows that $p_1(\delta,0,P')$ can only decrease for increasing δ. Consequently, it is necessary to take the least possible δ. But, for $\sigma > 0$, it is impossible to do this since then $p_1(\delta,\sigma,P') = 0$ for all sufficiently small δ. The question of determining an optimal δ when $\sigma > 0$, remains open.

Let us take

(6.23) $$\delta^2 = \sigma.$$

We now assume that there is no point $P'' \in \Pi_n$ for which $\Delta(P') - \Delta(P'') \geq \sigma > 0$. Denoting by $\mathring{P}'$ the point which is closest to P' such that $\Delta(\mathring{P}') = \inf_P \Delta(P)$, and by P''' the point of intersection of the interval $[P', \mathring{P}']$ with the boundary of Π_n, we obtain

(6.24) $$\Delta(P') - \Delta(P''') < \sigma \ , \quad P''' \in \Pi_n \quad .$$

From the convexity of $\Delta(P)$ where $P \in [P', \mathring{P}']$ and from condition (6.23), we find that

(6.25) $$\Delta(P') - \Delta(\mathring{P}') \leq \frac{\rho(P', \mathring{P}')}{\rho(P', P''')}\sigma \leq \rho(P', \mathring{P}')\delta \ .$$

We now consider a global description of the algorithm. Putting $P_0 = P_0'$ (where P_0' is an arbitrary point), and assuming that the values of δ, $\sigma = \delta^2$ and N have been chosen, we can determine a descent direction for Δ at the point P_0'.

If as a result of the solution of this problem, we find the point $P_0'' \in \Pi_n$, $\Delta(P_0) - \Delta(P_0'') \geq \sigma$, then as the descent direction for $\Delta(P)$ at the point P_0, we take the ray $P = P_0 + t(P'' - P_0)$, $t \geq 0$. Determining the least value in this direction, we find the point P_1. We now take P_1 as an initial point instead of P_0, and obtain in an analogous way the points P_2, P_3 etc.

Since $\inf_P \Delta(P) > -\infty$, it is obvious that there exists a j^0 (≥ 0) such that when determining the descent direction we find a point $P''_{j^0} \in \Pi_n$ for which $\Delta(P) - \Delta(P''_{j^0}) \geq \sigma$. Then the corresponding point P_{j^0} is the required point such that $\Delta(P_{j^0}) \approx \inf_P \Delta(P)$. We summarize these results in the following theorem.

Theorem 6.4. *The steep descent algorithm X always converges in the sense that for given $\epsilon > 0$, sufficiently large N and sufficiently small σ, there exists a j^0 such that*

(6.26) $$\Delta(P_{j^0}) - \inf_P \Delta(P) < \epsilon$$

with probability

(6.27) $$p_N > 1 - \epsilon.$$

The proof follows from (6.20), (6.23) and (6.25) if we take into account that under the given conditions, the domain of P for which $\Delta(P) \leq \Delta(P_0)$ is bounded.

As is obvious from inequality (6.25), for a proof of the theorem it is sufficient that $\rho(P_{j^0}, \mathring{P}_{j^0})$ is bounded, where $\mathring{P}_{j^0}$ is the nearest point to P_{j^0} such that $\Delta(\mathring{P}_{j^0}) = \inf_P \Delta(P)$. For the minimization of functions of the form (6.8), this implies that method X is also convergent in the case when the elements y_k are linearly dependent. Theorem 6.4 also remains valid if in algorithm 1, we restrict attention to a finite number of steps.

It should be noted that in the special case when A is the unit operator, the rapid descent algorithm X allows the best approximation by linear combinations of elements in any metric (with the usual properties of a norm) to be found.

In the general case, the rapid descent algorithms ensure convergence of an approximate solution of an operator equation to the exact solution only with respect to the residual. In order to guarantee the simultaneous convergence of the elements $x_n = \sum_{k=1}^{n} \alpha_k e_k$, it is necessary to place the additional restriction that $x_n \in D(A) \cap M$ on the minimization of $\Delta_n(P)$, where M is a certain given set.

Let M be a compact set. We denote by N the range of the operator A on the set $D(A) \cap M$.

Lemma 2. *If a closed operator A maps the set $D(A) \cap M$ one-to-one and onto the set N, then the inverse operator A^{-1} exists. The operators A and A^{-1} are continuous on $D(A) \cap M$ and N, respectively.*

For the proof we note that the continuity of A follows from the compactness of N and the condition that A is closed. The proof is now similar to that of the proof of Lemma 2 (§2.3).

Sufficient conditions for the compactness of M and N respectively in C and L_p are given by the inequalities

$$\|x\| \leq C_1, \ \|x(t) - x(t+h)\| \leq \omega_1(h) \to 0, \ h \to 0 , \tag{6.28}$$

and

$$\|Ax\| \leq C_2, \ \|(Ax)(t) - (Ax)(t+h)\| \leq \omega_2(h) \to 0, \ h \to 0 . \tag{6.29}$$

Since these inequalities are uniformly valid with respect to all x, they are obviously convexity conditions. Therefore, in its approximate formulation, the problem of the minimization of $\Delta_n(P)$ under the condition that x_n satisfies (6.28) and (6.29), reduces to a convex programming problem. The steep descent algorithm which is analogous to algorithm X and for which the same convergence theorem applies (see also §6.3), can be used for its numerical solution.

Let $\mathring{x}_n \in M$ be the point at which $\Delta_n(P)$ takes its least value on M.

Theorem 6.5. *Under the conditions of Lemma 2,*

$$\|\mathring{x}_n - \overset{*}{x}_M\| \to 0 , \ n \to \infty , \tag{6.30}$$

where $\|A\overset{*}{x}_M - y\| = \inf_{x \in D(A) \cap M} \|Ax - y\|$, *(A and A^{-1} are closed on $\overline{D(A)} \cap M$ and $\overline{N}$).*

The proof easily follows from the inequality

(6.31) $\|\overset{\circ}{x}_n - \overset{*}{x}_M\| = \|A^{-1}A(\overset{\circ}{x}_n - \overset{*}{x}_M)\| \leq \text{const}\ \omega_{A^{-1}}(\|A\overset{\circ}{x}_n - A\overset{*}{x}_M\|)$,

where $\omega_{A^{-1}}(h)$ denotes the modulus of continuity of the operator A^{-1} which tends to zero as $h \to 0$, and Theorem 3.2 which can be readily generalized to the case when $\Delta_n(P)$ is to be minimized under the additional restriction that $x_n \in D(A) \cap M$.

We note that in the cases when we are solving the equation $Ax = y$ in the spaces L_p or C, and it is known that $\|x'(t)\| < \infty$ and $\|y'(t)\| < \infty$, then it is natural to replace (6.28) and (6.29) by the simpler inequalities

(6.32) $$\|x'(t)\| \leq C_3 \quad , \quad \|(Ax)'(t)\| \leq C_4 \ ,$$

where the constants C_3 and C_4 do not depend on x. It should also be noted that the realisation of (6.32) in L_p and the uniform boundedness of $x(t)$ and $(Ax)(t)$ at the fixed point t_1, are sufficient conditions for the compactness of M and N in C.

§6.3 Some Algorithms for the Solution of Non-Linear Programming Problems

Let us examine the general programming problem of finding the minimum value of an objective function

(6.33) $$Q = Q(P) \quad (P \in D) \ ,$$

where D is a given domain of points $P(x_1, x_2, \ldots, x_n)$ in an n-dimensional Euclidean space.[24]

If Q is a linear function so that

(6.34) $$Q = \sum_{k=1}^{n} a_k x_k \ ,$$

and the domain D is bounded on all sides by hyperplanes, i.e. D is defined by a system of linear inequalities of the form

(6.35) $$\sum_{k=1}^{n} a_k^{(j)} x_k - b_j \leq 0 \quad (j = 1, 2, \ldots, m) \ ,$$

then we obtain a *linear programming* problem (see [52]).

If Q is convex, i.e. $Q[(P_1+P_2)/2] \leq [Q(P_1)+Q(P_2)]/2$ for any points P_1, $P_2 \in D$, and D is convex i.e. all lines joining any two of its points lie entirely in D, then we have a *convex programming* problem (see [52]).
The convex domain D is often defined by inequalities of the form

---

(24) The determination of the *maximum* value of an objective function Q is easily reduced to the above problem by considering the function $-Q$.

(6.36) $$H_j(P) \leq 0 \qquad (j = 1,2,\dots,m) \ ,$$

where $H_j(P)$ are convex functions.

If Q is non-linear, and (or) D cannot be defined by a system of linear inequalities, then we have a *non-linear programming* problem (see [165]).

It is clear that convex programming is a particular case of non-linear programming, and linear programming is a particular case of convex programming. In the case of convex programming, only one local minimum value of Q exists and this coincides with the global minimum of Q.

Problems already discussed in this chapter relating to the calculation of approximate solutions of equations of the form $Ax = y$, can be interpreted as non-linear programming problems.

1. (See §3.1 and §6.2). Minimize

(6.37) $$\Delta_n(P) = \|y - \sum_{k=1}^{n} \alpha_k A e_k\|^2 \ , \ P(\alpha_1,\alpha_2,\dots,\alpha_n) \ .$$

2. Minimize $\Delta_n(P)$ defined by (6.37) under the conditions

(6.38) $$\| \sum_{k=1}^{n} \alpha_k A_j e_k - y_j\|^2 \leq b_j \ , \ (j = 1,2,\dots,m) \ .$$

3. Determine

(6.39) $$m_n = \inf_{\alpha_k} \| \sum_{k=1}^{n} \alpha_k A e_k\| \ , \ \| \sum_{k=1}^{n} \alpha_k e_k\| = 1 \ ,$$

and

(6.40) $$M_n = \sup_{\alpha_k} \| \sum_{k=1}^{n} \alpha_k A e_k\| \ , \ \| \sum_{k=1}^{n} \alpha_k e_k\| = 1 \ ,$$

which play an important role in the determination of effective estimates in algorithms associated with the method of least squares (§3).

4. Minimize $\| \sum_{k=1}^{n} \alpha_k A e_k\|$ under the condition $\| \sum_{k=1}^{n} \alpha_k e_k\| = 1$ and constraints of the form (6.38).

It is readily seen that Problems 1 and 2 are convex programming problems.

Generalizations of the above examples to the case when A is a non-linear operator, will give rise to non-linear programming problems. For example, instead of Problem 2 we now have the problem of minimizing

(6.41) $$\Delta_n(P) = \|y - A(\sum_{k=1}^{n} \alpha_k e_k)\|^2$$

under the constraints

(6.42) $$\| A_j(\sum_{k=1}^{n} \alpha_k e_k) - y_j \| \leq b_j \quad (j = 1,2,\dots,m) .$$

It is clear that the problem of the minimization of $I(\sum_{k=1}^{n} \alpha_k e_k)$ under the conditions that $I_j(\sum_{k=1}^{n} \alpha_k e_k) \leq 0$, $j = 1,2,\dots,m$, where $I,I_1,\dots,I_m$ are functionals, is even more general.

Let us now examine some of the more general algorithms for the solution of non-linear programming problems, which do not require differentiability of the objective function and the constraints.(25)

Suppose we require the minimum value of $Q(P)$ in a domain D defined by the system of constraints

$$H_j(P) \leq 0 \quad (j = 1,2,\dots,m) ,$$

(6.43) $$R_s(P) = 0 \quad (s = 1,2,\dots,r) , \quad P = P(x_1,\dots,x_n) .$$

We shall assume that Q, H_j and R_s are continuous functions in the whole space. Without loss of generality we can assume that D is bounded, since it is always possible to add to the functions H_j, a function $\rho(0,P)-R$ with sufficiently large R, where $\rho(0,P)$ denotes the distance between the origin of coordinates and the point P. We shall also assume that Q attains its minimum in the region defined by the sphere $\rho(0,P) \leq R$. To satisfy this condition, it is sufficient to minimize, instead of Q, the function Q_1 defined by

(6.44) $$Q_1(P) = \begin{cases} Q(P) , & \rho(0,P) \leq R , \\ Q(P_1) , & \rho(0,P) \geq R , \end{cases}$$

where P_1 is the point of intersection of the ray from the origin through P with the sphere $\rho(0,P) = R$;

(6.45) $$x_i^{(1)} = \frac{R}{\rho(0,P)} x_i , \quad P_1(x_1^{(1)},\dots,x_n^{(1)}) , \; P(x_1,\dots,x_n) .$$

We denote by $D_\alpha(\alpha > 0)$, the domain defined by the system of inequalities

(6.46) $$\begin{aligned} H_j(P) &\leq \alpha , \; j = 1,2,\dots,m , \\ -\alpha \leq R_s(P) &\leq \alpha , \; s = 1,2,\dots,r . \end{aligned}$$

For the minimization of $Q(P)$ in D_α, we can apply the steep descent algorithm X(described in the previous section) but with the difference that we now have

(25) A collection of different algorithms is given in [37, 101, 165, 60, 52, 92].

(6.47) $$p_1(\delta,\sigma,P') = V(P')/(2\delta)^n \quad ,$$

where $V(P')$ denotes the volume of that part of Π_n which belongs to D_α and in which $Q(P)$ is less than $Q(P')$ by an amount of σ.(26)

The approximate solution of the problem of the minimization of $Q(P)$ in D_α for sufficiently small α, is simultaneously an approximate solution of the original problem.

Lemma 3. *If the domain D is not empty, then as* $\alpha \to 0$

(6.48) $$0 \le \inf_{P\in D} Q(P) - \inf_{P\in D_\alpha} Q(P) \le \eta(\alpha) \to 0 \ .$$

In addition any limit point $\mathring{P}$ *of the sequence* $\{\mathring{P}_\alpha\}$ *where* $Q(\mathring{P}_\alpha) = \inf_{P\in D_\alpha} Q(P)$, *has the property that* $Q(\mathring{P}) = \inf_{P\in D} Q(P)$.

Proof. Let L_α and L denote those points of D_α and D respectively, where $Q(P)$ attains its minimum. Let us assume that there exists a sequence $\{\alpha_i\}$ tending to zero as $i \to \infty$, such that $\inf_{P\in D} Q(P) - \inf_{P\in D_{\alpha_i}} Q(P) \ge d > 0$. From the continuity of $Q(P)$ and the boundedness of D_{α_i}, there exist points $\mathring{P}_{\alpha_i}$ for which $Q(\mathring{P}_{\alpha_i}) = \inf_{P\in D_{\alpha_i}} Q(P)$. Let $\mathring{P}$ be one of the limit points of the sequence $\mathring{P}_{\alpha_i}$. Then obviously $\inf_{P\in D} Q(P) - Q(\mathring{P}) \ge d > 0$. Consequently $\mathring{P} \notin D$, and there exists at least one integer j_0 or s_0 such that $H_{j_0}(\mathring{P}) > 0$ or $R_{s_0}(\mathring{P}) \ne 0$. But the last condition is not possible for sufficiently small α_i. In a similar way, we can readily show that any limit point $\mathring{P}$ of the sequence $\mathring{P}_\alpha \in L_\alpha$ cannot be outside of D. If we assume that it does not belong to L we obtain a contradiction since $\inf_{P\in D} Q(P) - Q(\mathring{P}) = 0$. #

Instead of Theorem 6.4 we have the following result.

Theorem 6.6. *If the domain D is not empty, then the steep descent algorithm X always converges to one of the points* $\mathring{P}$ *of the local minimum of Q in* D_α, *in the sense that given* $\epsilon > 0$, *there exists sufficiently small* δ,σ *and sufficiently large N, such that after a finite number of steps a point* P_s *is obtained for which the problem of the estimation of a descent direction is not solvable and*

(6.49) $$Q(P_s) - Q(\mathring{P}) \ge \epsilon \ .$$

with a probability $p \le \epsilon$.

---

(26) We now assume that the initial point P_0 belongs to D_α. We note that the problem of the determination of a point, which belongs to D_α, can be reduced to a finite sequence of problems of the type examined in [60].

Proof. We first note that we cannot have an infinite number of steps since this implies $\inf_{P \in D_\alpha} Q(P) = -\infty$. Let M_ϵ denote the set of points $P \in D_\alpha$ for which $Q(P) - Q(\overset{0}{P}) \geq \epsilon$, where $\overset{0}{P}$ is any local (interior or boundary) minimum of Q in D_α. From the conditions of the theorem and the property of uniform continuity, it is not difficult to show that

$$\inf_{P \in M_\epsilon} V(P) > 0 , \tag{6.50}$$

for any fixed $\delta > 0$ and sufficiently small $\sigma(\delta)$. Hence, for sufficiently large N, the probability that $P \in M_\epsilon$, and that at P the problem of the determination of a descent direction for Q in D_α is not solvable, is less than ϵ. This establishes the theorem. #

To determine the point $\overset{\circ}{P}$ where $Q(P)$ attains its minimum in D, we repeatedly select an initial point P_0 in D at random, and each time apply the steep descent algorithm. We choose $\overset{\circ}{P}$ to be one of the local minima at which $Q(\overset{\circ}{P})$ is least (see [13]).

It is known that in its approximate formulation, the minimization problem with constraints can always be reduced to a minimization problem without constraints (see [92]). Let us introduce the auxiliary function

$$\begin{aligned} \tilde{Q}(P) &= Q(P) + \sum_{j=1}^{m} \lambda_j \tilde{H}_j^2(P) + \sum_{s=1}^{r} \mu_s R_s^2(P) \\ &= Q(P) + H(P) , \end{aligned} \tag{6.51}$$

where

$$\tilde{H}_j(P) = \begin{cases} H_j(P) , & H_j(P) \geq 0 , \\ 0 , & H_j(P) \leq 0 , \end{cases} \tag{6.52}$$

and λ_j, μ_s are non-negative numbers.

Theorem 6.7. *If the domain D is not empty, then given $\epsilon > 0$ there exist arbitrarily large $\lambda_j(\epsilon)$ and $\mu_s(\epsilon)$ such that $\tilde{Q}(P)$ attains its minimum value at those points $\tilde{P}$ for which*

$$0 \leq \inf_{P \in D} Q(P) - Q(\tilde{P}) \leq \epsilon . \tag{6.53}$$

Proof. If $P \notin D_\alpha$, then

$$\tilde{Q}(P) \geq Q(P) + \alpha^2 \Big(\sum_{j=1}^{m} \lambda_j + \sum_{s=1}^{r} \mu_s \Big) \tag{6.54}$$

so that, for sufficiently large λ_j and μ_s, $\tilde{Q}(P)(P \notin D_\alpha)$ can be made arbitrarily large. This implies that $\inf \tilde{Q}(P)$, for sufficiently large λ_j and μ_s, is attained on D_α. Hence, from the definition of $\tilde{Q}(P)$, we obtain

(6.55) $$\inf_{P\varepsilon D} Q(P) = \inf_{P\varepsilon D} \tilde{Q}(P) \geq \inf_{P} \tilde{Q}(P) = \inf_{P\varepsilon D_\alpha} \tilde{Q}(P) \geq \inf_{P\varepsilon D_\alpha} Q(P) \ .$$

But from Lemma 3

$$0 \leq \inf_{P\varepsilon D} Q(P) - \inf_{P\varepsilon D_\alpha} Q(P) \leq \eta(\alpha) \leq \epsilon \ ,$$

for sufficiently small $\alpha(\epsilon)$. Since $\inf \tilde{Q}(P) = \tilde{Q}(\tilde{P}) = Q(\tilde{P}) + H(\tilde{P})$ and $H(\tilde{P}) \geq 0$ then

(6.56) $$\inf_{P\varepsilon D_\alpha} Q(P) \leq Q(\tilde{P}) \leq \inf_{P\varepsilon D_\alpha} \tilde{Q}(P) \leq \inf_{P\varepsilon D} Q(P)$$

and

(6.57) $$\inf_{P\varepsilon D} Q(P) - Q(\tilde{P}) \leq \inf_{P\varepsilon D} Q(P) - \inf_{P\varepsilon D_\alpha} Q(P) \leq \eta(\alpha) \leq \epsilon \ . \quad \#$$

From the proof of the theorem it follows that for sufficiently small ϵ, $\tilde{P}$ will be arbitrarily close to the set of points L where $Q(P)$ attains its minimum value in D. If L consists of interior points of D, then for sufficiently large λ_j and μ_s, $\tilde{P} \in L$.

After reducing the initial non-linear programming problem to the problem of the minimization of $\tilde{Q}(P)$, it is not difficult, by the replacement of independent variables, to reduce the latter to the standard problem of the minimization of the function $\phi(x_1,x_2,\ldots,x_n)$ in the unit n-dimensional cube $\Pi_n (0 \leq x_i \leq 1,\ i = 1,2,\ldots,n)$.

Let us now examine some general approaches to the solution of this standard problem. First of all, let us compare the Monte-Carlo method with the method of simple sorting. We assume that there exists a sequence of uniformly distributed random points in the n-dimensional cube Π_n. We divide each edge of this cube into $1/\delta$ equal parts, and then divide the whole cube Π_n into $(1/\delta)^n$ small cubes $\Pi_{n,\delta}$ each of volume δ^n, by hyperplanes which are perpendicular to the coordinate planes. The probability that any point of a sequence $T_1,T_2,\ldots$ falls into any one of the fixed small cubes $\Pi_{n,\delta}$ is obviously equal to

(6.58) $$p_1 = \delta^n \ .$$

The probability that only one of N points of the sequence $T_1,T_2,\ldots$ falls into $\Pi_{n,\delta}$ is equal to (see [120])

(6.59) $$p_N = 1 - (1 - \delta^n)^N.$$

From the relation that $(1+u)^{1/u} < e \approx 2.7$, we find

(6.60) $$p_N \le 1 - (1 - \delta^n)^{(1/\delta)^n} < 1 + (\delta^n - 1)/e \approx 0.63,\ 10^2 \le N \le (1/\delta)^n.$$

This result implies that a direct application of the Monte-Carlo method for the determination of the minimum value of ϕ in Π_n, requires a smaller number of operations than simple sorting, except when the result is required only to low accuracy†.

We obtain a different situation when the random search is applied only for the determination of the minimum of ϕ in the small cube $\Pi_{n,\delta}$, as in the steep descent algorithm X. If ϕ is differentiable then, for sufficiently small δ, the probability that any point of a sequence $T_1, T_2, \ldots$ falls into the required part of the small cube $\Pi_{n,\delta}$ will be arbitrarily close to 1/2, provided grad $\phi \neq 0$. Observing that

(6.61) $$\frac{\delta^n}{V_n(\delta)} = \left(\frac{4}{\pi}\right)^{n/2} \Gamma\left(\frac{n}{2} + 1\right) ,$$

where $V_n(\delta)$ is the volume of an n-dimensional sphere $\chi_{n,\delta}$ which is a refinement of the cube $\Pi_{n,\delta}$, and Γ is the gamma function which increases rapidly with n, it can be seen that it is more advantageous in the general case to construct a sequence of random points $T_1, T_2, \ldots$ which are uniformly distributed on the surface of the n-dimensional sphere $\chi_{n,\delta}$.

We can obtain this sequence by selecting from the uniformly distributed sequence in $\Pi_{n,\delta}$, only those points which lie inside the sphere $\chi_{n,\delta}$, and projecting them on to the surface of this sphere. A more effective algorithm (see [15]) is given by the formation of a uniformly distributed sequence with centre of distribution at the centre of the sphere $\chi_{n,\delta}$, and the projection of these points on to the surface of the sphere $\chi_{n,\delta}$.

A deterministic method which can be used for the evaluation of the global minimum of ϕ in Π_n, is given in [60]. This method is based on the use of the recurrence relation

(6.62) $$\inf_{P_n \in \Pi_n} \phi(P_n) = \inf_{0 \le x_n \le 1} \left[\inf_{P_{n-1} \in \Pi_{n-1}} \phi(P_{n-1}, x_n)\right] ,$$

where $P_n = (x_1, x_2, \ldots, x_n)$, $P_{n-1} = (x_1, x_2, \ldots, x_{n-1})$, and on the application of algorithms for the evaluation of the global minimum of a function of a single variable on [0,1]. These are more economical than simple sorting algorithms.

The following algorithm has been proposed in [7], for the evaluation of the global minimum of a function f defined on [0,1]. Starting with the point $x = z_0 = 0$, we define the first local minimum x_1 by the steepest descent algorithm. Then, by the known algorithm (see [121])

† But see R.S. Anderssen and P. Bloomfield, *J. Optimization Theory Appl.*, 16(1975)383-398. (Eds.)

(6.63) $$y_{j+1} = y_j + |f(y_j) - f(x_1) + \epsilon|/M, \quad (j = 0,1,2,\ldots)$$

where $M \leq \sup\limits_{x_1,x_2 \in [0,1]} |f(x_1) - f(x_2)|/|x_1 - x_2|$, we take z_1 to be the root of the equation

(6.64) $$f(y) = f(x_1) - \epsilon ,$$

where ϵ is a given small positive number. Starting with z_1, we again use the steepest descent algorithm to evaluate the local minimum x_2 which then takes the role of x_1 etc.

§6.4 Criteria for the Comparison of Different Algorithms

The result of an approximate solution of any problem in a discrete form is given by a finite-dimensional numerical vector $R^{(\epsilon)}(r_1^{(\epsilon)}, r_2^{(\epsilon)}, \ldots, r_p^{(\epsilon)})$, which is obtained as a result of the application of a finite number of operations $\tilde{A}$ to a finite-dimensional numerical input vector $I^{(\epsilon)}(i_1^{(\epsilon)}, i_2^{(\epsilon)}, \ldots, i_s^{(\epsilon)})$,

(6.65) $$R^{(\epsilon)} = \tilde{A} I^{(\epsilon)} .$$

We denote the exact value of the input data by $I(i_1, i_2, \ldots, i_s)$, the desired result in the discrete form by $R(r_1, r_2, \ldots, r_p)$ and the transformation algorithm from I to R by A, where

(6.66) $$R = AI .$$

The values of the components of the vectors I and R are usually unknown, but on the other hand it is known that

(6.67) $$\|I - I^{(\epsilon)}\|^2 = \sum_{j=1}^{s} |i_j - i_j^{(\epsilon)}|^2 \leq \epsilon ,$$

and we require

(6.68) $$\|R - R^{(\epsilon)}\|^2 = \sum_{j=1}^{p} |r_j - r_j^{(\epsilon)}|^2 \leq \delta_p .$$

Here we have chosen one of many possible norms, but it is well known that for finite-dimensional vectors all of the commonly used norms are equivalent, (see [8]).

Let an element $\overset{*}{\chi}$ of a certain metric space S be the required exact solution of a problem, and let ϕ_p be an operator which maps the p-dimensional Euclidean space E_p onto a certain subspace $S_p \subseteq S$. We will assume that the approximate solution of the problem is defined by the condition

(6.69) $$\rho(\overset{*}{\chi}, \phi_p \tilde{A} I^{(\epsilon)}) \leq \delta ,$$

where ρ denotes distance in the space S.

Definition. *Let $\tilde{N}$ denote the set of all possible algorithms $\tilde{A}$ for which* (6.69) *is valid. Any subset $\tilde{M} \subset \tilde{N}$ we shall call a method of solution for the problem* (6.67)-(6.69).

We shall denote by $\tau(\tilde{A})$ and $\Pi(\tilde{A})$ the computer time and computer storage respectively, which are necessary for the solution of the problem (6.67)-(6.69) by means of the algorithm $\tilde{A}$. One of possible criteria for selecting an algorithm from the subset $\tilde{M}$ is given by

(6.70) $$\min_{\tilde{A} \in \tilde{M}} \tau(\tilde{A}) \ , \ \Pi(\tilde{A}) \leq \Pi_{\mathrm{Re}} \ , \ \tau(\tilde{A}) \leq \tau_{\mathrm{Re}} \ ,$$

where τ_{Re} and Π_{Re} are limits with respect to computer time and memory storage respectively.

In practice many criteria may be used for the comparison of algorithms which are different from the one given. Not infrequently, algorithms are compared with respect to their programming difficulties, and to their reliability in the presence of random inconsistencies in the working of the computer.

Programming difficulties are determined by the specific character of the collection of commands and by the capability of the programmer. As a rule, the smaller the number of commands (operations) required to represent the algorithm, the easier it is to write the program. A decrease in the number of commands however, does not necessarily imply a decrease in the computational time of the algorithm. If, for example, we compare a direct algorithm with an iterative one, then the number of commands of the corresponding program may decrease considerably but for a large number of iterations, the computational time may be greatly increased.

In order to account for this, instead of minimizing $\tau(\tilde{A})$, let us minimize

(6.71) $$T(\tilde{A}) = t(\tilde{A}) + q\tau(\tilde{A})$$

under the restriction

(6.72) $$T(\tilde{A}) \leq \tau_{\mathrm{Re}} \ , \ \Pi(\tilde{A}) \leq \Pi_{\mathrm{Re}} \ ,$$

where $t(\tilde{A})$ denotes the programming time for the algorithm $\tilde{A}$, and q is a weight which may be defined in terms of the number of times the algorithm will be used.

Let us assume that the average time the computer works without stopping is given by $\bar{\tau}$. It is desirable that $\tau(\tilde{A}) < \bar{\tau}$. If this condition is not satisfied then $\tilde{A}$ should be modified, even at the expense of an increase in $\tau(\tilde{A})$, so that in a period of time less than but close to $\bar{\tau}$, intermediate results are obtained and stored. If, after stopping, the algorithm $\tilde{A}$ can be restarted using the last intermediate results, it is possible to reduce the computational time.

In some situations it is not possible to allow for the stopping of the computer. Then different methods (doubling, testing on simple partial examples, testing of a check sum, the solution by different methods, etc.), which lead to an increase of $\tau(\tilde{A})$ but on the other hand increase the reliability of the results, should be used.

In this way for the realization of each algorithm $\tilde{A}$ on a computer, the probability $p(\tilde{A})$ of the determination of a correct result, should be established. This leads to an alternative formulation of the problem for determining an optimum algorithm $\tilde{A}^*$. We now choose $\tilde{A}^*$ to be the algorithm which solves the given problem and minimizes (6.71) under the conditions

$$(6.73) \qquad T(\tilde{A}) \leq \tau_{\mathrm{Re}} \; , \; \Pi(\tilde{A}) \leq \Pi_{\mathrm{Re}} \; , \; 1 \geq p(\tilde{A}) \geq p_{\mathrm{Re}} > 0 \; .$$

It should be noted that the minimization of $T(\tilde{A})$ is not always the most appropriate condition. In certain cases it may be more appropriate to minimize $\Pi(\tilde{A})$ by placing restrictions on $T(\tilde{A})$ and $p(\tilde{A})$. Other criteria are possible (see [88, 146]).

Let us examine a technico-economic criterion for the choice of an optimum algorithm. Let the solution of the problem (6.67)-(6.69) yield an income $D(\epsilon,\delta)$, and let this problem be solved by the algorithm $\tilde{A}$ which incurs an expenditure $E(\tilde{A})$. For an optimum solution, we choose an algorithm $\tilde{A}^*$ and values ϵ^* and δ^* which maximize the profit

$$(6.74) \qquad \Pi(\epsilon,\delta,\tilde{A}) = D(\epsilon,\delta) - E(\tilde{A}) \quad .$$

Until now we have considered algorithms for the solution of a particular problem. In practice, we are more often interested in algorithms for the solution of a class of problems which are characterized by a vector of parameters $\omega \in \Omega$. In this case, the algorithms $\tilde{A}$ are functions of ω,

$$(6.75) \qquad \tilde{A} = \tilde{A}(\omega) \; , \quad \omega \in \Omega \; ,$$

and are called *standard algorithms*. Criteria for the comparison of different algorithms to solve the given class of problems, will be similar to those discussed above.

If we extend the class of problems, the number of suitable algorithms and methods necessarily decreases, and difficulties in the development of effective methods quickly increase. Therefore in practice we consider the direct problem: for a given method $\tilde{M}$, we define the maximum domain $\Omega(\tilde{M})$ of values of the parameters ω for which the method has the required effectiveness (for example, each of the components of its algorithms satisfies conditions (6.73) for every $\omega \in \Omega(\tilde{M})$).

In the general case when it is required to define the optimum algorithm on a given set of algorithms (not necessarily consisting of algorithms for the solution only of the particular problem), it is necessary to have criteria for the comparison of any two algorithms of the given set. Such criteria can be obtained on the basis of decision theory (see [146]), and the theory of discrete self-organizing systems (see [39]).

Chapter 2

BASIC RESULTS FROM THE THEORY OF ONE-DIMENSIONAL LINEAR SINGULAR INTEGRAL EQUATIONS †

§7 SINGULAR INTEGRALS AS OPERATORS

We first introduce the concept of a singular integral.(1)

Definition. *Let a function f be defined almost everywhere on a curve Γ, and let t_0 be a point of Γ. We delete from Γ those points which are within a distance ϵ from t_0, and assume that f is integrable on the remaining part Γ_ϵ of the curve Γ, for all $\epsilon > 0$. If $\lim_{\epsilon\to 0}\int_{\Gamma_\epsilon} f(t)dt$ exists, then this limit is called the singular integral of the function f along Γ.*

Let $\Gamma \equiv \gamma$, the unit circle with $t = e^{i\theta}$, and choose $f(t) = \phi(t)/[\pi i(t-t_0)]$ where $t_0 = e^{i\theta_0} \in \gamma$. We examine the singular integral

$$S\phi = \frac{1}{\pi i}\int_\gamma \frac{\phi(t)dt}{t-t_0} = \lim_{\epsilon\to 0}\frac{1}{\pi i}\int_{\gamma_\epsilon}\frac{\phi(t)dt}{t-t_0}, \tag{7.1}$$

where γ_ϵ denotes γ with the arc within the circle centre t_0 and radius ϵ deleted. The function ϕ is called the *density* of the singular integral. It is readily verified that

$$S\phi = i\, S_1\phi_1 + \frac{1}{2\pi}\int_{-\pi}^{\pi}\phi(e^{i\theta})d\theta , \tag{7.2}$$

where

$$\begin{aligned} S_1\phi_1 &= -\frac{1}{\pi}\int_{-\pi}^{\pi}\frac{\phi_1(\theta+\theta_0) - \phi_1(\theta_0)}{2\tan(\theta/2)}d\theta , \\ &= -\lim_{\substack{h\to 0\\ h>0}}\frac{1}{\pi}\left(\int_{-\pi}^{-h} + \int_h^{\pi}\right)\frac{\phi_1(\theta+\theta_0) - \phi_1(\theta_0)}{2\tan(\theta/2)}d\theta, \quad \phi_1(\theta) = \phi(e^{i\theta}). \end{aligned} \tag{7.3}$$

N.N. Luzin [94] was the first to show that if $\phi_1 \in L_2([-\pi, \pi])$ then $S_1\phi_1 \in L_2([-\pi, \pi])$, and also(2) $\|S_1\phi_1\|_{L_2} < C_1\|\phi_1\|_{L_2}$. From this it follows

(1) In this section we will examine one-dimensional singular integrals. For the general definition of a singular integral and some properties of multi-dimensional singular integrals, see §23.

(2) Here and subsequently, $C_1, C_2, \ldots$ are positive constants which do not depend on values appearing in the corresponding estimates.

† For an up to date summary of the theory see Chapter VI of P.P. Zabreyko et al., *Integral Equations - a reference text* (a translation from the Russian) Noordhoff International Publishing, Leyden (1975). (Eds.)

that if $\phi \in L_2(\gamma)$, and its Laurent expansion is given by

$$\phi(t) = \sum_{k=-\infty}^{\infty} \phi_k t^k , \quad \text{where } \phi_k = \frac{1}{2\pi i} \int_\gamma \frac{\phi(t)}{t^{k+1}}\, dt , \tag{7.4}$$

then

$$S\phi = \sum_{k=0}^{\infty} \phi_k t_0{}^k - \sum_{k=-\infty}^{-1} \phi_k t_0{}^k , \tag{7.5}$$

and

$$\|S\phi\|^2_{L_2(\gamma)} = \sum_{k=-\infty}^{\infty} |\phi_k|^2 = \|\phi\|^2_{L_2(\gamma)} , \quad \text{i.e. } \|S\|_{L_2(\gamma)} = 1 . \tag{7.6}$$

Following N.N. Luzin, many investigations of the properties of the integrals $S_1\phi_1$ and $S\phi$ have been made.[3] More recently, these results have been extended to a larger class of contours of integration. Some of these results have been summarized in [142] where questions concerning properties of singular integrals when their densities are integrable with arbitrary weight, and integrable to other orders (greater than 1), have been examined in detail.

Theorem 7.1. (B.V. Khvedelidze [142]). *Let Γ be a closed or open Lyapunov curve*[4], *and let $\phi \in L_p(\Gamma, \rho)$, i.e. $|\phi(t)|^p \rho(t)$ is integrable on Γ where $p > 1$ and the weight function ρ is defined by*

$$\rho(t) = \prod_{k=1}^{m_1} |t-d_k|^{\delta_k(p-1)} \prod_{k=m_1+1}^{m} |t-d_k|^{-\delta_k} \tag{7.7}$$

$$(0 \le \delta_k < 1, \quad d_k \in \Gamma, \quad k = 1,2,\dots,m) .$$

Then

$$S\phi = \frac{1}{\pi i} \int_\Gamma \frac{\phi(t)dt}{t-t_0} \in L_p(\Gamma, \rho) , \tag{7.8}$$

and S is a bounded operator on $L_p(\Gamma, \rho)$.

An estimate of $\|S\|$ is of practical interest, but in general such estimates are difficult to obtain. We shall restrict attention to an examination of one case.

---

(3) For a list of results up to 1965, see [51].

(4) A simple smooth curve is called a *Lyapunov curve* if the angle of inclination θ of the tangent to the x-axis, considered as a function of arc length s, satisfies a Hölder condition,

$$|\theta(s_1) - \theta(s_2)| \le C_2|s_1 - s_2|^\alpha , \quad 0 < \alpha \le 1 .$$

Theorem 7.2. *If*

$$\rho(x) = (b-x)^{\frac{1}{2}} (x-a)^{\frac{1}{2}} (b-x)^{\epsilon_1} (x-a)^{\epsilon_2} ; \tag{7.9}$$

$$\epsilon_1 = \epsilon_2 = 0, -1; \quad \epsilon_1 = 0, \epsilon_2 = -1; \quad \epsilon_2 = 0, \epsilon_1 = -1 ;$$

then

$$\|S\|_{L_2([a,b],\rho)} = 1 . \tag{7.10}$$

This theorem is proved in [1]. More precisely, it is proved there that the operator σ, where

$$\sigma\psi = \frac{1}{\pi i} \int_a^b \frac{\rho^{-1}(x)\ \psi(x)}{x - y}\, dx \quad , \tag{7.11}$$

maps the space $L_2(\rho^{-1}) = L_2([a,b], \rho^{-1})$ into the space $L_2(\rho) = L_2([a,b], \rho)$, and $\|\sigma\| = 1$. But if $\psi \in L_2(\rho^{-1})$ then obviously $\rho^{-1}\psi \in L_2(\rho)$ and conversely, and $\|\psi\|_{L_2(\rho^{-1})} = \|\rho^{-1}\psi\|_{L_2(\rho)}$, from which the required result follows.

In the proof given in [1], the following important formulae are obtained

$$\begin{aligned}
&\frac{1}{\pi} \int_a^b \frac{\rho^{-1}(x)\ T_k^*(x)dx}{x - y} = T_k(y) \qquad (\epsilon_1 = 0,\ \epsilon_2 = -1),\\
&\frac{1}{\pi} \int_a^b \frac{\rho^{-1}(x)\ T_k^*(x)dx}{x - y} = -\ T_k(y) \qquad (\epsilon_1 = -1,\ \epsilon_2 = 0),\\
&\frac{1}{\pi} \int_a^b \frac{\rho^{-1}(x)\ T_k^*(x)dx}{x - y} = T_{k-1}(y) \qquad (\epsilon_1 = \epsilon_2 = 0),\\
&\frac{1}{\pi} \int_a^b \frac{\rho^{-1}(x)\ T_k^*(x)dx}{x - y} = -\ T_{k+1}(y) \qquad (\epsilon_1 = \epsilon_2 = -1),
\end{aligned} \tag{7.12}$$

where T_k^* and T_k are polynomials of degree k which are orthonormalized with respect to the weights ρ^{-1} and ρ respectively, on the interval $[a,b]$.

In the same article it is also proved that $\|S\|_{L_2(\Gamma,\rho)} = 1$ in the case when Γ consists of a finite number of intervals $(a_1, b_1), (a_2, b_2),\dots,(a_N, b_N)$, $(a_k < b_k < a_{k+1} < \infty)$ and

$$\rho(x) = \left(-\prod_{k=1}^{N} \frac{x-b_k}{x-a_k}\right)^{\frac{1}{2}} (x-a_1)^{\epsilon_1} \left(\frac{x-a_2}{x-b_1}\right)^{\epsilon_2} \cdots \left(\frac{x-a_N}{x-b_{N-1}}\right)^{\epsilon_N} \left(\frac{1}{b_N-x}\right)^{\epsilon_{N+1}} \tag{7.13}$$

for $x \in \Gamma$, where each ϵ_k is equal to either 0 or 1.

As is obvious from Theorem 7.1, except in the case when $p = 1$, the integrability of the functions ϕ and $S\phi$ is the same. An analogous situation occurs when ϕ and $S\phi$ are Hölder continuous.

The properties in the latter case were derived by I.I. Privalov [118], and have received their most detailed examination in the book by

N.I. Muskhelishvili [108].

Let Γ be a closed piecewise-smooth contour. We denote by $H(\alpha, \Gamma)$, the normed linear space of functions ϕ which are Hölder continuous on Γ with index α, and

$$\|\phi\|_H^\alpha = \max_{t\in\Gamma} |\phi(t)| + \sup_{\substack{t_1,t_2\in\Gamma \\ t_1\neq t_2}} \frac{|\phi(t_1) - \phi(t_2)|}{|t_1 - t_2|^\alpha} . \tag{7.14}$$

Theorem 7.3. (N.I. Muskhelishvili [108]). *The space $H(\alpha, \Gamma)$ is a Banach space. In addition if $\phi \in H(\alpha, \Gamma)$ $(0 < \alpha < 1)$, then $S\phi \in H(\alpha, \Gamma)$ and S is a bounded operator on $H(\alpha, \Gamma)$.*

One of the shortcomings of the space $H(\alpha, \Gamma)$ is that it is very difficult to estimate $\|S\|$ on it. In [54, 57] a space has been introduced which overcomes this difficulty.

It is known that under fairly general assumptions concerning Γ and ϕ (see [108,118]) the Sokhotski-Plemelj formulae are valid:

$$\phi = \phi^+ - \phi^- , \qquad S\phi = \phi^+ + \phi^- , \tag{7.15}$$

where ϕ^+ and ϕ^- are the boundary values of the Cauchy integral $\frac{1}{2\pi i}\int_\Gamma \frac{\phi(t)dt}{t-z}$ on the left-hand and right-hand sides respectively of Γ, described in the positive direction. We denote by $W(\Gamma)$ the normed linear space of functions ϕ for which ϕ^+ and ϕ^- are continuous, and define

$$\|\phi\|_{W(\Gamma)} = \max_{t\in\Gamma} |\phi^+(t)| + \max_{t\in\Gamma} |\phi^-(t)| . \tag{7.16}$$

Theorem 7.4. *If Γ is a closed simple smooth contour then $W(\Gamma)$ is a Banach space and*

$$\|S\|_{W(\Gamma)} = 1 . \tag{7.17}$$

Proof. Equation (7.17) follows from the fact that in the case of a closed contour $(S\phi)^+ = \phi^+$, $(S\phi)^- = -\phi^-$ and so $\|S\phi\|_{W(\Gamma)} = \|\phi\|_{W(\Gamma)}$.

For the proof of the completeness of $W(\Gamma)$, we assume that $\|\phi_n-\phi_m\|_{W(\Gamma)} \to 0$, $\phi_n, \phi_m \in W(\Gamma)$, and show that as a result ϕ_n converges to $\phi \in W(\Gamma)$. In fact, from $\|\phi_n-\phi_m\|_{W(\Gamma)} \to 0$ and the completeness of $C(\Gamma)$, there exist continuous functions ϕ_+ and ϕ_- such that ϕ_n^+ and ϕ_n^- converge uniformly to ϕ_+ and ϕ_-, respectively. Since (see [118, p.202]) the condition

(7.18) $$F(z) = \frac{1}{2\pi i} \int_\Gamma \frac{f(t)dt}{t - z} = 0 \quad \text{for } z \text{ outside } \Gamma$$

is necessary and sufficient for the boundary value of F, taken along any non-tangential path within Γ, to coincide with f almost everywhere on Γ, then $\int_\Gamma [\phi_n^+(t)/(t-z)]dt = 0$ for z outside Γ and so $\int_\Gamma [\phi_+(t)/(t-z)]dt = 0$ for z outside Γ. i.e. ϕ_+ is the boundary value of its own Cauchy integral almost everywhere on Γ. But for a continuous function ϕ_+ (see [118, p.67]) such that $\int_\Gamma [\phi_+(t)/(t-z)]dt = 0$ for z outside Γ, the limiting value of its Cauchy integral from within Γ, along any non-tangential path, always exists and coincides with ϕ_+. Thus the subscript "+" on the function ϕ_+ becomes a superscript.

By the change of variable $u = 1/(t-z_1)$ where z_1 is any point inside Γ, we reduce the examination of the properties of ϕ_- to the previous case. Putting $\phi = \phi^+ - \phi^-$, we conclude the proof of the theorem. #

In conclusion we note that S is not a bounded operator in $C(\Gamma)$, since N.N. Luzin [94] has constructed a function ϕ_1 which is continuous on $[-\pi,\pi]$, periodic with period 2π, and such that $S_1\phi_1$ has an infinity of discontinuities on a set which is dense in $[-\pi,\pi]$.(5)

§8 SINGULAR INTEGRAL EQUATIONS WHICH ARE SOLVABLE IN CLOSED FORM

§8.1 The Dominant Equation in the Case of a Closed Contour

Let us examine the dominant equation

(8.1) $$K^0\phi \equiv a(t)\,\phi(t) + \frac{b(t)}{\pi i} \int_\Gamma \frac{\phi(\tau)d\tau}{\tau - t} = f(t) \ , \quad t \in \Gamma \ ,$$

in which Γ is a smooth, closed, simple curve; a, b, $f \in H(\alpha, \Gamma)$ i.e. they are Hölder continuous on Γ with index α; and $a^2-b^2 \neq 0$ on Γ. Without loss of generality, we shall assume that the coordinate origin 0 lies inside Γ.

The general solution of this equation in closed form was first given by I.N. Vekua [16] (see also [33]).

Applying the Sokhotski-Plemelj formulae (7.15) we transform (8.1) to the form

(8.2) $$K^0\phi \equiv (a + b)\phi^+ - (a - b)\phi^- = f \ ,$$

(5) It is shown in [40] that $\exp(S_1\phi_1) \in L_p$ for any $0 < p < \infty$ if ϕ_1 is continuous.

from which

$$\phi^+ = G\phi^- + g, \quad G = \frac{a-b}{a+b}, \quad g = \frac{f}{a+b}. \tag{8.3}$$

The problem of the determination of the function $\Phi(z) = \frac{1}{2\pi i}\int_\Gamma \frac{\phi(t)dt}{t - z}$ and its boundary values $\phi^\pm$ from (8.3) is called the *Riemann boundary value problem*. The general solution of the Riemann boundary value problem in a closed form was first given by F.D. Gakhov [28]. (Usually a Riemann problem is taken to be a more general problem in which Φ can have arbitrary singularities outside of Γ. We shall restrict attention to the special case when Φ can be represented by a Cauchy integral, since (8.3) is only considered in connection with equation (8.1)).

Definition. *The quantity*

$$\varkappa = \frac{1}{2\pi i}\int_\Gamma d[\ln G(t)] = \frac{1}{2\pi}[\arg G(t)]_\Gamma \tag{8.4}$$

is called the index of equation (8.1) *or of the operator* K^0, *and also the index of the Riemann boundary value problem as defined by* (8.3).

From the definition of index, it follows that $\psi(t) = \ln[G(t)t^{-\varkappa}]$ is single-valued on Γ. By applying to it the Sokhotski-Plemelj formulae, we obtain

$$G(t) = t^{\varkappa}\frac{\exp\psi^+(t)}{\exp\psi^-(t)},$$

$$\psi^\pm(t) = \pm\frac{1}{2}\ln[G(t)t^{-\varkappa}] + \frac{1}{2\pi i}\int_\Gamma \frac{\ln[G(\tau)\tau^{-\varkappa}]}{\tau - t}\,d\tau. \tag{8.5}$$

We substitute into (8.3) this expression for G. On dividing both sides of (8.3) by $\exp\psi^+$, and writing $g_1 = g/\exp\psi^+$, we obtain

$$\frac{\phi^+(t)}{\exp\psi^+(t)} - g_1^+(t) = t^{\varkappa}\frac{\phi^-(t)}{\exp\psi^-(t)} - g_1^-(t). \tag{8.6}$$

From the theorem on analytic continuation [33, §13], equation (8.6) implies that the function

$$F(z) = \begin{cases} \dfrac{\Phi(z)}{\exp\Psi(z)} - G_1(z), & z \text{ inside } \Gamma, \\[2ex] z^{\varkappa}\dfrac{\Phi(z)}{\exp\Psi(z)} - G_1(z), & z \text{ outside } \Gamma, \end{cases} \tag{8.7}$$

is analytic in the entire plane, where

$$\Phi(z) = \frac{1}{2\pi i} \int_\Gamma \frac{\phi(\tau)d\tau}{\tau - z} \;, \quad \Psi(z) = \frac{1}{2\pi i} \int_\Gamma \frac{\psi(\tau)}{\tau - z} d\tau \text{ and } G_1(z) = \frac{1}{2\pi i} \int_\Gamma \frac{g_1(\tau)d\tau}{\tau - z} \;.$$

If $\varkappa \leq 0$, then F is zero at infinity; if $\varkappa > 0$, it has a pole of order $\varkappa$-1 at infinity. From Liouville's theorem it follows that $F(z) \equiv P_{\varkappa-1}(z)$ (a polynomial of degree $\varkappa$-1), where $P_{\varkappa-1}(z) \equiv 0$ for $\varkappa \leq 0$.

Thus both sides of (8.6) are equal to $P_{\varkappa-1}(t)$ and we have

$$\phi^+(t) = g_1^+(t) \exp \psi^+(t) + P_{\varkappa-1}(t) \exp \psi^+(t) \;,$$

$$(8.8) \qquad \phi^-(t) = g_1^-(t)\, t^{-\varkappa} \exp \psi^-(t) + P_{\varkappa-1}(t) t^{-\varkappa} \exp \psi^-(t) \;,$$

$$\text{and} \quad \phi(t) = \phi^+(t) - \phi^-(t) \;.$$

In addition $\phi(t) = \phi^+(t) - \phi^-(t)$ is obviously a solution of the original equation, (8.1).

Since Φ must be zero at infinity, and $\exp \Psi(z)$ is equal to 1 at infinity when $\varkappa < 0$, (8.8) will be the required solution provided that the condition that $G_1(z)$ has a zero of order greater than $-\varkappa$+1 at infinity is satisfied.

For $\varkappa \geq 0$ it is obvious from (8.8) that both the Riemann boundary value problem and the corresponding dominant equation (8.1) have $\varkappa$ linearly independent solutions.

It is easily shown that equations (8.8) determine all the required solutions under fairly general assumptions concerning the class of functions in which these solutions are to be determined (see [33]). It has been proved (see [47, 142])that (8.8) determine the required solutions for more general assumptions on a, b, f and Γ. In particular, in [55] it has been shown that the same method of solution is applicable when G is an arbitrary continuous function. It should also be noted that (8.8) determine the solution of (8.1) in the case when Γ consists of a finite number of non-overlapping closed curves.

From (8.8) it follows that in order to calculate a solution, it is necessary to calculate the index $\varkappa$ and singular integrals. A discussion of the calculation of the index has been given by Gakhov [33]. For the calculation of singular integrals we can apply known methods (see [53, 61] and also §11).

§8.2 The Dominant Equation in Exceptional Cases

In connection with problems from the theory of automatic control which are examined in Chapter 4, the solution of (8.1)(and the corresponding Riemann problem (8.3)) in the exceptional case when $a^2 - b^2$ has zeros of integer order at isolated points of Γ, is important. We derive here, with only minor changes, the analysis given by Gakhov [33, §15 and §25].

Let the functions a and b be such that

$$G(t) = \frac{\prod_{k=1}^{\mu} (t - \alpha_k)^{m_k}}{\prod_{j=1}^{\nu} (t - \beta_j)^{p_j}} \quad G_1(t) , \tag{8.9}$$

where $\alpha_k \neq \beta_j$, α_k and $\beta_j \in \Gamma$; m_k, p_j are positive integers; and G_1 is a non-vanishing function satisfying a Hölder condition on Γ. We call the points β_j the *poles* of G, and introduce the notation

$$\varkappa = \frac{1}{2\pi} [\arg G_1(t)]_\Gamma , \quad \sum_{j=1}^{\nu} p_j = p, \quad \sum_{k=1}^{\mu} m_k = m. \tag{8.10}$$

We shall seek solutions in the class of functions bounded on Γ. It is easily seen that the boundary value problem (8.3) with G given by (8.9), cannot be satisfied by finite ϕ^+ and ϕ^- if we allow g to have poles at points different from β_j, or if at these points the order of the pole of g exceeds p_j. Hence we assume that g may have poles at the points β_j, with orders not exceeding p_j. In addition, to apply the following theory it is essential to assume that the functions $G_1(t)$ and $\prod_{j=1}^{\nu} (t - \beta_j)^{p_j} g(t)$ are sufficiently differentiable at the singular points. The order of the required highest derivative is determined from the following analysis.

Writing G_1 in the form

$$\begin{aligned} G_1(t) &= t^{\varkappa} \frac{\exp \psi_1^+(t)}{\exp \psi_1^-(t)} , \\ \psi_1^{\pm}(t) &= \pm \frac{1}{2} \ln [t^{-\varkappa} G_1(t)] + \frac{1}{2\pi i} \int_\Gamma \frac{\ln[G_1(\tau)\, \tau^{-\varkappa}]}{\tau - t} d\tau , \end{aligned} \tag{8.11}$$

and substituting in (8.3) the expression for G as given by (8.9), we obtain

$$\prod_{j=1}^{\nu} (t-\beta_j)^{p_j} \frac{\phi^+(t)}{\exp \psi_1^+(t)} - g_2^+(t) = \prod_{k=1}^{\mu} (t-\alpha_k)^{m_k} t^{\varkappa} \frac{\phi^-(t)}{\exp \psi_1^-(t)} - g_2^-(t) , \tag{8.12}$$

where

$$g_2^{\pm}(t) = \pm \frac{1}{2} \prod_{j=1}^{\nu} (t-\beta_j)^{p_j} \frac{g(t)}{\exp \psi_1^+(t)} + \frac{1}{2\pi i} \int_\Gamma \frac{\prod_{j=1}^{\nu} (\tau-\beta_j)^{p_j} g(\tau)}{\exp \psi_1^+(\tau)\, (\tau-t)} d\tau . \tag{8.13}$$

From (8.12), properties of analytic continuation, and the

generalized Liouville theorem, it follows that

$$\phi^+(t) = \frac{\exp \psi_1^+(t)}{\prod_{j=1}^{\nu} (t-\beta_j)^{p_j}} [g_2^+(t) + P_{\varkappa+m-1}(t)] ,$$

(8.14) $$\phi^-(t) = \frac{\exp \psi_1^-(t)}{t^{\varkappa} \prod_{k=1}^{\mu} (t-\alpha_k)^{m_k}} [g_2^-(t) + P_{\varkappa+m-1}(t)] ,$$

and $\phi(t) = \phi^+(t) - \phi^-(t)$.

Here $P_{\varkappa+m-1}$ is a polynomial of degree $\varkappa+m-1$, and $P_{\varkappa+m-1}(t) \equiv 0$ if $\varkappa+m-1 < 0$. Equations (8.14) give solutions which in general will become infinite at the points α_k, β_j; in addition ϕ^- will not be zero at infinity. In order that the solutions have the required properties it is necessary that the function $g_2^+ + P_{\varkappa+m-1}$ has zeros of order p_j at the points β_j, and that the function $g_2^- + P_{\varkappa+m-1}$ has zeros of order m_k at the points α_k. If $\varkappa+m-1 < 0$, then in addition ϕ^- must have a zero of order greater than $-\varkappa-m+1$ at infinity.

Suppose $\varkappa \geq p$.(6) Obviously in this case, the problem is reduced to the construction of Hermite polynomials $Q_{\varkappa+m-1}$ satisfying the conditions

(8.15)
$$Q_{\varkappa+m-1}^{(i)}(\beta_j) = - g_2^{+(i)}(\beta_j) \quad (i = 0,1,\dots,p_j-1,\ j = 1,2,\dots,\nu),$$
$$Q_{\varkappa+m-1}^{(\ell)}(\alpha_k) = - g_2^{-(\ell)}(\alpha_k) \quad (\ell = 0,1,\dots,m_k-1,\ k = 1,2,\dots,\mu) .$$

It is known (see [109]) that there exists a unique polynomial Q_{p+m-1} of degree $p+m-1$, which satisfies equations (8.15). Hence it is not difficult to see that

(8.16) $$Q_{\varkappa+m-1}(t) = Q_{p+m-1}(t) + \prod_{j=1}^{\nu} (t-\beta_j)^{p_j} \prod_{k=1}^{\mu} (t-\alpha_k)^{m_k} P_{\varkappa-p-1}(t) .$$

Substituting the polynomial $Q_{\varkappa+m-1}$ in (8.14) instead of $P_{\varkappa+m-1}$ we obtain, using (8.16),

(6) An examination of the remaining cases is given in [33, §15].

$$\phi^+(t) = \frac{\exp \psi_1^+(t)}{\prod_{j=1}^{\nu} (t-\beta_j)^{\nu_j}} [g_2^+(t) + Q_{p+m-1}(t)] + \exp \psi_1^+(t) \prod_{k=1}^{\mu} (t-\alpha_k)^{m_k} P_{\varkappa-p-1}(t),$$

$$(8.17)\quad \phi^-(t) = \frac{\exp \psi_1^-(t)}{t^{\varkappa} \prod_{k=1}^{\mu} (t-\alpha_k)^{m_k}} [g_2^-(t) + Q_{p+m-1}(t)] + \frac{\exp \psi_1^-(t)}{t^{\varkappa}} \prod_{j=1}^{\nu} (t-\beta_j)^{p_j} P_{\varkappa-p-1}$$

and $\phi(t) = \phi^+(t) - \phi^-(t)$,

where $P_{\varkappa-p-1}$ is an arbitrary polynomial of degree $\varkappa$-p-1, $P_{-1}(t) \equiv 0$. Thus for $\varkappa \geq p > 0$, the dominant equation (8.1) always has $\varkappa$-p linearly independent solutions.

It is obvious that from the computational point of view, the tabulation of the functions ϕ^+, ϕ^- and ϕ using equations (8.17), reduces to the problem of the calculation of index, singular integrals, and the construction of Hermite interpolation polynomials (see [41]).

§8.3 The Dominant Equation in the Case of an Open Contour

Let us consider the equation

$$(8.18)\qquad K^0\phi \equiv a(t)\,\phi(t) + \frac{b(t)}{\pi i} \int_{\Gamma} \frac{\phi(\tau)d\tau}{\tau - t} = f(t)\ ,\quad t \in \Gamma\ ,$$

under the assumption that Γ is the union of a finite set of non-overlapping simple smooth arcs $\Gamma_1, \Gamma_2, \ldots, \Gamma_s$ with end-points d_1, d_2; d_3, d_4; .. ..; d_{2s-1}, d_{2s} respectively; $a, b, f \in H(\alpha,\Gamma)$, $0 < \alpha < 1$; and $a^2-b^2 \neq 0$ on Γ.

We shall briefly explain the method of solution of (8.18) based on the method which is described in more detail in the books by N.I. Muskhelishvili [108] and F.D. Gakhov [33]. We shall look for solutions of (8.18) in the class of functions which are Hölder continuous everywhere on Γ except at the points $d_1, d_2, \ldots, d_{2s}$ where integrable singularities are allowed.(7) Applying the Sokhotski-Plemelj formulae we obtain

$$(8.19)\qquad K^0\phi \equiv (a(t) + b(t))\phi^+(t) - (a(t) - b(t))\phi^-(t) = f(t)\ ,$$

whence

$$(8.20)\qquad \phi^+ = G\phi^- + g,\quad G = \frac{a-b}{a+b}\ ,\quad g = \frac{f}{a+b}\ .$$

(7) If we seek ϕ in the class of integrable functions for which $\phi^\pm$ are integrable, we arrive at the same result.

Solving the Riemann boundary value problem (8.20), we obtain all the solutions of (8.18) by using the equation $\phi = \phi^+ - \phi^-$.

We introduce the piecewise analytic function

$$\Psi(z) = \frac{1}{2\pi i}\int_\Gamma \frac{\psi(\tau)d\tau}{\tau - z}, \quad \psi(t) = \ln G(t) \,. \tag{8.21}$$

From the Sokhotski-Plemelj formulae, $\ln G(t) = \psi^+(t) - \psi^-(t)$ and

$$G(t) = \frac{\exp \psi^+(t)}{\exp \psi^-(t)} \,. \tag{8.22}$$

Unlike in the case of a closed contour, we cannot substitute the representation (8.22) for G into (8.20), because on dividing each side of this equation by $\exp \psi^+$, we will obtain the function $g_1 = g/\exp \psi^+$ (compare with equation (8.6)), which may not be integrable on Γ.

Let us then rewrite $\Psi(z)$ in the form

$$\Psi(z) = \frac{1}{2\pi i}\int_\Gamma \frac{\psi(\tau) - \psi(d_k)}{\tau - z}\,d\tau + \frac{1}{2\pi i}\psi(d_k)\sum_{j=1}^{s} \ln\left(\frac{d_{2j} - z}{d_{2j-1} - z}\right) . \tag{8.23}$$

Then obviously

$$\exp \Psi(z) = \prod_{j=1}^{s}\left(\frac{d_{2j} - z}{d_{2j-1} - z}\right)^{\psi(d_k)/2\pi i} \Omega(z) \quad , \tag{8.24}$$

where Ω is Hölder continuous in a neighbourhood of d_k, for $k = 1,2,\ldots,2s$.

Again from the Sokhotski-Plemelj formulae,

$$\begin{aligned}\psi^\pm(t) = \pm\frac{1}{2}\psi(t) &+ \frac{1}{2\pi i}\int_\Gamma \frac{\psi(\tau) - \psi(t)}{\tau - t}\,d\tau + \\ &+ \frac{1}{2\pi i}\sum_{k=1}^{2s}(-1)^k\,[\psi(t) - \psi(d_k)]\ln(d_k - t) + \\ &+ \frac{1}{2\pi i}\sum_{k=1}^{2s}(-1)^k\,\psi(d_k)\ln(d_k - t) \,,\end{aligned} \tag{8.25}$$

so that

$$\exp \psi^\pm(t) = \prod_{k=1}^{2s}(d_k - t)^{(-1)^k\psi(d_k)/2\pi i}\,\omega_\pm(t) \quad , \tag{8.26}$$

where the functions $\omega_\pm$ are Hölder continuous on Γ.

Now let

$$\Pi(z) = \prod_{k=1}^{2s}(d_k - z)^{\lambda_k} \,, \tag{8.27}$$

where λ_k are integers defined by the conditions

$$(8.28)\qquad -1 < \alpha_k + \lambda_k < 1, \quad \alpha_k + i\beta_k = (-1)^k \psi(d_k)/2\pi i$$

for $k = 1,2,\ldots,2s$. The function X where

$$(8.29)\qquad X(z) = \Pi(z) \exp \Psi(z) ,$$

is called a *canonical solution* of the homogeneous Riemann problem corresponding to (8.20). It is readily seen that X does not become zero outside of Γ, and as a consequence of the inequalities (8.28) its boundary values $x^{\pm}(t)$ are integrable on Γ. It is also obvious that, except at the points d_k, the functions $x^{\pm}$ are Hölder continuous, do not become zero, and satisfy the equation

$$(8.30)\qquad G(t) = \frac{x^+(t)}{x^-(t)} .$$

Substituting this representation for G into (8.20) we obtain

$$(8.31)\qquad \frac{\phi^+(t)}{x^+(t)} - \tilde{g}^+(t) = \frac{\phi^-(t)}{x^-(t)} - \tilde{g}^-(t) , \quad \tilde{g}(t) = \frac{g(t)}{x^+(t)} .$$

Equation (8.31) implies that the function

$$(8.32)\qquad F(z) = \frac{\Phi(z)}{X(z)} - \tilde{G}(z) , \quad z \notin \Gamma ,$$

where

$$\Phi(z) = \frac{1}{2\pi i} \int_\Gamma \frac{\phi(\tau)}{\tau - z} d\tau \quad \text{and} \quad \tilde{G}(z) = \frac{1}{2\pi i} \int_\Gamma \frac{\tilde{g}(\tau)}{\tau - z} d\tau ,$$

is analytic at points of the plane not on Γ, and that in a neighbourhood of infinity it has the form

$$(8.33)\qquad F(z) = \frac{A_1}{z^{1-\varkappa}} + \frac{A_2}{z^{2-\varkappa}} + \ldots + \frac{B_1}{z} + \frac{B_2}{z^2} + \ldots ,$$

where A_ν, B_ν are constants and

$$(8.34)\qquad \varkappa = -\sum_{k=1}^{2s} \lambda_k .$$

Since in a neighbourhood of d_k we have

$$(8.35)\qquad |F(z)| < \frac{\text{constant}}{|z - d_k|^\alpha} , \quad \alpha < 1 ,$$

it follows that F has removable singularities at d_k, so that it is analytic in the plane. Again, using Liouville's theorem, we have

(8.36) $$F(z) = \frac{\Phi(z)}{X(z)} - \tilde{G}(z) \equiv P_{\mathcal{K}-1}(z) \quad ,$$

where $P_{\mathcal{K}-1}$ $\mathcal{K}$ is a polynomial of degree $\mathcal{K}$-1 for $\mathcal{K} > 0$, and $P_{\mathcal{K}-1}(z) \equiv 0$ for $\mathcal{K} \leq 0$. We readily obtain

$$\phi^+(t) = \tilde{g}^+(t)x^+(t) + P_{\mathcal{K}-1}(t)x^+(t) \ ,$$

(8.37) $$\phi^-(t) = \tilde{g}^-(t)x^-(t) + P_{\mathcal{K}-1}(t)x^-(t) \ ,$$

and $$\phi(t) = \phi^+(t) - \phi^-(t) \ ,$$

where

$$x^{\pm}(t) = \prod_{k=1}^{2s} (d_k - t)^{\frac{(-1)^k \psi(d_k)}{2\pi i} + \lambda_k} \times$$

(8.38) $$\times \exp \{\pm \frac{1}{2} \psi(t) + \frac{1}{2\pi i} \int_\Gamma \frac{\psi(\tau) - \psi(t)}{\tau - t} d\tau + \frac{1}{2\pi i} \sum_{k=1}^{2s} (-1)^k [\psi(t) - \psi(d_k)] \times$$
$$\times \ln(d_k - t)\} \ ,$$

$$\tilde{g}(t) = g(t)/x^+(t) \ , \ \psi(t) = \ln G(t) \ .$$

For $\mathcal{K} < 0$, equations (8.37) determine a solution of the original equation (8.18) provided that at infinity $\tilde{G}(z)$ has a zero of order greater than $-\mathcal{K}+1$, which imposes an additional $-\mathcal{K}$ conditions on the right-hand side, f.

A canonical function x and a solution ϕ are, in general, not uniquely determined by the conditions (8.28). A quantity λ_k, which corresponds to a point d_k, is determined uniquely only in the case when α_k is an integer (in this case $\lambda_k = -\alpha_k$). The *ends* d_k for which α_k are integers are called *special*; the remaining ends are called *non-special*. For non-special **ends**, the quantities λ_k are determined only to within ±1 i.e. we can select λ_k so that either $\alpha_k + \lambda_k < 0$, or $\alpha_k + \lambda_k > 0$.

Definition. *We shall say that the solution* ϕ *which is determined by* (8.37) *belongs to the class* $h(d_1, d_2, \ldots, d_q)$, *where* $d_1, d_2, \ldots, d_q$ *are non-special ends, if integers* λ_k *are chosen so that* $\alpha_k + \lambda_k > 0$ *for the ends* $d_1, d_2, \ldots, d_q$ *and* $\alpha_k + \lambda_k < 0$ *for all the remaining non-special ends.. The integer* $\mathcal{K}$ *determined by* (8.34) *is called the index of the given class* $h(d_1, d_2, \ldots, d_q)$.

It is not difficult to verify (see (11.24)) that solutions of the class $h(d_1, d_2, \ldots, d_q)$ are Hölder continuous in neighbourhoods of the non-special ends $d_1, d_2, \ldots, d_q$. In addition they are bounded in neighbourhoods of those special ends for which Re $\psi(d_k) \neq 0$, and can become unbounded (like

$\ell n(t - d_k))$ in neighbourhoods of the special ends d_k for which Re $\psi(d_k) = 0$.

We note that the problem of the approximate calculation of (8.37) reduces to the calculation of the index, and of singular integrals. This problem is examined in detail in §11.3.

§8.4 A System of Dominant Equations

Let us examine the system of singular integral equations

$$K^0\phi \equiv a(t)\,\phi(t) + \frac{b(t)}{\pi i}\int_\Gamma \frac{\phi(\tau)d\tau}{\tau - t} = f(t), \quad t \in \Gamma, \tag{8.39}$$

where Γ is a closed smooth contour with the origin of the complex plane inside Γ; a, b are matrices (of functions) of order m which satisfy the conditions $\det(a\pm b) \neq 0$, $t \in \Gamma$; $f(f_1, f_2,\dots,f_m)$ is a known vector; and ϕ is an unknown vector. We assume that a, b, $f \in H(\alpha,\Gamma)$ i.e. their components are Hölder continuous on Γ with index α.

In the case when $m > 1$ equation (8.39) does not, in general, have a solution in closed form (see [21]).

The Sokhotski-Plemelj formulae

$$S\phi \equiv \frac{1}{\pi i}\int_\Gamma \frac{\phi(\tau)d\tau}{\tau - t} = \phi^+(t) + \phi^-(t),\ \phi(t) = \phi^+(t) - \phi^-(t) \tag{8.40}$$

relate the vector ϕ (which is Hölder continuous on Γ) to the vectors ϕ^+ and ϕ^- (which are also Hölder continuous on Γ and are boundary values from the inside and outside of Γ respectively of the piecewise analytic vector function

$$\Phi(z) = \frac{1}{2\pi i}\int_\Gamma \frac{\phi(\tau)d\tau}{\tau - z}\Big) .$$

Applying this we obtain

$$K^0\phi \equiv (a + b)\phi^+ - (a - b)\phi^- = f , \tag{8.41}$$

from which

$$\phi^+ = G\phi^- + g, \quad G = (a+b)^{-1}(a-b), \quad g = (a+b)^{-1}f . \tag{8.42}$$

The problem of constructing the functions $\phi^\pm$ satisfying (8.42) is called, as in the case of a single function, the *Riemann boundary value problem*. For $g = 0$, it is known as the *homogeneous* Riemann problem.

If the matrix G can be represented in the form

$$G(t) = \chi^+(t)\,[\chi^-(t)]^{-1} , \tag{8.43}$$

where χ^+ and χ^- are matrices analytic[8] in the interior and exterior respectively of Γ, det $\chi^{\pm} \neq 0$, and χ^- has finite order at infinity, then the inhomogeneous Riemann problem is easily solved by analogy with the case when $m = 1$.

On substituting (8.43) into (8.42) and using the Sokhotski-Plemelj formulae, we obtain

$$(\chi^+)^{-1}\,\phi^+ - (\chi^-)^{-1}\,\phi^- = [(\chi^+)^{-1}\,g]^+ - [(\chi^+)^{-1}\,g]^- \quad ,$$

or

$$(\chi^+)^{-1}\,\phi^+ - [(\chi^+)^{-1}\,g]^+ = (\chi^-)^{-1}\,\phi^- - [(\chi^+)^{-1}\,g]^- \quad . \tag{8.44}$$

From analytic continuation and Liouville's theorem, it follows from (8.44) that each side of this equation is equal to a polynomial vector $P(p_1, p_2, \ldots, p_m)$, so that

$$\begin{aligned} \phi^+ &= \chi^+P + \chi^+[(\chi^+)^{-1}\,g]^+ \quad , \\ \phi^- &= \chi^-P + \chi^-[(\chi^+)^{-1}\,g]^- \quad . \end{aligned} \tag{8.45}$$

If we know the structure of the matrix χ^- at infinity, those solutions for which ϕ^- becomes zero at infinity are readily determined from (8.45). With these conditions satisfied, all solutions of (8.39) will then be given by

$$\phi = (\chi^+ - \chi^-)P + \chi^+[(\chi^+)^{-1}\,g]^+ - \chi^-[(\chi^+)^{-1}\,g]^- \quad . \tag{8.46}$$

It is obvious that we require a solution of the so-called *factorization problem* for G i.e. the determination of the expression (8.43). Although in the case when $m = 1$ this problem has a comparatively simple solution, an analogous method for $m > 1$ cannot, in general, be used.

Following [145], we shall define a set of sufficient conditions on G for which the factorization problem can be solved in closed form. We introduce the homogeneous Riemann problem

$$\chi^+(t) = G(t)\,\chi^-(t) \quad , \tag{8.47}$$

and consider the case when the matrix

$$X(z) = \exp\left(\frac{1}{2\pi i}\int_\Gamma \frac{\ln G(\tau)d\tau}{\tau - z}\right) \tag{8.48}$$

satisfies the boundary condition (8.47). Letting z tend to $t \in \Gamma$ from the interior and the exterior of Γ, we obtain on using the Sokhotski-Plemelj

---

(8) A matrix is said to be *analytic*, if all its elements are analytic.

formulae that

$$(8.49)\qquad \chi^{\pm} = \exp(\pm h/2 + Sh/2)\ , \qquad h = \ln G.$$

If we substitute these expressions for $\chi^{\pm}$ into (8.47), then we must have

$$(8.50)\qquad \exp(h/2 + Sh/2) = \exp(h)\,\exp(-h/2 + Sh/2)\ .$$

In order for equation (8.50) to be valid, it is necessary and sufficient that the matrices

$$(8.51)\qquad B = h \quad\text{and}\quad C = -h/2 + Sh/2$$

satisfy the equation

$$(8.52)\qquad \exp(B)\,.\,\exp(C) = \exp(B + C)\ .$$

This equation is, in general, not valid for arbitrary matrices B and C of order m, see [158].

We define

$$(8.53)\qquad [B, C] = BC - CB\ ,$$

and we shall say that the matrices B and C *commute* if $[B, C] \equiv 0$.

Lemma 1. *The condition $[B, C] \equiv 0$ is sufficient for the validity of* (8.52).

The proof is elementary, using known properties of matrix functions (see [98]).

Lemma 2. *The matrices B and C defined by* (8.51) *commute if, for any t and τ, $[h(t), h(\tau)] \equiv 0$.*

In fact the commutativity of B and C is equivalent to the commutativity of h and Sh, but

$$(8.54)\qquad \begin{aligned}[h(t), Sh(t)] &= h(t)\left(\frac{1}{\pi i}\int_\Gamma \frac{h(\tau)d\tau}{\tau - t}\right) - \left(\frac{1}{\pi i}\int_\Gamma \frac{h(\tau)d\tau}{\tau - t}\right)h(t) \\ &= \frac{1}{\pi i}\int_\Gamma \frac{[h(t), h(\tau)]d\tau}{\tau - t}\ .\end{aligned}$$

Definition. *A functional matrix h such that $[h(t), h(\tau)] = 0$ for all t,τ is said to be function-commuting.*

A class of function-commuting matrices has been examined by V.V. Morozov [106], from an algebraic point of view.

Lemma 3. *The matrix $h = \ln G$ is function-commuting if and only if the matrix G is function-commuting.*

The proof of this lemma, like that of lemma 1, readily follows from the properties of matrix functions.

Theorem 8.1. (V.V. Morozov [106]). *In order that an arbitrary matrix A be function-commuting it is necessary and sufficient that it have a representation of the form*

(8.55) $$A(t) = \phi_1(t)\, A_1 + \dots + \phi_r(t)\, A_r \quad ,$$

where $\phi_1,\dots,\phi_r$ are linearly independent functions, and $A_1,\dots,A_r$ are linearly independent, pairwise commuting constant matrices.

We note that in the case when $m = 2$, in addition to the results given above, there is given in [145] a series of sufficient conditions for the validity of (8.52) when B and C are non-commuting, or when h does not commute with Sh.

In conclusion we observe that for some different singular integrals, equations which are not considered above are solvable in closed form (see [33,11]).

§9 THE THEORY OF SINGULAR INTEGRAL EQUATIONS

§9.1 Basic Concepts

Let us consider the equation

$$K\phi \equiv a(t)\,\phi(t) + \frac{b(t)}{\pi i}\int_\Gamma \frac{\phi(\tau)d\tau}{\tau - t} + \lambda\int_\Gamma k(\tau,t)\,\phi(\tau)d\tau = f(t)\ , \tag{9.1}$$

$t \in \Gamma$, under the following assumptions: EITHER

1. Γ is a closed Lyapunov contour[9]; $a,b \in C(\Gamma;\, d_1,d_2,\ldots,d_m)$ i.e. continuous everywhere on Γ except at the points $d_k \in \Gamma$ at which finite discontinuities may occur; $f \in L_p(\Gamma,\rho)$, $p > 1$, with

$$\rho(t) = \prod_{k=1}^{m} |t - d_k|^{\delta_k}\ ,\quad -1 < \delta_k < 1\ ; \tag{9.2}$$

the kernel $k(\tau,t)$ is such that $\int_\Gamma k(\tau,t)\,\phi(\tau)d\tau$ is a completely continuous operator in $L_p(\Gamma,\rho)$; λ is a complex parameter; either (a) $a^2-b^2 \neq 0$ on Γ, or (b) a^2-b^2 becomes zero on Γ; OR

2. Γ is a closed piecewise smooth contour; a, b, f and $\int_\Gamma k(\tau,t)\,\phi(\tau)d\tau$ (where ϕ is any function from the class of functions in which a solution of (9.1) is sought) belong to $H(\alpha,\Gamma)$, $0 < \alpha \leq 1$; λ is a complex parameter; either (a) $a^2-b^2 \neq 0$ on Γ, or (b) a^2-b^2 becomes zero on Γ.

We note that under the assumption that a, b and f have a finite number of discontinuities, the problem of an open contour readily reduces to that of a closed contour. To do this, it is sufficient to make the original contour a closed one, and to assume on the additional contour that $a=1$, $b=k=f=0$. We also note that a sufficient condition for $\int_\Gamma k(\tau,t)\phi(\tau)d\tau$ to be completely continuous on $L_2(\Gamma,\rho)$ is that $k(\tau,t) \in L_2(\Gamma,(\rho(\tau))^{-1}\rho(t))$ i.e.

$$\int_\Gamma\int_\Gamma |k(\tau,t)|^2(\rho(\tau))^{-1}\rho(t)\,|dt|\ |d\tau| < \infty\ . \tag{9.3}$$

We shall call the equation of the form (8.1) with the same $a(t)$ and $b(t)$ as given in (9.1) the *dominant equation* corresponding to the *complete equation* (9.1), and the term $\int_\Gamma k(\tau,t)\,\phi(\tau)d\tau$ the *regular part* of (9.1). If for the regular part we introduce the notation

(9) By a closed Lyapunov contour we shall mean a finite number of closed Lyapunov curves $\Gamma_0, \Gamma_1,\ldots,\Gamma_m$ where Γ_0 contains all the other curves within it.

(9.4) $$k\phi = \int_\Gamma k(\tau,t)\ \phi(\tau)d\tau\ ,$$

we can write the complete equation as

(9.5) $$K\phi \equiv K^0\phi + \lambda k\phi = f(t)\ .$$

We shall call the equation

(9.6) $$K'\psi \equiv a(t)\ \psi(t) - \frac{1}{\pi i}\int_\Gamma \frac{b(\tau)\ \psi(\tau)d\tau}{\tau - t} + \lambda\int_\Gamma k(t,\tau)\ \psi(\tau)d\tau = 0\ ,$$

which is obtained from the homogeneous equation $K\phi = 0$ by a transposition of variables in the kernel, the *adjoint equation*, or *transposed equation*.

We shall call the operators K^0 and K' the *dominant* and *adjoint* (*transposed*) *operators* respectively of K.

Let $t = t(s)$, $(0 \leq s \leq L)$ be the parametric equation of the contour Γ where s is the arc length. We shall denote functions of the real variables s and s_0, which are defined by the transformations $\tau = \tau(s)$, $t = t(s_0)$, by the same letters which denoted the functions of the original complex variables τ and t. Then the operator K as defined by the left hand side of (9.1) can be written as

(9.7) $$K\phi \equiv a(s_0)\ \phi(s_0) + \frac{b(s_0)}{\pi i}\int_0^L \frac{\phi(s)\ \tau'(s)ds}{\tau(s) - t(s_0)} + \lambda\int_0^L k(s,s_0)\ \phi(s)\tau'(s)ds\ .$$

The operator K^* defined by

(9.8) $$K^*\chi \equiv \overline{a(s_0)}\ \chi(s_0) + \frac{\overline{t'(s_0)}}{\pi i}\int_0^L \frac{\overline{b(s)}\ \chi(s)ds}{\overline{\tau(s)} - \overline{t(s_0)}} + \bar{\lambda}\int_0^L \overline{k(s_0,s)}\ \overline{t'(s_0)}\chi(s)ds\ ,$$

will be called the operator *conjugate to* K.

In the case when $a^2 - b^2 \neq 0$, we shall call the integer $\varkappa$, where

(9.9) $$\varkappa = \frac{1}{2\pi}\left[\arg\frac{a-b}{a+b}\right]_\Gamma = \frac{1}{2\pi i}\int_\Gamma d\left[\ln\frac{a-b}{a+b}\right]\ ,$$

the *index of the equation* (9.1), or the *index of the operator* K.

It follows at once that the index of the operators K' and K^* is given by $-\varkappa$.

For equation (9.1), under either of the assumptions 1(a) or 2(a), the following statements are valid (Noether's theorems):

1. The number of linearly independent solutions is finite.
2. A necessary and sufficient condition for solvability is

$$\int_\Gamma f(t)\,\psi_j(t)dt = 0\ , \quad (j = 1,2,\dots,n')\ , \tag{9.10}$$

where $\{\psi_j\}$ is a complete system of linearly independent solutions of the adjoint homogeneous equation (9.6).

3. The difference between the number n of linearly independent solutions of the equation $K\phi = 0$, and the number n' of linearly independent solutions of the adjoint equation $K'\psi = 0$, depends only on the dominant part of the operator K and is equal to its index, i.e.

$$n - n' = \varkappa. \tag{9.11}$$

It should be noted that condition (9.10) is equivalent to the orthogonality of f with respect to all solutions of the equation $K'\psi = 0$.

As shown in [29, 30, 140], under assumptions 1(b) and 2(b), Noether's theorems do not apply to (9.1). In these cases we have an improperly posed problem (see §2).

From the work of I.N. Vekua [18] and B.V. Khvedelidze [142] it follows that, in cases 1(a) and 2(a), the inverse operator $K^{-1}(\lambda)$ is a meromorphic function of λ. The poles of this function are called the *characteristic values* of the operator K.

§9.2 Methods of Regularization

The regularization of an equation implies the construction of a Fredholm integral equation of the second kind with the property that the solution of the original equation is contained among its solutions. The generalized problem of the equivalent regularization of equation (9.1) was first formulated and solved by I.N. Vekua ([18]). The formulation of the problem as given in [18] is as follows: determine an operator M of the same form as K such that

1) the operators MK and KM are Fredholm operators;
2) either the equations $K\phi = f$ and $MK\phi = Mf$ are equivalent in the usual sense, or the equations $K\phi = f$ and $KM\psi = f$ are equivalent in the sense that if one of them is solvable then so is the other, and their solutions are connected by the equation $\phi = M\psi$.

A solution of this problem under condition 2(a) of §9.1 is given, for example, by the operator $M = K_0^{0\prime}$ where

$$K_0^{0\prime}\psi \equiv a_0(t)\,\psi(t) - \frac{1}{\pi i}\int_\Gamma \frac{b_0(\tau)\,\psi(\tau)d\tau}{\tau - t} \quad ,$$

(9.12)

$$a_0(t) = \frac{a(t)}{a^2(t) - b^2(t)} \quad , \quad b_0(t) = \frac{b(t)}{a^2(t) - b^2(t)} \quad .$$

Straightforward calculations give (see [143]),

$$K_0^{0\prime}K\phi \equiv \phi + T\phi \quad , \qquad KK_0^{0\prime}\phi \equiv \phi + \tilde{T}\phi \; ,$$

(9.13)

$$T = -\frac{1}{\pi i}T_{ab_0} - \frac{1}{\pi i}T_{bb_0}\,S + K_0^{0\prime}k \; ,$$

$$\text{and} \quad \tilde{T} = -\frac{a}{\pi i}T_{b_0} + \frac{b}{\pi i}T_{a_0} + kK_0^{0\prime} \; ,$$

where the operator T_ω is defined by

(9.14)

$$T_\omega\phi \equiv \int_\Gamma \frac{\omega(\tau) - \omega(t)}{\tau - t}\,\phi(\tau)d\tau \; .$$

As usual, S denotes the singular integral operator $S\phi = \frac{1}{\pi i}\int_\Gamma \frac{\phi(\tau)d\tau}{\tau - t}$, and k is defined by (9.4).

B.V. Khvedelidze [143] has shown that the operator $K_0^{0\prime}$ also solves the problem of equivalent regularization under conditions 1(a) of §9.1.

Here it is relevant to stress that although, as a result of regularization, we reduce the problem of the approximate solution of a singular integral equation to a corresponding problem for Fredholm integral equations, the problem of finding direct methods for the approximate solution of singular integral equations remains. The basic reasons for this situation are the following: firstly, the development and analysis of methods for the approximate solution of singular integral equations without regularization is necessary, since the process of regularization can be computationally complicated; secondly, a direct solution of the problem will allow one to consider the application of such methods to a wider class of equations than is possible with the above procedure of reduction to a Fredholm integral equation of the second kind. In particular, we can consider singular integral equations for which the problem of equivalent regularization has not yet been solved (for example, some systems of singular integral equations) or equations for which regularization is not known or is not possible (for example, some forms of singular integral equations of convolution type, or the special case of singular integral equations with a^2-b^2 equal to zero on some set of points).

§9.3 The Integrability and Smoothness of Solutions

It is not difficult to see from the inequalities

$$\int_\Gamma \left|\int_\Gamma k(\tau,t)\phi(\tau)d\tau\right|^2 \rho(t)|dt| \le \int_\Gamma \left(\int_\Gamma |k(\tau,t)\rho^{-\frac{1}{2}}(\tau)\phi(\tau)\rho^{\frac{1}{2}}(\tau)|\ |d\tau|\right)^2 \rho(t)|dt|$$

(9.15)

$$\le \int_\Gamma\int_\Gamma |k(\tau,t)|^2\rho^{-1}(\tau)\rho(t)|d\tau dt| \int_\Gamma |\phi(\tau)|^2\rho(\tau)|d\tau| ,$$

that, if (9.3) is satisfied then the operator K of (9.1) will be a bounded linear operator in $L_2(\rho,\Gamma)$ provided that a and b are bounded everywhere on Γ except perhaps on a set of points of measure zero.

If the conditions 1(a) of §9.1 are satisfied, then this operator can be regularized. If $|k(\tau,t)\rho^{-1}(\tau)|$ and $|f(t)|$ are bounded on Γ then, assuming that equation (9.1) under conditions 1(a) has a solution which is integrable on Γ with weight ρ, it follows that this solution is p^{th} order integrable for any $p > 1$.

Under the conditions 2(a) of §9.1, any integrable solution automatically belongs to $H(\alpha,\Gamma)$, $\alpha < 1$.[10] If a,b and f possess derivatives of rth order, $a^{(r)}$, $b^{(r)}$, $f^{(r)} \in H(\alpha,\Gamma)$ ($a^2-b^2 \neq 0$ on Γ), and $k_t^{(r)}(\tau,t) \in H(\alpha,\Gamma)$ (where differentiation is with respect to t and the Hölder condition holds uniformly with respect to τ), then any solution ϕ of equation (9.1) which is integrable on Γ, will automatically have its rth derivative $\phi^{(r)} \in H(\alpha,\Gamma)$, $\alpha < 1$. Finally, if a,b ($a^2-b^2 \neq 0$ on Γ) and f are analytic inside an annulus $\tilde{D}$ which contains Γ, and $k(\tau,t)$ is an analytic function of t inside $\tilde{D}$ (uniformly with respect to τ), then any integrable solution on Γ will be analytic inside $\tilde{D}$.

A proof of these facts follows almost trivially from the representation

$$K^0\phi = f(t) - \lambda\int_\Gamma k(\tau,t)\ \phi(\tau)d\tau , \tag{9.16}$$

the properties of the solution of the dominant equation as given by (8.8), and the corresponding properties of singular integrals (see §7 and [108, 142, 56]).

An alternative proof can be given making use of equations (9.13)-(9.14) and the corresponding properties of Fredholm integral equations of the second kind.

---

(10) If $\alpha = 1$ and Γ is a Lyapunov curve, the solution will be contained in the Zygmund space $Z(\Gamma)$ (see [173, 56]).

As in the above case, when the solution ϕ possesses an rth order derivative $\phi^{(r)}$, both proofs yield the important inequality

$$(9.17) \qquad |\phi^{(r)}(t)| < C_1 M_{f^{(r)}} + C_2 H_{f^{(r)}} + C_3 \int_\Gamma |\phi(\tau)|\,|d\tau| \,.$$

There is no point in deriving general explicit expressions for C_1, C_2 and C_3, since they are complicated and give a conservative bound. It is better to obtain them in each particular case, as required.

The first method of proof has the advantage that for a class of integrable solutions, certain conclusions can be made about their smoothness when the condition $a^2-b^2 \neq 0$ on Γ has been violated at a set of points. We now apply equations (8.14) rather than (8.8). In particular, if a, b, f and k possess r^{th} order derivatives and $k_t^{(r)}(\tau,t) \in H(\alpha,\Gamma)$ (uniformly with respect to τ) then, for equation (9.1) when $(a-b)/(a+b)$ is of the form (8.9), each integrable solution will have an rth order derivative (which belongs to $H(\alpha,\Gamma)$ for $\alpha < 1$) at all points except α_k and β_j where it will have derivatives of order $r-m_k-1$ and $r-p_j-1$ respectively.

Let us make the substitution

$$(9.18) \qquad t = t(u),\ u = \exp(2\pi i\ell/L),\ \tau = \tau(v),\ v = \exp(2\pi is/L)\ ,\ 0 \leq \ell,\ s \leq L,$$

where ℓ, s are arc lengths along Γ. We shall also denote functions of the variables u,v which are obtained from this substitution, by the same letters which denoted these functions in the original variables t, τ. Then equation (9.1) will be transformed to

$$(9.19) \qquad K\phi \equiv a(u)\phi(u) + \frac{b(u)}{\pi i}\int_\gamma \frac{\phi(v)dv}{v-u} + \int_\gamma k_1(v,u)\phi(v)dv = f(u)\ ,$$

where

$$(9.20) \qquad k_1(v,u) = \lambda k(v,u)\tau'(v) + \frac{b(u)}{\pi i}\left\{\frac{\tau'(v)(v-u) - [\tau(v)-t(u)]}{[\tau(v)-t(u)](v-u)}\right\} .$$

Hence it follows that a transformation of equation (9.1) to one where the contour of integration is the unit circle (which is a convenient form for its approximate solution) can lead to a decrease in the order of differentiability of its coefficients, kernel and solution when compared with the original equation.

§9.4 Stability of Solutions

For the implementation of some of the methods for the approximate solution of equation (9.1), which will be described later, it will be necessary to represent the functions a, b, f and k in a particular form. Since in general the representation of functions and their transformation from one form to another can only be done approximately, it is natural to pose the following problem.

Suppose (9.1) is an approximate equation, and assume (from physical or other considerations) that the "exact" equation

$$(9.21)\qquad K_1\phi_1 \equiv a_1(t)\phi_1(t) + \frac{b_1(t)}{\pi i}\int_\Gamma \frac{\phi_1(\tau)d\tau}{\tau - t} + \lambda_1\int_\Gamma k_1(\tau,t)\phi_1(\tau)d\tau = f_1(t) ,$$

is solvable and possesses $m > 0$ linearly independent solutions. Suppose further that the coefficients and kernels of equations (9.1) and (9.21) are known only to satisfy the inequalities

$$(9.22)\qquad \begin{gathered} |a - a_1| < \epsilon, \quad |b - b_1| < \epsilon, \quad |f - f_1| < \epsilon, \\ |\lambda k(\tau,t) - \lambda_1 k_1(\tau,t)| < \epsilon \quad (\tau,t \in \Gamma,\ \epsilon < 1) . \end{gathered}$$

The problem is now to determine whether equation (9.1) is solvable and, if it is, to find the number of its linearly independent solutions. In addition we shall require bounds on the errors $|\phi_j - \phi_{1j}|$, where ϕ_j and ϕ_{1j} are the solutions of (9.1) and (9.21) respectively which satisfy certain given conditions.

If nothing is known concerning the solvability of (9.21) then, after solving (9.1), it is natural to ask what conclusions may be drawn concerning the solvability of (9.21), and the properties of its solution using the assumptions of (9.22).

Let us examine these questions, assuming for simplicitv that $\Gamma \equiv \gamma$, $(|t| = |\tau| = 1,\ t = e^{i\theta})$, $a(e^{i\theta}) = a(\theta)$, $b(e^{i\theta}) = b(\theta)$ etc. We assume that $|a^{(\nu)}(\theta)|$, $|b^{(\nu)}(\theta)|$, $|f^{(\nu)}(\theta)|$ and $|k_\theta^{(\nu)}|$ for $\nu = 0,1,\dots,r$ remain bounded for decreasing ϵ, where the rth order derivatives of a, b, f and k are in $H(\alpha,\gamma)$, $(0 < \alpha \leq 1)$ with Hölder constants that remain bounded as $\epsilon \to 0$. Then the corresponding functions of (9.21) have r^{th} order derivatives that are in $H(\alpha,\gamma)$.

In fact, if $r = 0$ then

$$\begin{aligned} |a_1(\theta+h) - a_1(\theta)| &\leq |a_1(\theta+h) - a(\theta+h)| + |a_1(\theta) - a(\theta)| + \\ &\quad + |a(\theta+h) - a(\theta)| \\ &< C_4 h^\alpha . \end{aligned}$$

If $r > 0$ then

$$\left| \frac{a_1(\theta+h_1) - a_1(\theta)}{h_1} - \frac{a_1(\theta+h_2) - a_1(\theta)}{h_2} \right|$$

$$\leq \left| \frac{a_1(\theta+h_1) - a(\theta+h_1)}{h_1} \right| + \left| \frac{a_1(\theta) - a(\theta)}{h_1} \right| + \left| \frac{a_1(\theta+h_2) - a(\theta+h_2)}{h_2} \right| +$$

$$+ \left| \frac{a_1(\theta) - a(\theta)}{h_2} \right| + \left| \frac{a(\theta+h_1) - a(\theta)}{h_1} - a'(\theta) \right| + \left| a'(\theta) - \frac{a(\theta+h_2) - a(\theta)}{h_2} \right| ,$$

which can be made less than any preassigned positive number for all sufficiently small h_1 and h_2 which are not equal to zero. This implies that a_1 has a derivative. Furthermore

$$|a_1'(\theta) - a'(\theta)| \leq \left| a_1'(\theta) - \frac{a_1(\theta+h) - a_1(\theta)}{h} \right| +$$

$$+ \left| \frac{a_1(\theta+h) - a_1(\theta) - a(\theta+h) + a(\theta)}{h} \right| +$$

$$+ \left| \frac{a(\theta+h) - a(\theta)}{h} - a'(\theta) \right| \to 0 ,$$

as $h,\epsilon \to 0$. The proof is now completed by induction. Similar results hold for b_1, f_1 and k_1.

If $a(\theta)$, $b(\theta)$, $f(\theta)$, $k(\theta_0,\theta)$ are trigonometric polynomials, then whether or not the above assumptions are satisfied will depend upon the rate of decrease of ϵ. For example, if $\epsilon = \text{constant} \times n^{-(r+\alpha)}$, where n is the order of the polynomials a, b, f and k, and (9.22) holds for any n, then from the inverse theorem of S.N. Bernstein [9], $a_1(\theta)$, $b_1(\theta)$, $f_1(\theta)$ and $k_1(\theta_0,\theta)$ have r^{th} order derivatives which are in $H(\alpha,\gamma)$.(11) In addition (see [27]) $a^{(\nu)}$, $b^{(\nu)}$, $f^{(\nu)}$ and $k^{(\nu)}$ ($\nu = 0,1,2,\ldots,r$) have their moduli bounded uniformly with respect to the order of the polynomials, and are also contained in $H(\alpha,\gamma)$.

Let a, b, f, k and a_1, b_1, f_1, k_1 be Hölder continuous with index $\alpha(0 < \alpha < 1)$, uniformly with respect to τ and ϵ. Let us consider (9.1) and (9.21) as linear equations in the Banach space $W(\gamma)$ (see §7):

(9.23) $$\|\phi\|_{W(\gamma)} = \max_{t\in\gamma} |\phi^+(t)| + \max_{t\in\gamma} |\phi^-(t)| ,$$

(11) For $\alpha = 1$, $H(1,\gamma)$ should be replaced by the Zygmund space $Z(\gamma)$, see [173].

where $\phi^{\pm}$ are related to ϕ by the Sokhotski-Plemelj formulae. We estimate the norm of the operator $(K_1-K) = \Delta K$:

$$(9.24)\qquad \Delta K\phi \equiv (a_1(t)-a(t))\phi(t) + \frac{b_1(t)-b(t)}{\pi i}\int_{\gamma}\frac{\phi(\tau)d\tau}{\tau - t} + \int_{\gamma}(\lambda_1 k_1(\tau,t)-\lambda k(\tau,t))\phi(\tau)d\tau\ .$$

From the equation

$$(9.25)\qquad \int_{\gamma}\frac{a(\tau)-a_1(\tau)}{\tau - t}\phi(\tau)d\tau = \int_{\gamma_\epsilon}\frac{a(\tau)-a(t)+a_1(t)-a_1(\tau)}{\tau - t}\phi(\tau)d\tau + $$
$$+\int_{\gamma-\gamma_\epsilon}\frac{a(\tau)-a_1(\tau)}{\tau - t}\phi(\tau)d\tau +$$
$$+[a_1(t)-a(t)]\int_{\gamma-\gamma_\epsilon}\frac{\phi(\tau)d\tau}{\tau - t} - [a_1(t)-a(t)]\int_{\gamma}\frac{\phi(\tau)d\tau}{\tau - t}\ ,$$

where γ_ϵ is that part of γ within a circle with centre at t and radius $\epsilon^{1/\alpha}$; and also from similar equations for $b-b_1$, $f-f_1$ and $k-k_1$; we find after some algebra that

$$(9.26)\qquad \|\Delta K\|_{W(\gamma)} \leq C_5\,\epsilon\,\ell n(1/\epsilon)\ ,$$

where C_5 does not depend on ϵ. We shall state this result as a theorem.

<u>Theorem 9.1</u>. *If, in equations* (9.1) *and* (9.21), *a, b, f, k and a_1, b_1, f_1, k_1 satisfy inequalities* (9.22) *and are Hölder continuous uniformly with respect to τ and ϵ, then* (9.26) *is valid.*

It is not difficult to see that in the space $L_2(\Gamma,\rho)$ if the inequalities (9.22) are satisfied then, without additional restrictions,

$$(9.27)\qquad \|\Delta K\|_{L_2(\Gamma,\rho)} \leq C_6\epsilon\ ,$$

where C_6 does not depend on ϵ.

The inequalities (9.26) and (9.27) imply that the operators K and K_1 differ from each other by a perturbation which is arbitrarily small in norm. On the other hand, the relation between the solutions of equations (9.1) and (9.22) will be determined by the general theory of perturbations of operators (some of these results will be given below).

Finally, we note that a very detailed analysis of the stability of the Riemann boundary value problem, with regard to deformations of the contour Γ, has been developed by L.A. Chikin, [150].

§9.5 A General Functional Approach

The operator K as defined by the left hand side of equation (9.1), under the conditions 1(a) or 2(a), is a specific example of the so-called "normally solvable" operators in a Banach space, which have been investigated by M.G. Krein, S.M. Nikolskiy, F.V. Atkinson, M.A. Krasnosel'skiy, S.G. Mikhlin, I.Ts. Gokhberg etc. We shall state here some of the results that have been obtained.

Let E be a Banach space, and A a closed linear operator on E. We shall let $D(A)$ denote the domain of A, $R(A)$ the range of A, and $L(A)$ the subspace of solutions for which $Ax = 0$ (null space). Let $\alpha(A)$ be the dimension of $L(A)$ and $\beta(A)$ the deficiency index of A i.e., the dimension of the factor space $E/\overline{R(A)}$.

Definition 1. *The operator A is called normally solvable if the equation*

$$\ell(y) = 0 \,, \tag{9.28}$$

where ℓ is an arbitrary functional of the conjugate space E^* for which $\ell A = 0$, is a necessary and sufficient condition for the solvability of $Ax = y$.

The normal solvability of an operator A is equivalent to the closure of $R(A)$.

Definition 2. *The number pair $(\alpha(A), \beta(A))$ is called the defect characteristic of the operator A, or d-characteristic of A. If $\alpha(A)$ and $\beta(A)$ are finite, then the d-characteristic of A is also said to be finite. If the d-characteristic of A is finite, the difference $\alpha(A) - \beta(A)$ is called the index of the operator A, and is denoted by $\mathcal{H}(A)$.*

If a normally solvable operator A has a finite d-characteristic, then the perturbation of A by a linear operator B which is either completely continuous or has a sufficiently small norm does not affect the normal solvability and finiteness of the d-characteristic of the perturbed operator; i.e. the operator $A+B$ is normally solvable and has a finite d-characteristic. In addition, the index of the operator is stable with respect to these perturbations, i.e.

(9.29) $$\mathcal{H}(A + B) = \mathcal{H}(A) .$$

Definition 3. *A bounded linear operator M is a regularizing operator for the operator A if, for all $x \in D(A)$,*

(9.30) $$MAx = x + Tx ,$$

where T is a completely continuous operator. If, for the operator A, there exists a regularizing operator then we say that A admits a regularization.

If the operator A admits a regularization then A is normally solvable and $\alpha(A)$ is finite. If the operator A is normally solvable and its d-characteristic is finite, then the operator A admits a regularization.

Definition 4. *Let B be a bounded linear operator in E. The set of all complex numbers λ for which an operator $A-\lambda B$ is normally solvable and has a finite d-characteristic is called the domain of Noether for the operator $A-\lambda B$.*

The domain of Noether is an open set. Consequently it can be represented as the union of a finite or countable number of connected components. Let N be any one of these components. For all $\lambda \in N$, the index of the operator $A-\lambda B$ will take the same value. For all $\lambda \in N$ (except perhaps at an isolated set of points in N) $\alpha(A-\lambda B)$ takes one and the same value. At the points λ of this isolated set, $\alpha(A-\lambda B)$ will have a larger value.

Under the conditions 1(b) or 2(b), the operator K is not normally solvable. In this case the stability of the solutions can be ensured for perturbations which do not remove the solutions from a given compact set, see [128]. Difficulties which can arise with an arbitrary perturbation may be circumvented by the methods of regularization [129], and of quasi-solutions [69]; for more details see §2.

It should be noted that a general functional approach to the analysis of singular integral equations which is more specific than that given in [43], has been developed by Z.I. Khalilov [139] and by Yu. I. Cherskiy [148].

An important advantage of the general functional approach is that the results obtained are valid not only with respect to the operator K of (9.1), but also with respect to other operators. This does not imply, however, that the necessity to study subclasses of operators or particular operators is eliminated. The contraction of a class of operators being studied will often lead to an enrichment of their properties, although valuable results obtained from the study of subclasses of operators will not necessarily be valid for all operators in a larger class.

For example the results quoted above from perturbation theory do not completely include the results of the previous section since the original assumptions (9.22) in general are not equivalent to a perturbation by an operator with a sufficiently small norm or which is completely continuous (in the space $W(\Gamma)$).

Let us consider one more example which illustrates these general ideas. I. Ts. Gokhberg [42] has proved the following theorem(12): In order that the operator K, as defined by the left-hand side of (9.1), should admit a regularization, it is necessary and sufficient that $a^2 - b^2$ should never become zero. In this case a general functional approach to the solution of the regularization of singular integral equations in exceptional cases gives a negative result. But practical considerations require a more positive result. A more detailed study of a series of exceptional cases for singular integral equations has led (see [33]) to their regularization in a more general sense (the corresponding Fredholm equation of the second kind is obtained by the solution of the equation $a\phi + bS\phi = f - k\phi$, making use of (8.14)).

(12) In $L_p(\rho,\Gamma)$ the corresponding result has been proved by D.F. Kharazov and B.V. Khvedelidze [140].

Chapter 3

THE NUMERICAL SOLUTION OF SINGULAR INTEGRAL EQUATIONS WITH CAUCHY KERNEL

§10 RESULTS FROM THE CONSTRUCTIVE THEORY OF FUNCTIONS

§10.1 Approximation on a Circle

Let γ denote the unit circle, and let the complex-valued function $\phi(t)$ be defined at the points of γ, with $\phi \in L_p(\gamma)$, $p \geq 1$ and $t = e^{i\theta}$. We define the *measure of the deviation* of the best approximation to ϕ on γ to be

$$(10.1) \qquad I_n(\phi) = \inf_{T_n \in H_n} \|\phi - T_n\|_{L_p} \,,$$

where H_n denotes the space of all polynomials T_n of the form $\sum_{k=-n}^{n} a_k t^k$. It should be noted that since $t^k = e^{ik\theta} = \cos k\theta + i \sin k\theta$, $-\pi \leq \theta \leq \pi$, the problem of polynomial approximation on a circle is equivalent to that of approximation to functions defined on $[-\pi,\pi]$ by trigonometric functions of order not exceeding n.

In the book by N.I. Akhieser [2, §89], it has been proved that

$$(10.2) \qquad I_n(\phi) \leq \frac{\pi\|\phi_\theta^{(r)}\|_{L_p}}{2(n+1)^r} \,, \quad (r = 0,1,2,\ldots)$$

and

$$(10.3) \qquad I_n(\phi) < \left(\frac{\pi}{2} + \frac{4}{\pi}\right) \omega_p\left(\frac{1}{n+1} \,;\, \phi_\theta^{(r)}\right) (n+1)^{-r} \,,$$

for $r = 0,1,2,\ldots$, where, for any $\delta > 0$ and any function $f \in L_p(\gamma)$, the modulus of continuity is defined by

$$(10.4) \qquad \omega_p(\delta;\, f) = \sup_{0<h\leq\delta} \left\{ \frac{1}{2\pi} \int_{-\pi}^{\pi} |f[e^{i(\theta+h)}] - f(e^{i\theta})|^p d\theta \right\}^{1/p} .$$

Equations (10.1)-(10.3) also hold when $p = \infty$, where

$$(10.5) \qquad \|f\|_{L_\infty} = \|f\|_{C(\gamma)} = \operatorname*{ess\,sup}_{t\in\gamma} |f(t)|$$

and

$$\omega_\infty(\delta;\, f) = \omega(\delta;\, f) = \sup_{0<h\leq\delta}\ \operatorname{ess\,sup}_{t\varepsilon\gamma} |f(te^{ih}) - f(t)| \quad . \tag{10.6}$$

Since, from (7.2) and (7.3), the conjugate function $\tilde{\phi}$ (see [2, §69]) is given by

$$\tilde{\phi} = -S_1\phi_1 = i\left(S\phi - (2\pi)^{-1}\int_{-\pi}^{\pi}\phi(e^{i\theta})d\theta\right),$$

it follows that

$$I_n(S\phi) \leq \frac{\pi}{2}\cdot\frac{\|\phi_\theta^{(r)}\|_{L_p}}{(n+1)^r}, \quad (r = 1,2,\ldots), \tag{10.7}$$

$$\text{and}\qquad I_n(S\phi) < \left(\frac{\pi}{2} + \frac{4}{\pi}\right)\omega_p\left(\frac{1}{n+1};\, \phi_\theta^{(r)}\right)(n+1)^{-r}, \quad (r = 1,2,\ldots). \tag{10.8}$$

If, for the polynomial of best approximation $T_n^* = \sum_{k=-n}^{n} \alpha_k^* t^k$, we replace the coefficients α_k^* by the corresponding Fourier coefficients, it follows that

$$\rho_n^+(\phi^+) \leq \frac{\pi}{2}\cdot\frac{\|\phi_\theta^{+(r)}\|_{L_p}}{(n+1)^r} \qquad (r = 0,1,2,\ldots) \tag{10.9}$$

and

$$\rho_n^-(\phi^-) \leq \frac{\pi}{2}\cdot\frac{\|\phi_\theta^{-(r)}\|_{L_p}}{(n+1)^r} \qquad (r = 0,1,2,\ldots). \tag{10.10}$$

Here we have defined

$$\rho_n^+(\phi^+) = \inf_{T_n^+\varepsilon H_n^+} \|\phi^+ - T_n^+\|_{L_p} \tag{10.11}$$

and

$$\rho_n^-(\phi^-) = \inf_{T_n^-\varepsilon H_n^-} \|\phi^- - T_n^-\|_{L_p}, \tag{10.12}$$

where ϕ^+ and ϕ^- denote the boundary values of a function which is analytic on the inside and outside respectively of γ, and H_n^+, H_n^- denote polynomials of the form $\sum_{k=0}^{n} \alpha_k t^k$, $\sum_{k=-n}^{-1} \alpha_k t^k$ respectively.

The relations (10.7)-(10.12) are also valid when $p = \infty$. Hence, in particular, it follows that

(10.13) $$\rho_n(\phi) \le \frac{\pi}{2} \cdot \frac{\|\phi_\theta^{(r)}\|_{W(\gamma)}}{(n+1)^r} \quad , \quad (r = 0,1,2,\dots) \quad ,$$

where

(10.14) $$\rho_n(\phi) = \inf_{H_n} \|\phi - T_n\|_{W(\gamma)} \quad ,$$

the norm in the space $W(\gamma)$ being defined by (7.16).

We shall now consider some particular polynomials T_n which approximate an arbitrary function ϕ. First we consider the interpolation polynomial

(10.15) $$u_n(\phi,t) = \sum_{k=-n}^{n} a_k t^k, \quad a_k = \frac{1}{2n+1} \sum_{j=-n}^{n} \phi(\tau_j)\, \tau_j^{-k} \quad ,$$

$$\tau_j = \exp\,[2\pi i j/(2n+1)] \; .$$

It is not difficult to verify that

(10.16) $$u_n(\phi,t) = \frac{1}{2n+1} \sum_{j=-n}^{n} \phi(\tau_j) \left(\frac{t}{\tau_j}\right)^{-n} \left\{ \frac{1 - (t/\tau_j)^{2n+1}}{1 - (t/\tau_j)} \right\} \quad ,$$

and $$u_n(\phi,\tau_j) = \phi(\tau_j) \; .$$

Theorem 10.1. *The following inequalities are valid*

(10.17) $$|u_n(\phi,t)| \le \left\{ \frac{4}{\pi} + \frac{2}{\pi} \ln\,[\frac{2}{\pi}(2n+1)] \right\} \max_j |\phi(\tau_j)| ,$$

(10.18) $$|u_n^+(\phi,t)| \le \left\{ \frac{n+1}{2n+1} + \frac{2}{\pi} + \frac{2}{\pi} \ln\,[\frac{2}{\pi}(2n+1)] \right\} \max_j |\phi(\tau_j)| ,$$

(10.19) $$|u_n^-(\phi,t)| \le \left\{ \frac{n}{2n+1} + \frac{2}{\pi} + \frac{2}{\pi} \ln\,[\frac{2}{\pi}(2n+1)] \right\} \max_j |\phi(\tau_j)| ,$$

where $u_n^+(\phi,t) = \sum_{k=0}^{n} a_k t^k$ *and* $u_n^-(\phi,t) = \sum_{k=-n}^{-1} a_k t^k$.

The inequality (10.17) *cannot be improved since there exists a function such that*

$$\max_t |u_n(\phi,t)| = \left\{ \frac{4}{\pi} + \frac{2}{\pi} \ln\,[\frac{2}{\pi}(2n+1)] + O(\frac{1}{n}) \right\} \max_j |\phi(\tau_j)| \; .$$

A proof of this theorem will not be given here; it may be found in [63].

From (10.18) and (10.19) we obtain

$$(10.20)\qquad \|u_n(\phi,t)\|_{W(\gamma)} \leq \left\{1 + \frac{4}{\pi} + \frac{4}{\pi}\ln\left[\frac{2}{\pi}(2n+1)\right]\right\} \max_j |\phi(\tau_j)| \ .$$

Let $\phi \in C(\gamma)$ have on γ continuous derivatives $\phi_\theta^{(r)}$, $r \geq 0$. Let $T_n^* = \sum_{k=-n}^{n} \alpha_k^* t^k$ be the polynomial of best approximation to ϕ on γ i.e.

$$(10.21)\qquad \max_t |\phi(t) - T_n^*(t)| = E_n(\phi,\gamma) = \inf_{T_n \in H_n} \max_t |\phi(t) - T_n(t)| \ .$$

Then obviously $\max_j |\phi(\tau_j)| \leq M_\phi = \max_t |\phi(t)| = \|\phi\|_{C(\gamma)}$, $u_n(T_n^*, \tau) = T_n^*$ so that from (10.17)

$$(10.22)\qquad \begin{aligned} |\phi(t) - u_n(\phi,t)| &\leq |\phi(t) - T_n^*(t)| + |u_n(T_n^* - \phi,t)| \\ &\leq \left\{1 + \frac{4}{\pi} + \frac{2}{\pi}\ln\left[\frac{2}{\pi}(2n+1)\right]\right\} E_n(\phi,\gamma) \ . \end{aligned}$$

Similarly from (10.20) we obtain

$$(10.23)\qquad \|\phi(t) - u_n(\phi,t)\|_{W(\gamma)} \leq \left\{2 + \frac{4}{\pi} + \frac{4}{\pi}\ln\left[\frac{2}{\pi}(2n+1)\right]\right\} \rho_n(\phi) \ .$$

Theorem 10.2. *The inequality*

$$(10.24)\qquad \|u_n(\phi,t)\|_{L_2(\gamma)} \leq \max_j |\phi(\tau_j)| \ ,$$

is valid.

It is not difficult to show that

$$(10.25)\qquad \frac{1}{2\pi}\int_{-\pi}^{\pi} |u_n(\phi,t)|^2 dt = \frac{1}{2\pi i}\int_\gamma t^{-1} u_n \bar{u}_n \, dt = \frac{1}{(2n+1)} \sum_{j=-n}^{n} |\phi(\tau_j)|^2 \ ,$$

from which (10.24) follows at once.

From (10.24) we obtain by a similar method to that used for the derivation of (10.22), that

$$(10.26)\qquad \|\phi(t) - u_n(\phi,t)\|_{L_2(\gamma)} \leq 2\, E_n(\phi,\gamma) \ ,$$

and in particular

(10.27) $$\|\phi(t) - u_n(\phi,t)\|_{L_2(\gamma)} \le \pi \frac{\max_\theta |\phi_\theta^{(r)}(t)|}{(n+1)^r} .$$

Let us introduce the generalized Fejér polynomials

$$\sigma_{n+1}(\phi,t;\mu) = \sum_{|k|\le n} \lambda\left(\frac{k}{n+1}\right) c_k t^k ,$$

where $$c_k = \frac{1}{2\pi i}\int_\gamma \frac{\phi(t)dt}{t^{k+1}} , \quad t = e^{i\theta} ,$$

(10.28)

and $$\lambda(\xi) = \begin{cases} 1 , & \text{for } |\xi| \le 1-\mu , \\ \frac{1}{\mu}(1-|\xi|) , & \text{for } 1-\mu \le |\xi| \le 1 . \end{cases}$$

It is known (see [2, §90]) that

(10.29) $$|\phi(t) - \sigma_{n+1}(\phi,t;\mu)| \le \left\{1 + \frac{4}{\pi} + \frac{2}{\pi}\ln\left(\frac{2-\mu}{\mu}\right)\right\} E_{(n+1)(1-\mu)}(\phi,\gamma) .$$

If we put $\mu = 1/(r+1)$ and recalling that $(1+1/r)^r \le e \approx 2.7$, we obtain on using (10.2) that

(10.30) $$|\phi(t) - \sigma_{n+1}(\phi,t;(r+1)^{-1})| \le e[\pi/2 + 2 + \ln(2r+1)] \frac{\max_\theta |\phi_\theta^{(r)}(t)|}{(n+1)^r} .$$

If we put $\mu = 1/(n+1)$, then $\sigma_{n+1}(\phi,t;\mu) = s_n(\phi,t)$ where $s_n(\phi,t)$ is n^{th} partial sum of the Fourier series for $\phi(t)$. Then

(10.31) $$|\phi(t) - s_n(\phi,t)| \le \frac{D_1 + 1 + D_2 \ln(n+1)}{n^r} \max_\theta |\phi_\theta^{(r)}(t)| ,$$

where D_1 and D_2 are constants such that $D_1 < 2$ and $D_2 < 1$.

If ϕ is analytic in the annulus between the circles $|z| = R_1 < 1$ and $|z| = R_2 > 1$, then we can represent ϕ as $\phi = \phi^+ - \phi^-$ where

(10.32) $$\phi^+(z) = \sum_{k=0}^{\infty} c_k z^k , \quad \phi^-(z) = -\sum_{k=-\infty}^{-1} c_k z^k,$$

$$c_k = \frac{1}{2\pi i}\int_\gamma \frac{\phi(t)dt}{t^{k+1}} ,$$

so that ϕ^+ is analytic inside the circle $|z| = R_2$, and ϕ^- is analytic outside the circle $|z| = R_1$.

If we assume that $|\phi|$ is integrable on the circles $|z| = R_1$ and $|z| = R_2$, then for $k \geq 0$,

$$|c_k| \leq R_2^{-k} \frac{1}{2\pi} \int_{-\pi}^{\pi} |\phi(R_2 e^{i\theta})| d\theta \,, \tag{10.33}$$

and for $k < 0$,

$$|c_k| \leq R_1^{-k} \frac{1}{2\pi} \int_{-\pi}^{\pi} |\phi(R_1 e^{i\theta})| d\theta \,. \tag{10.34}$$

Hence we obtain

$$\begin{aligned} E_n(\phi) \leq \rho_n(\phi) &\leq \max_t |\phi^+(t) - \sum_{k=0}^{n} c_k t^k| + \max_t |\phi^-(t) - \sum_{k=-n}^{-1} c_k t^k| \\ &\leq \sum_{k=n+1}^{\infty} |c_k| + \sum_{k=-\infty}^{-n-1} |c_k| \\ &\leq \frac{R_2^{-n-1}}{1-R_2^{-1}} \cdot \frac{1}{2\pi} \int_{-\pi}^{\pi} |\phi(R_2 e^{i\theta})| d\theta + \frac{R_1^{n+1}}{1-R_1} \cdot \frac{1}{2\pi} \int_{-\pi}^{\pi} |\phi(R_1 e^{i\theta})| d\theta \,. \end{aligned} \tag{10.35}$$

We shall now state three important theorems from the constructive theory of functions.

Theorem 10.3. (S.N. Bernstein [9]). *For every polynomial* $T_n = \sum_{k=-n}^{n} \alpha_k t^k$, $t = e^{i\theta}$, *the inequality*

$$|T_n'(t)| \leq n \max_\theta |T_n(t)| \tag{10.36}$$

holds.

Theorem 10.4. (S.N. Bernstein [9]). *If, for all* n, $E_n(\phi,\gamma) \leq$ constant $n^{-(p+\alpha)}$, *where* p *is a non-negative integer and* $0 < \alpha < 1$, *then* ϕ *has a* p^{th} *order derivative and* $\phi^{(p)} \in H(\alpha,\gamma)$.(1)

Theorem 10.5. (A.L. Garkavi [27]). *For every function* $\phi(t)$, $t = e^{i\theta}$, *which possesses an* r^{th} *order derivative* $\phi_\theta^{(r)}(t)$, *and for every polynomial* $T_n(t) = \sum_{k=-n}^{n} \alpha_k t^k$ *we have*

$$|\phi_\theta^{(r)}(t) - T_{n,\theta}^{(r)}(t)| \leq \max_\theta |\phi(t) - T_n(t)| n^r + (1+\pi/2) c_{n,r} E_n(\phi_\theta^{(r)},\gamma) \,, \tag{10.37}$$

where $c_{n,r} \leq \frac{4}{\pi^2} \ln (p+1) + \pi e + 4$, $p = \min \{r,n\}$.

---

(1) If $\alpha = 1$, then $\phi^{(p)}$ is in the Zygmund space $Z(\gamma)$, (see [173]) i.e.

$$|\phi^{(p)}(e^{i(\theta+h)}) + \phi^{(p)}(e^{i(\theta-h)}) - 2\phi^{(p)}(e^{i\theta})| < C_1 h \,.$$

From (10.37) and (10.22) it follows that

$$|\phi_\theta^{(\nu)}(t) - u_{n,\theta}^{(\nu)}(\phi,t)| \le \left\{\frac{\pi}{2} + 2 + \ln\left[\frac{2}{\pi}(2n+1)\right] + \frac{\pi}{2}\left(1+\frac{\pi}{2}\right)c_{n,\nu}\right\} \times \tag{10.38}$$
$$\times \frac{\max_\theta |\phi_\theta^{(r)}(t)|}{(n+1)^{r-\nu}} .$$

§10.2 Approximation on Other Sets

Let us now assume that $\phi(x)$ is defined for all $x \in (-\infty, \infty)$. If we make the change of variable

$$t = \frac{x-i}{x+i} = e^{i\theta} , \quad x = i\,\frac{1+t}{1-t} = -\cot\frac{\theta}{2} , \tag{10.39}$$

then we obtain the function $\psi(t) = \phi[i(1+t)/(1-t)]$, which is defined on γ.

We shall now state sufficient properties on ϕ, so that ψ may have derivatives $\psi_\theta^{(r)}(t)$, $r \ge 0$, with bounded modulus for all $t \in \gamma$. Formal differentiation gives

$$\psi_\theta^{(r)} = \phi_x^{(r)}\,[(1+x^2)/2]^r + \phi_x^{(r-1)}P_{2r-1}(x) + \ldots + \phi_x' P_{r+1}(x) , \tag{10.40}$$

where P_j are polynomials of degree j. Therefore it seems that a necessary and sufficient condition for the boundedness of $|\psi_\theta^{(r)}|$ is given by the boundedness of $|\phi_x^{(r)}|$ everywhere on the real axis, together with conditions such as $|\phi_x^{(\nu)}| < C_\nu |x|^{-r-\nu}$, $(\nu = 1,2,\ldots,r)$ being valid for sufficiently large $|x|$. But if $\phi(x)$ is analytic at infinity i.e. if it can be represented by

$$\phi(x) = A_0 + \frac{A_1}{x} + \frac{A_2}{x^2} + \ldots , \tag{10.41}$$

then obviously ψ is analytic in a neighbourhood of $t = 1$ and consequently all derivatives of ψ with respect to t and θ are bounded in a neighbourhood of $t = 1$. This suggests that for sufficiently large $|x|$, we put $\phi(x) = \chi(u)$ where $u = 1/x$, and instead of (10.40), consider the equation

$$\psi_\theta^{(r)}(t) = \chi_u^{(r)}\left(\frac{1+x^2}{2}\right)^r\left(-\frac{1}{x^2}\right)^r + \chi_u^{(r-1)}\cdot\theta_1(x) + \ldots + \chi_u'\,\theta_{r-1}(x). \tag{10.42}$$

It is readily seen that the functions $\theta_1, \theta_2, \ldots, \theta_{r-1}$ are bounded for sufficiently large $|x|$. Consequently, a sufficient condition for the boundedness of $|\psi_\theta^{(r)}(t)|$ is given by the boundedness of $|\phi_x^{(r)}(x)|$ everywhere, and the boundedness of $|\phi_u^{(r)}(x)|$ for sufficiently large $|x|$.

From (10.2), (10.22) and (10.39) it follows that

$$|\phi(x) - \sum_{k=-n}^{n} a_k \left(\frac{x-i}{x+i}\right)^k| \le \left\{ 2 + \frac{\pi}{2} + \ln\left[\frac{2}{\pi}(2n+1)\right]\right\} \frac{\max_\theta |\psi_\theta^{(r)}(t)|}{(n+1)^r} , \tag{10.43}$$

where $\quad a_k = \frac{1}{2n+1} \sum_{j=-n}^{n} \phi(x_j) \exp[-2\pi i jk/(2n+1)]$

and $\quad x_j = -\cot[\pi j/(2n+1)]$.

If $\phi(\pm\infty) = 0$ then $\sum_{k=-n}^{n} a_k = 0$,

$$\sum_{k=-n}^{n} a_k \left(\frac{x-i}{x+i}\right)^k = \sum_{k=-n}^{n-1} \gamma_k \left(\frac{x-i}{x+i}\right)^k \left(\frac{2i}{x+i}\right) , \tag{10.44}$$

and $\quad \gamma_k = a_{-n} + a_{-n+1} + \ldots + a_k = -(a_{k+1} + a_{k+2} + \ldots + a_n)$.

If ϕ is a real function, then $\bar{a}_k = a_{-k}$ so that

$$\sum_{k=-n}^{n} a_k \left(\frac{x-i}{x+i}\right)^k = a_0 + 2\sum_{k=1}^{n} [a_k^{(1)} \cos k\theta - a_k^{(2)} \sin k\theta] , \tag{10.45}$$

$$a_k^{(1)} = \frac{1}{2n+1} \sum_{j=-n}^{n} \phi(x_j) \cos\left(\frac{2\pi jk}{2n+1}\right), \quad a_k^{(2)} = \frac{(-1)}{2n+1} \sum_{j=-n}^{n} \phi(x_j) \sin\left(\frac{2\pi jk}{2n+1}\right) .$$

Suppose now that ϕ is defined on the interval [-1,1]. We make the change of variables

$$x = \cos\theta, \; t = e^{i\theta} , \tag{10.46}$$

and write $\psi_1(t) = \phi(\cos\theta)$. Since $\psi_1(\bar{t}) = \psi_1(t)$, it readily follows that

$$u_n(\psi_1, t) \equiv \sum_{k=-n}^{n} a_k t^k = a_0 + 2\sum_{k=1}^{n} a_k T_k^{(1)}(x) \equiv u_n(\phi, x) , \tag{10.47}$$

$$a_k = \frac{1}{(2n+1)} [\phi(1) + 2\sum_{j=1}^{n} \phi(x_j) \cos\left(\frac{2\pi jk}{2n+1}\right)] , \; x_j = \cos\left(\frac{2\pi j}{2n+1}\right) ,$$

where

$$T_k^{(1)} = \cos k\theta = \left\{[x + i(1-x^2)^{\frac{1}{2}}]^k + [x - i(1-x^2)^{\frac{1}{2}}]^k\right\} /2 , \tag{10.48}$$

are the Chebyshev polynomials of the first kind which are orthogonal with respect to the weight function $(1-x^2)^{-\frac{1}{2}}$ on the interval [-1,1].

Differentiating ψ_1, we see that if the modulus of the rth derivative of ϕ is bounded, then the rth derivative $\psi_{1,\theta}^{(r)}$ is also bounded in the modulus.

As before we obtain

$$|\phi(x) - u_n(\phi,x)| \leq \left\{2 + \frac{\pi}{2} + \ln\left[\frac{2}{\pi}(2n+1)\right]\right\} \frac{\max_\theta |\psi_{1,\theta}^{(r)}(t)|}{(n+1)^r} . \tag{10.49}$$

If we had written $\psi_2(t) = \sin\theta\, \phi(\cos\theta)$, $t = e^{i\theta}$ then, since $\psi_2(\bar{t}) = -\psi_2(t)$, we would have obtained

$$\begin{aligned} &u_n(\psi_2,t) \equiv \sum_{k=-n}^{n} a_k t^k = \sum_{k=1}^{n} b_k \sin k\theta \equiv v_n(\phi,x) \sin\theta \\ &\text{and} \quad v_n(\phi,x) = \sum_{k=1}^{n} b_k T_{k-1}^{(2)}, \; b_k = \frac{4}{(2n+1)} \sum_{j=1}^{n} \phi(x_j) \sin\left(\frac{2\pi j}{2n+1}\right) \sin\left(\frac{2\pi jk}{2n+1}\right) , \end{aligned} \tag{10.50}$$

where

$$T_{k-1}^{(2)} = \frac{\sin k\theta}{\sin\theta} = \frac{[x + i(1-x^2)^{\frac{1}{2}}]^k - [x - i(1-x^2)^{\frac{1}{2}}]^k}{2i(1 - x^2)^{\frac{1}{2}}} , \tag{10.51}$$

are the Chebyshev polynomials of the second kind which are orthogonal with respect to the weight function $(1-x^2)^{\frac{1}{2}}$ on the interval $[-1,1]$.

By the change of variables

$$u = \frac{1+x}{1-x} , \quad x = \frac{u-1}{u+1} , \tag{10.52}$$

we reduce the function ϕ defined on the semi-axis $[0,\infty)$ to the function $\phi_1(x) = \phi[(1+x)/(1-x)]$ which is defined on $[-1,1]$, and as a consequence we obtain results concerning the approximation of ϕ, analogous to those given above. In a similar way we may examine other domains of definition for ϕ.

§10.3 Examples of Complete Systems of Functions

Previously we have examined the system of functions $1, t, t^{-1}, t^2, t^{-2}, \ldots, t^n, t^{-n}, \ldots$, which is complete in both $C(\gamma)$ and $L_2(\gamma)$. But for particular functions ϕ, approximation by other complete systems may be more effective.

On the unit circle γ the system of functions

$$\frac{1}{t-z_1^+} , \frac{1}{t-z_1^-} , \ldots , \frac{1}{t-z_k^+} , \frac{1}{t-z_k^-} , \ldots \tag{10.53}$$

is complete in $C(\gamma)$ and $L_2(\gamma)$ if and only if

$$\sum_{k=1}^{\infty} (1 - |z_k^-|) = \infty, \quad \sum_{k=1}^{\infty} (1 - 1/|z_k^+|) = \infty , \tag{10.54}$$

where $|z_k^-| < 1$, $|z_k^+| > 1$.

A proof of this statement can be found in [2, App.A].

The following system of rational functions has been constructed by Walsh [135]:

$$\begin{aligned} \Phi_k^+(z) &= \left(1 - \frac{1}{|z_k^+|^2}\right)^{\frac{1}{2}} \left(\frac{z_k^+}{z_k^+ - z}\right) \prod_{s=1}^{k} \frac{z_{s-1}^+ z - \overline{(z_{s-1}^+)}^{-1}}{z_{s-1}^+ - z} ; \\ \Phi_k^-(z) &= (1 - |z_k^-|^2)^{\frac{1}{2}} \left(\frac{1}{z - z_k^-}\right) \prod_{s=1}^{k} \frac{1 - z|z_{s-1}^-|^2}{z - z_{s-1}^-} , \end{aligned} \tag{10.55}$$

$$\text{where} \quad \frac{z_0^+ z - (\bar{z}_0^+)^{-1}}{z_0^+ - z} = \frac{1 - |z_0^-|^2 z}{z - z_0^-} = 1 .$$

This system has the properties:

$$\begin{aligned} &\frac{1}{2\pi i} \int_\gamma t^{-1} \Phi_k^+(t) \overline{\Phi_j^+(t)} \, dt = \begin{cases} 0 , & k \neq j , \\ 1 , & k = j ; \end{cases} \\ &\frac{1}{2\pi i} \int_\gamma t^{-1} \Phi_k^-(t) \overline{\Phi_j^-(t)} \, dt = \begin{cases} 0 , & k \neq j , \\ 1 , & k = j ; \end{cases} \\ &\int_\gamma t^{-1} \Phi_k^+(t) \overline{\Phi_j^-(t)} \, dt = \int_\gamma t^{-1} \Phi_k^-(t) \overline{\Phi_j^+(t)} \, dt = 0. \end{aligned} \tag{10.56}$$

In Akhieser [2, App.A], it has been proved that for the completeness of the sequence of functions

$$\frac{1}{x-c_1^+} , \frac{1}{x-c_1^-} , \ldots , \frac{1}{x-c_k^+} , \frac{1}{x-c_k^-} , \ldots \tag{10.57}$$

$$(\text{Im } c_k^+ < 0 , \quad \text{Im } c_k^- > 0) ,$$

in $L_2(-\infty,\infty)$, it is necessary and sufficient that

$$(10.58)\qquad \sum_{k=1}^{\infty} \frac{\operatorname{Im} c_k^-}{1+|c_k^-|^2} = \infty, \quad \sum_{k=1}^{\infty} \frac{\operatorname{Im} c_k^+}{1+|c_k^+|^2} = -\infty .$$

From elementary properties of the Fourier transform we have that

$$\int_0^{\infty} e^{-ic_k^+ u}\, e^{iux}\, du = \frac{i}{x-c_k^+}, \quad \int_{-\infty}^{0} e^{-ic_k^- u}\, e^{iux}\, du = \frac{-i}{x-c_k^-},$$

$$(10.59)\qquad \int_{-\infty}^{\infty} \frac{e^{-iux}\, dx}{x-c_k^+} = \psi_k^+(u) = \begin{cases} -2\pi i \exp(-ic_k^+ u), & u > 0 \\ 0, & u < 0; \end{cases}$$

$$\int_{-\infty}^{\infty} \frac{e^{-iux}\, dx}{x-c_k^-} = \psi_k^-(u) = \begin{cases} 0, & u > 0, \\ 2\pi i \exp(-ic_k^- u), & u < 0. \end{cases}$$

Hence, from the well known results (see [133]),

$$\|\phi(x)\|_{L_2(-\infty,\infty)} = \|f(u)\|_{L_2(-\infty,\infty)},$$

$$(10.60)\qquad \text{where } \phi(x) = \int_{-\infty}^{\infty} f(u)\, e^{iux}\, du$$

$$\text{and} \quad f(u) = \frac{1}{2\pi} \int_{-\infty}^{\infty} \phi(x)\, e^{-iux}\, dx,$$

we conclude that the system of functions $\psi_1^+, \psi_1^-, \ldots, \psi_k^+, \psi_k^-, \ldots$ is complete in $L_2(-\infty,\infty)$ if and only if the conditions (10.58) are satisfied.

Assuming in particular that $c_k^+ = -ik$, we have that $e^{-u}, e^{-2u}, \ldots, e^{-ku}, \ldots$ is complete in $L_2(0,\infty)$. A complete system in $L_2(-\infty,\infty)$ is given by $e^{-u^2}, ue^{-u^2}, \ldots, u^k e^{-u^2}, \ldots$; and in $L_2(0,\infty)$, another complete system is given by the functions $e^{-u}, ue^{-u}, \ldots, u^k e^{-u}, \ldots$; see [133].

We note from the geometric theory of functions (see [40]) that the system of functions $1, t, t^{-1}, t^2, t^{-2}, \ldots, t^n, t^{-n}, \ldots$ is complete in $L_2(\Gamma)$, where Γ is an arbitrary rectifiable Jordan curve which does not pass through the coordinate origin.

If Γ is an arbitrary closed bounded set of the complex plane then in $C(\Gamma)$ a complete system of functions is given by $1, t, \bar{t}, t^2, \bar{t}^2, \ldots, t^n, \bar{t}^n, \ldots$. This follows from Weierstrass' theorem [110] that any continuous function on a bounded plane rectangle can be approximated arbitrarily closely by polynomials in x and y, and also from the fact that in Γ, $x = (t+\bar{t})/2$ and $y = (t-\bar{t})/2i$.

§11 THE APPROXIMATE EVALUATION OF SINGULAR INTEGRALS AND CAUCHY INTEGRALS

§11.1 Methods Based on the Approximation of the Density Function

Let γ denote the unit circle, and let ϕ be a sufficiently smooth[2] function defined on γ. We construct the interpolation polynomial $u_n(\phi,t) = \sum_{k=-n}^{n} a_k t^k$ (see (10.15)), and assume that

$$S\phi \equiv \frac{1}{\pi i}\int_\gamma \frac{\phi(\tau)d\tau}{\tau - t} \approx \frac{1}{\pi i}\int_\gamma \frac{u_n(\phi,\tau)d\tau}{\tau - t} \equiv Su_n(\phi) , \tag{11.1}$$

(see [61, 25]).

From the definition of a singular integral, we have that

$$S\tau^k = \frac{1}{\pi i}\int_\gamma \frac{\tau^k d\tau}{\tau - t} = \begin{cases} t^k, & k \geq 0 , \\ -\,t^k, & k < 0 . \end{cases} \tag{11.2}$$

Hence

$$Su_n(\phi) = \sum_{k=0}^{n} a_k t^k - \sum_{k=-n}^{-1} a_k t^k . \tag{11.3}$$

If in equation (11.3) we substitute the values of a_k as given by equation (10.15), we find

$$Su_n(\phi) = \frac{1}{(2n+1)} \sum_{j=-n}^{n} \phi_j \left\{ 1 + \frac{2i\,\sin[n(\theta-\theta_j)/2]\sin[(n+1)(\theta-\theta_j)/2]}{\sin[(\theta-\theta_j)/2]} \right\} \tag{11.4}$$

where

$$\phi_j = \phi(t_j),\quad t_j = \exp(i\theta_j),\quad \theta_j = 2\pi j/(2n+1),\quad t = \exp(i\theta) .$$

Theorem 11.1. *We have*

$$|S\phi - Su_n(\phi)| \leq \|S\phi - Su_n(\phi)\|_{W(\gamma)} \leq \left\{ 2 + \frac{4}{\pi} + \frac{4}{\pi}\,\ln\left[\frac{2}{\pi}(2n+1)\right] \right\} \rho_n(\phi), \tag{11.5}$$

where $\rho_n(\phi)$ *is defined by* (10.14).

Proof. From the definition of the space $W(\gamma)$ (see §7), we have

$$|S\phi - Su_n(\phi)| \leq \|S\phi - Su_n(\phi)\|_{W(\gamma)} = \|\phi - u_n(\phi)\|_{W(\gamma)} ,$$

[2] The required degree of smoothness will be as indicated in the subsequent arguments.

and the result follows immediately from (10.23). #

From (11.5) and (10.13) it follows in particular that

$$(11.6) \qquad |S\phi - Su_n(\phi)| \le \left\{\pi + 2 + 2\ \ln\left[\tfrac{2}{\pi}(2n+1)\right]\right\} (n+1)^{-r} \|\phi_\theta^{(r)}\|_{W(\gamma)} .$$

If ϕ is analytic in the annulus between the circles $|t| = R_1 < 1$ and $|t| = R_2 > 1$, then from (11.5) and (10.35) we obtain

$$(11.7) \qquad |S\phi - Su_n(\phi)| \le \left\{2 + \frac{4}{\pi} + \frac{4}{\pi}\ln\left[\frac{2}{\pi}(2n+1)\right]\right\} \times$$

$$\times \left\{\frac{R_2^{-n-1}}{1-R_2^{-1}} \cdot \frac{1}{2\pi}\int_{-\pi}^{\pi} |\phi(R_2 e^{i\theta})|\,d\theta + \frac{R_1^{n+1}}{1-R_1} \cdot \frac{1}{2\pi}\int_{-\pi}^{\pi} |\phi(R_1 e^{i\theta})|\,d\theta\right\} .$$

From (10.37), (11.6) and (11.7), we may also obtain estimates for $|(S\phi)^{(\nu)} - (Su_n(\phi))^{(\nu)}|$.

For the generalized Fejér polynomials, it is obvious that $S\sigma_{n+1}(\phi) = \sigma_{n+1}(S\phi)$, and hence (see (10.29)),

$$(11.8) \qquad |S\phi - S\sigma_{n+1}(\phi)| \le \left[1 + \frac{4}{\pi} + \frac{2}{\pi}\ln\left(\frac{2-\mu}{\mu}\right)\right] E_{(n+1)(1-\mu)}(S\phi,\gamma) .$$

From (10.7), on putting $\mu = 1/(r+1)$, we obtain

$$(11.9) \qquad |S\phi - S\sigma_{n+1}(\phi)| \le e[\pi/2 + 2 + \ln(2r+1)](n+1)^{-r} \max_\theta |\phi_\theta^{(r)}(t)| ,$$

$(r = 1,2,\ldots)$. Again, if we put $\mu = 1/(n+1)$, we obtain

$$(11.10) \qquad |S\phi - Ss_n(\phi)| \le [D_1 + 1 + D_2 \ln(n+1)]n^{-r} \max_\theta |\phi_\theta^{(r)}(t)| ,$$

$(r = 1,2,\ldots)$, $D_1 < 2$, $D_2 < 1$, (cf. equn. 10.31).

Since

$$\rho_n(\phi) \le \|\phi - \sigma_{n+1}(\phi)\|_{W(\gamma)} \le \|\phi - \sigma_{n+1}(\phi)\|_{C(\gamma)} + \|S\phi - S\sigma_{n+1}(\phi)\|_{C(\gamma)} ,$$

then

$$(11.11) \qquad |S\phi - Su_n(\phi)| \le \left\{2 + \frac{4}{\pi} + \frac{4}{\pi}\ln\left[\frac{2}{\pi}(2n+1)\right]\right\} \times$$

$$\times\ 2e[\pi/2 + 2 + \ln(2r+1)](n+1)^{-r} \max_\theta |\phi_\theta^{(r)}(t)| .$$

Theorem 11.2. *The estimate*

$$(11.12) \qquad \|S\phi - Su_n(\phi)\|_{L_2(\gamma)} = \|\phi - u_n(\phi)\|_{L_2(\gamma)} \le 2\ E_n(\phi,\gamma) ,$$

is valid.

The proof easily follows from (10.26) and the fact that $\|S\phi\|_{L_2(\gamma)} = \|\phi\|_{L_2(\gamma)}$. Applying known estimates for $E_n(\phi,\gamma)$ we can, as above, obtain a series of useful estimates for $\|S\phi - Su_n(\phi)\|_{L_2(\gamma)}$.

For the Cauchy integral $\frac{1}{2\pi i}\int_\gamma \frac{\phi(\tau)d\tau}{\tau - z}$, we assume that

$$\frac{1}{2\pi i}\int_\gamma \frac{\phi(\tau)d\tau}{\tau-z} = \begin{cases} \Phi^+(z) \approx \sum_{k=0}^{n} a_k z^k, & |z| \le 1, \\ \Phi^-(z) \approx -\sum_{k=-n}^{-1} a_k z^k, & |z| \ge 1, \end{cases} \tag{11.13}$$

where the a_k are determined by (10.15).

Theorem 11.3. *The estimate*

$$\max_{|z|\le 1} |\Phi^+(z) - \sum_{k=0}^{n} a_k z^k| + \max_{|z|\ge 1} |\Phi^-(z) - \sum_{k=-n}^{-1} a_k z^k| \le \le \left\{ 2 + \frac{4}{\pi} + \frac{4}{\pi} \ln\left[\frac{2}{\pi}(2n+1)\right]\right\} \rho_n(\phi), \tag{11.14}$$

is valid.

Proof. The result follows from the maximum-modulus theorem for analytic functions and Theorem 11.1. #

From the approximation to ϕ by a finite number of terms of the Fourier-Laurent series $\sum_{k=-\infty}^{\infty} c_k t^k$, where $c_k = \frac{1}{2\pi i}\int_\gamma \tau^{-k-1}\phi(\tau)d\tau$, we obtain the following result.

Theorem 11.4. *The estimates*

$$\begin{aligned} &|\Phi^+(z) - \sum_{k=0}^{n} c_k z^k| \le r^{n+1} \max_{t\in\gamma} |\Phi^+(t) - \sum_{k=0}^{n} c_k t^k|, \quad |z| \le r \le 1, \\ &|\Phi^-(z) - \sum_{k=-n}^{-1} c_k z^k| \le r^{n+1} \max_{t\in\gamma} |\Phi^-(t) - \sum_{k=-n}^{-1} c_k t^k|, \quad |z| \ge \frac{1}{r} \ge 1, \end{aligned} \tag{11.15}$$

are valid.

Proof. From the maximum-modulus theorem, we obtain

$$\left| \frac{\Phi^+(z) - \sum_{k=0}^{n} c_k z^k}{z^{n+1}} \right| \le \max_{t\in\gamma} |\Phi^+(t) - \sum_{k=0}^{n} c_k t^k|, \quad |z| \le 1,$$

so that for $|z| \leq r$, the first inequality of (11.15) follows. The second inequality may be proved similarly on writing $z = 1/u$.

Finally we consider the case in which $\phi^{(\nu)} \in C(\gamma; d_1, d_2, \ldots, d_m)$, $(\nu = 0,1,\ldots,r)$, i.e. we consider functions whose derivatives of order ν are continuous everywhere on γ except at the points $d_1, d_2, \ldots, d_m \in \gamma$ at which $\phi^{(\nu)}$ can have discontinuities of the first kind. We construct the Hermite polynomials R_k using the conditions

(11.16)
$$R_k^{(\nu)}(c) = \phi^{(\nu)}(c) \quad (\nu = 0,1,\ldots,r) \quad ;$$
$$c = d_k,\ d_{k+1},\ (d_{m+1} = d_1),\ (k = 1,2,\ldots,m) .$$

It is known, (see [41]) that

(11.17)
$$R_k(t) = (t-d_k)^{r+1} \sum_{\nu=0}^{r} \phi^{(\nu)}(d_{k+1}) \frac{(t-d_{k+1})^\nu}{\nu!} \left\{\frac{1}{(t-d_k)^{r+1}}\right\}^{(r-\nu)}_{(d_{k+1})} +$$
$$+ (t-d_{k+1})^{r+1} \sum_{\nu=0}^{r} \phi^{(\nu)}(d_k) \frac{(t-d_k)^\nu}{\nu!} \left\{\frac{1}{(t-d_{k+1})^{r+1}}\right\}^{(r-\nu)}_{(d_k)} ,$$

where $\left\{\frac{1}{(t-c)^{r+1}}\right\}^{(r-\nu)}_{(d)}$ denotes the first $r-\nu+1$ terms in the Taylor series expansion of the function $1/(t-c)^{r+1}$ in a neighbourhood of the point d.

The functions

(11.18)
$$\phi_k(t) = \begin{cases} \phi(t) - R_k(t) , & t \in \widetilde{d_k d}_{k+1} , \\ 0 , & t \notin \widetilde{d_k d}_{k+1} , \end{cases}$$

have derivatives with respect to t, (and with respect to θ), up to the rth order which are bounded in modulus at all points of γ. Hence we can represent $S\phi$ in the form

(11.19)
$$S\phi = \frac{1}{\pi i} \sum_{k=1}^{m} \int_{\widetilde{d_k d}_{k+1}} \frac{\phi(\tau)d\tau}{\tau - t}$$
$$= \sum_{k=1}^{m} S\phi_k + \frac{1}{\pi i} \sum_{k=1}^{m} \int_{\widetilde{d_k d}_{k+1}} \frac{R_k(\tau)d\tau}{\tau - t} ,$$

and reduce the problem of the evaluation of $S\phi$ to the evaluation of a singular integral for sufficiently smooth functions ϕ_k, and to the calculation of integrals which can be evaluated exactly.

Singular integrals along curves which are different from γ can, in many cases, be reduced to singular integrals along γ by means of a change of variables. We shall illustrate this by considering the integrals

$$\sigma\phi = \frac{(1-x^2)^{\frac{1}{2}}}{\pi}\int_{-1}^{1}\frac{\phi(y)dy}{(1-y^2)^{\frac{1}{2}}(y-x)} \quad \text{and} \quad S\phi = \frac{1}{\pi i}\int_{-1}^{1}\frac{\phi(y)dy}{y-x}, \tag{11.20}$$

which are important in applications (see [119]). A straightforward calculation shows that

$$\sigma\phi = \frac{1}{i}\left\{\frac{1}{\pi i}\int_{\gamma}\frac{\psi_1(\tau)d\tau}{\tau - t} - \frac{1}{2\pi i}\int_{\gamma}\frac{\psi_1(\tau)d\tau}{\tau}\right\},$$

where $t = x+i(1-x^2)^{\frac{1}{2}}$; $\tau = e^{i\theta}$, $-\pi < \theta \leq \pi$; $\psi_1(\tau) = \phi(\cos\theta)$; and

$$S\phi = \frac{1}{(1-x^2)^{\frac{1}{2}}}\cdot\left\{-\frac{1}{\pi i}\int_{\gamma}\frac{\psi_2(\tau)d\tau}{\tau - t} + \frac{1}{2\pi i}\int_{\gamma}\frac{\psi_2(\tau)d\tau}{\tau}\right\}, \tag{11.21}$$

where
$$\psi_2(t) = \begin{cases} \phi(\cos\theta)\sin\theta, & \theta \in [0,\pi], \\ -\phi(\cos\theta)\sin\theta, & \theta \in [-\pi,0]. \end{cases}$$

Let us summarize the proposed method. If ϕ is piecewise differentiable on γ, the problem of an approximate calculation of $S\phi$ reduces to:

1) the construction of $u_n(\phi,t)$ (see 10.15) from the given values $\phi(t_j)$, where $t_j = \exp[2\pi ij/(2n+1)]$, for $j = 0,\pm1,\pm2,\ldots,\pm n$;
2) the construction of the Hermite polynomial (11.17);
3) numerical differentiation for the estimation of $\|\phi_\theta^{(r)}\|_{C(\gamma)}$.

In addition, on the compact set where $\|\phi_\theta^{(r)}\|_{C(\gamma)} \leq$ constant for every given $\epsilon > 0$, it is not difficult to obtain an $n(\epsilon)$ such that $|S\phi - Su_n(\phi)| \leq \epsilon$ for all $n > n(\epsilon)$. Although the set of functions $\{Su_n(\phi,t)\}$ is not the most economized ϵ-net (see [23]), it has the significant advantage of being readily calculated from a table of values $\{\phi(t_j)\}$, and allows the value of $S\phi$ at any point $t \in \gamma$ to be obtained to within any prescribed accuracy.

§11.2 Quadrature Formulae

We shall now consider the approximate calculation of a singular integral when the only assumption we shall make concerning the density function of the integrand is that it belongs to a certain Hölder class $H(\alpha,\Gamma)$, where $0 < \alpha \leq 1$.

Suppose we are given an arc Γ, of length L, and end points a and b. We divide Γ into segments by the points $a = t_1, t_2, \ldots, t_{n+1}, t_{n+2} = b$, so that $|t_{j+1} - t_j| = |t_j - t_{j-1}|$, $j \neq n+1$, $|t_{n+2} - t_{n+1}| \leq |t_{n+1} - t_n|$. We let s denote the arc abscissa of the point t; we assume $M = \max\limits_{t,t_0 \in \Gamma} |s-s_0|/|t-t_0|$ and assume that n is chosen so that $|t_{j+1} - t_j| \geq L/4n$, $j \neq n+1$, $|t_{j+1} - t_j| \leq |t_k - t_j|$, $k \neq j$ and that on the arc (t_j, t_{j+1}) we have $|s_j - s| \leq 2|t_j - t| \leq 4|t_j - t_{j+1}|$.

<u>Theorem 11.5.</u> *Under the above assumptions, for every function $\phi \in H(\alpha,\Gamma)$, we have*

$$\Delta = \Bigg| \frac{1}{\pi i} \int_\Gamma \frac{\phi(t)}{t-t_k}\, dt - \frac{1}{\pi i} \phi(t_k) \ln\left(\frac{b-t_k}{t_k-a}\right) - \sum_{j=1}^{n+1}{}' \left(\frac{\phi(t_j) - \phi(t_k)}{t_j - t_k}\right) \frac{(t_{j+1} - t_j)}{\pi i} \Bigg| \tag{11.22}$$

$$\leq \frac{8}{\pi} H_\phi^\alpha \left(\frac{1}{\alpha} + M \ln(2n)\right) (L/n)^\alpha ,$$

where $\sum'$ denotes summation over all $j \neq k$, $k-1$; $k = 1,2,\ldots,n+1$.

<u>Proof.</u> We rewrite the left hand side of (11.22) in the form

$$\Delta = \frac{1}{\pi} \left| \sum_{j=1}^{n+1}{}' \int_{t_j}^{t_{j+1}} \left(\frac{\phi-\phi_k}{t-t_k} - \frac{\phi_j-\phi_k}{t_j-t_k} \right) dt + \int_{t_{k-1}}^{t_{k+1}} \frac{\phi-\phi_k}{t-t_k}\, dt \right| ,$$

and transform the expression in the round brackets as follows:

$$\frac{\phi-\phi_k}{t-t_k} - \frac{\phi_j-\phi_k}{t_j-t_k} = \frac{\phi-\phi_j}{t-t_k} + \frac{(\phi_j-\phi_k)(t_j-t)}{(t_j-t_k)(t-t_k)} .$$

We then find

$$\frac{1}{\pi} \left| \sum_{j=1}^{n+1}{}' \int_{t_j}^{t_{j+1}} \left(\frac{\phi-\phi_k}{t-t_k} - \frac{\phi_j-\phi_k}{t_j-t_k} \right) dt \right|$$

$$\leq \frac{1}{\pi} \sum_{j=1}^{n+1}{}' \left| \int_{t_j}^{t_{j+1}} \left(\frac{|\phi-\phi_j|}{|t-t_k|} + \frac{|\phi_j-\phi_k|\; 2|t_j-t_{j+1}|^\alpha}{|t_j-t_k|^\alpha\; |t-t_k|} \right) |dt| \right|$$

$$\leq \frac{8}{\pi} H_\phi^\alpha M\left(\frac{L}{n}\right)^\alpha \int_{s_k+L/4n}^{s_k+L/2} \frac{ds}{s-s_k} = \frac{8}{\pi} H_\phi^\alpha M\left(\frac{L}{n}\right)^\alpha \ln(2n) ;$$

and
$$\left| \frac{1}{\pi i} \int_{t_{k-1}}^{t_{k+1}} \frac{\phi-\phi_k}{t-t_k}\, dt \right| \leq \frac{8}{\pi} H_\phi^\alpha \frac{1}{\alpha} \left(\frac{L}{n}\right)^\alpha .$$

From these two inequalities, the theorem follows. #

It should be noted that for a given accuracy in the evaluation of $\phi(t_j)$, the accuracy in the calculation of $(\phi(t_j) - \phi(t_k))/(t_j - t_k)$ diminishes as n increases. In practice, n should be selected in such way that the accuracy of the estimate (11.22) is the same as the accuracy of the input data.

Let us now consider the evaluation of integrals of the form

$$\frac{1}{\pi i} \int_\Gamma \frac{\phi(t)dt}{(t-c)^\gamma\,(t-t_0)} ,$$

where c is either a or b, $\gamma = \lambda+i\mu \neq 0$, $0 \leq \lambda < 1$; $t_0 \neq c$; the branch of $(t-c)^\gamma$ is chosen appropriately. For simplicity let us restrict attention to the case when $\Gamma \equiv [-1,1]$, $t \equiv x$.

Theorem 11.6. *Let $\phi \in H(\alpha)$ on $[-1,1]$. Let $\sigma_n(x_0)$ denote an open neighbourhood of x_0, with centre at x_0 and length $2/n$, n being chosen so that $|x_0-c| > 3/n$ where $c = \pm 1$. Divide $[-1,1]$ into $2n$ equal parts by the points $-1 = x_1 < x_2 < \ldots < x_{2n+1} = 1$. Under these conditions we have,*

$$\int_{-1}^{1} \frac{\phi(x)dx}{(x-c)^\gamma\,(x-x_0)} = \phi(x_0) \int_{-1}^{1} \frac{dx}{(x-c)^\gamma\,(x-x_0)} + \tag{11.23}$$
$$+ \sum_{j=1}^{2n}{}' \left(\frac{\phi(x_j) - \phi(x_0)}{x_j - x_0}\right) \left(\frac{(x_{j+1}-c)^{1-\gamma} - (x_j-c)^{1-\gamma}}{1-\gamma}\right) + \Delta_n ,$$

and

$$\int_{-1}^{1} \frac{dx}{(x-c)^\gamma\,(x-x_0)} = \pm \frac{\pi \cot(\pi\gamma)}{(x_0-c)^\gamma} \pm \sum_{s=0}^{\infty} \frac{(\pm 1)^{s+\gamma}}{(s+\gamma)2^{s+\gamma}} (x_0-c)^s , \tag{11.24}$$

where $\sum'$ denotes summation over all j such that $x_j, x_{j+1} \notin \sigma_n(x_0)$. The symbol "+" or "-" corresponds to $c = -1$ or $+1$ respectively, and

$$|\Delta_n| < \frac{6H_\phi^\alpha}{|x_0-c|^\lambda} \left\{ \frac{1}{\alpha} + \frac{1}{1-\lambda} + \ln\left[\frac{n^2(1-x_0^2)}{2}\right] \right\} n^{-\alpha} . \tag{11.25}$$

Proof. We shall show first of all that, under the conditions of the theorem, the following inequalities are valid for any $f \in H(\alpha, [-1,1])$:

(11.26)
$$\Delta_1 = \frac{1}{\pi} \int_{\sigma_n(x_0)} \left| \frac{f(x) - f(x_0)}{(x-c)^\gamma (x-x_0)} \right| dx < \frac{4 H_f^\alpha}{\pi\alpha n^\alpha |x_0-c|^\lambda} ,$$

(11.27)
$$\Delta_2 = \frac{1}{\pi} \int_{[-1,1]-\sigma_n(x_0)} \left| \frac{f(x)}{(x-c)^\gamma (x-x_0)} \right| dx$$
$$< \frac{2 M_f}{\pi |x_0-c|^\lambda} \left\{ \frac{1}{1-\lambda} + \ln \left[\frac{n^2(1-x_0^2)}{2} \right] \right\} .$$

Let $c = -1$. For $x \in \sigma_n(x_0)$, $x+1 > x_0+1-1/n > (1+x_0)/2$ so that

$$\Delta_1 < \frac{2^\lambda H_f^\alpha}{\pi} \int_{x_0-1/n}^{x_0+1/n} \frac{|x-x_0|^{\alpha-1}}{(x_0+1)^\lambda} dx = \frac{2^{\lambda+1} H_f^\alpha}{\pi\alpha n^\alpha (x_0+1)^\lambda} ,$$

which proves (11.26) when $c = -1$. It is obvious that

$$\Delta_2 < \frac{M_f}{\pi} \left[\int_{-1}^{x_0-1/n} \frac{dx}{(x+1)^\lambda (x_0-x)} + \int_{x_0+1/n}^{1} \frac{dx}{(x+1)^\lambda (x-x_0)} \right] = \frac{M_f}{\pi} (G_1+G_2) ,$$

where $M_f = \max_{x\in[-1,1]} |f(x)|$. We rewrite G_1 as

$$G_1 = \int_{-1}^{\xi} \frac{dx}{(x+1)^\lambda (x_0-x)} + \int_{\xi}^{x_0-1/n} \frac{dx}{(x+1)^\lambda (x_0-x)} ,$$

and estimate the first integral using $x_0-x > x_0-\xi$ and the second integral using $x+1 > \xi+1$. Then

$$G_1 < \frac{1}{(x_0-\xi)} \cdot \frac{(\xi+1)^{1-\lambda}}{(1-\lambda)} + \frac{1}{(\xi+1)^\lambda} \ln \left[\frac{1}{n(x_0-\xi)} \right] .$$

Assuming $\xi+1 = x_0-\xi = (x_0+1)/2$, we find that

$$G_1 < \frac{2^\lambda}{(x_0+1)^\lambda} \left\{ \frac{1}{1-\lambda} + \ln \left[\frac{n(x_0+1)}{2} \right] \right\} .$$

G_2 can be easily estimated to give

$$G_2 < \int_{x_0+1/n}^{1} \frac{dx}{(x_0+1)^\lambda\ (x-x_0)} < \frac{1}{(x_0+1)^\lambda}\ \ln[n(1-x_0)]\ .$$

Combining these inequalities for G_1 and G_2, we obtain (11.27) when $c = -1$. For $c = 1$, the proof is analogous.

To prove the theorem, it is obvious that we shall also need estimates for the quantities

$$I_1 = \frac{1}{\pi i} \int_{\tilde{\sigma}_n(x_0)} \frac{\phi(x) - \phi(x_0)}{(x-c)^\gamma\ (x-x_0)}\ dx\ ,$$

$$I_2 = \frac{1}{\pi i} {\sum_{j=1}^{2n}}' \int_{x_j}^{x_{j+1}} \left[\frac{\phi(x) - \phi(x_0)}{(x-c)^\gamma\ (x-x_0)} - \frac{\phi(x_j) - \phi(x_0)}{(x-c)^\gamma\ (x_j-x_0)} \right] dx\ ,$$

where $\tilde{\sigma}_n(x_0)$ denotes that part of the interval $[-1,1]$ which remains if we delete all intervals $(x_j,\ x_{j+1})$ included under the summation $\sum'$.

Applying the same methods that were used in proving (11.26) and (11.27), we find

$$|I_1| < \frac{H_\phi^\alpha}{\pi} \int_{\tilde{\sigma}_n(x_0)} \frac{|x-x_0|^{\alpha-1}}{|x-c|^\lambda}\ dx < \frac{H_\phi^\alpha}{\pi} \int_{x_0-3/2n}^{x_0+3/2n} \frac{|x-x_0|^{\alpha-1}\ dx}{|x-c|^\lambda}$$

$$< \frac{6\ H_\phi^\alpha}{\pi\alpha n^\alpha\ |x_0-c|^\lambda}\ ,$$

$$|I_2| < \frac{1}{\pi} \sum_{j=1}^{2n} \frac{|\phi(x) - \phi(x_j)|}{|x-c|^\lambda\ |x-x_0|}\ dx + \frac{1}{\pi} \sum_{j=1}^{2n} \int_{x_j}^{x_{j+1}} \frac{|\phi(x_j)-\phi(x_0)|\,|x_j-x_0|}{|x-c|^\lambda\ |x_j-x_0|\ |x-x_0|}\ dx$$

$$< \frac{2\ H_\phi^\alpha}{\pi} \int_{[-1,1]-\sigma_n(x_0)} \frac{dx}{n^\alpha |x-c|^\lambda\ |x-x_0|}$$

$$< \frac{4\ H_\phi^\alpha}{\pi |x_0-c|^\lambda\ n^\alpha} \left\{ \frac{1}{1-\lambda} + \ln\left[\frac{n^2(1-x_0^2)}{2}\right] \right\}.$$

Hence, since $|\Delta_n| \le |I_1| + |I_2|$, (11.25) follows.

It now remains to prove the series representation (11.24). To do this we introduce the integrals

$$I(x_0) = \frac{1}{\pi i} \int_{-1}^{1} \frac{dx}{(x-c)^{\gamma}\,(x-x_0)} \;,\quad \Psi(z) = \frac{1}{2\pi i} \int_{-1}^{1} \frac{dx}{(x-c)^{\gamma}\,(x-z)} \;,$$

and the function $\chi(z) = (z-c)^{-\gamma}$ which takes the value of $(x-c)^{-\gamma}$ on the left-hand side of the interval [-1,1]. From the Sokhotski-Plemelj formulae, we obtain

$$\Psi^+(x_0) - \Psi^-(x_0) = (x_0-c)^{-\gamma},\ \Psi^+(x_0) + \Psi^-(x_0) = I(x_0).$$

But

$$\chi^+(x_0) - \chi^-(x_0) = (x_0-c)^{-\gamma} - e^{-2\pi i\gamma}\,(x_0-c)^{-\gamma} \;,$$

hence

$$\Psi^+(x_0) - \frac{\chi^+(x_0)}{1-\exp(-2\pi i\gamma)} = \Psi^-(x_0) - \frac{\chi^-(x_0)}{1-\exp(-2\pi i\gamma)} \;.$$

On using properties of analytic continuation, we have that

$$\text{(11.28)} \qquad \Psi(z) = \frac{(z-c)^{-\gamma}}{1-\exp(-2\pi i\gamma)} + \sum_{s=0}^{\infty} A_s (z-c)^s \;,\ |z-c| \leq 2 \;.$$

If C denotes the circle of radius 2 with centre at c, then

$$2\pi i A_s = \frac{1}{1-\exp(-2\pi i\gamma)} \int_C \frac{(z-c)^{-\gamma}\,dz}{(z-c)^{s+1}} +$$

$$+ \frac{1}{2\pi i} \int_{-1}^{1} \left[\frac{1}{(x-c)^{\gamma}} \int_C \frac{dz}{(x-z)(z-c)^{s+1}} \right] dx$$

$$= - \frac{1}{1-\exp(-2\pi i\gamma)} \left[\frac{(z-c)^{-\gamma-s}}{\gamma + s} \right]_C = \pm \frac{(\pm 1)^{s+\gamma}\, 2^{-s-\gamma}}{\gamma + s} \;,$$

where "+" corresponds to $c = -1$ and "-" corresponds to $c = +1$. Noting that

$$\text{(11.29)} \qquad I(x_0) = \Psi^+(x_0) + \Psi^-(x_0) = \frac{1+\exp(-2\pi i\gamma)}{1-\exp(-2\pi i\gamma)} (x_0-c)^{-\gamma} + $$

$$+ 2 \sum_{s=0}^{\infty} A_s (x_0-c)^s \;,$$

and substituting the above value for A_s, we obtain the required result. This completes the proof of the theorem. #

Similar methods have been given in [89, 162, 126, 167] for the evaluation of other forms of singular integrals. In [126], in addition to an estimate of the type (11.25) (with $\gamma = 0$), an estimate for the Hölder constant of the singular integral is given.

It is not difficult to see that by using similar techniques we can approximately evaluate Cauchy integrals, (see [89]).

§11.3 Some Other Methods

In the case of ordinary integrals higher accuracy quadrature rules can be obtained by using, as approximations to the integrand, piecewise polynomials of degree higher than one. In addition we assume that the integrand is sufficiently smooth. A similar method can be used for singular integrals, see [53].

A.A. Korneychuk has pointed out the essential advantage of using polygonal approximations for the density function, rather than piecewise-constant approximation, in the case when the density has a second derivative the modulus of which is bounded. Let Γ be a smooth simple closed curve, and ϕ be defined on Γ with $|\phi''| < C$. We denote by $'\phi$ the polynomial

$$'\phi(t) = \phi_k \frac{t_{k+1} - t}{t_{k+1} - t_k} + \phi_{k+1} \frac{t - t_k}{t_{k+1} - t_k} , \tag{11.30}$$

$$(t \in \widetilde{t_k, t_{k+1}},\ t_k \in \Gamma),\ \phi_k = \phi(t_k)\ (k = 1,2,\ldots,n,\ t_{n+1} = t_1) ,$$

and by $s(t_1, t_2)$ the length of the shortest arc of Γ joining the points t_1 and t_2. It is not difficult to show that

$$|\phi(t) - {}'\phi(t)| \leq C \max_{1\leq k\leq n} \left\{ \frac{s(t_k, t_{k+1})}{|t_{k+1} - t_k|}\, s^2(t_k, t_{k+1}) \right\} . \tag{11.31}$$

From

$$\begin{aligned} \phi(t) - \phi(t_0) &= \int_{t_0}^{t} \phi'(u)du = \phi'(t_0)(t-t_0) + \int_{t_0}^{t} [\phi'(u) - \phi'(t_0)]du \\ &= \phi'(t_0)(t-t_0) + \int_{t_0}^{t}\int_{t_0}^{u} \phi''(v)dv , \end{aligned} \tag{11.32}$$

where the integration is taken along the curve Γ, we find that for $t \in \widetilde{t_k, t_{k+1}}$,

$$|\phi(t) - {}'\phi(t)| = \left| \frac{\phi(t) - \phi_k}{t_{k+1} - t_k} (t_{k+1} - t) + \frac{\phi(t) - \phi_{k+1}}{t_{k+1} - t_k} (t - t_k) \right|$$

$$= \left| \frac{\phi'(t_{k+1}) - \phi'(t_k)}{t_{k+1} - t_k} (t - t_{k+1})(t - t_k) + \frac{\int_{t_k}^{t} \int_{t_k}^{u} \phi''(v)dv}{t_{k+1} - t_k} (t_{k+1} - t) + \right.$$

$$\left. + \frac{\int_{t_{k+1}}^{t} \int_{t_{k+1}}^{u} \phi''(v)dv}{t_{k+1} - t_k} (t - t_k) \right|$$

$$\leq C \frac{s(t_k, t_{k+1})}{|t_{k+1} - t_k|} \left| s(t_k, t)\, s(t, t_{k+1}) + s^2(t_k, t) + s^2(t, t_{k+1}) \right|$$

$$\leq C \max_k \frac{s(t_k, t_{k+1})}{|t_{k+1} - t_k|} s^2(t_k, t_{k+1}) .$$

On putting

$$(11.33) \qquad S\phi = \frac{1}{\pi i} \int_\Gamma \frac{\phi(\tau)d\tau}{\tau - t} \approx \frac{1}{\pi i} \int_\Gamma \frac{'\phi(\tau)d\tau}{\tau - t} ,$$

and using (11.31) we can obtain, by the same method as was used in the proof of Theorem 11.5, the estimate

$$(11.34) \qquad \left| S\phi - \frac{1}{\pi i} \int_\Gamma \frac{'\phi(\tau)d\tau}{\tau - t} \right| < \frac{\text{constant} \times \ln n}{n^2} ,$$

which has a higher power of n than does (11.22).

The method used for the construction of highest accuracy quadrature formulae [86], can also be applied to singular integrals, see [61, 82]. In fact it is known ([86]) that the formula

$$(11.35) \qquad \frac{1}{2\pi} \int_0^{2\pi} f(\tau)d\theta \approx \frac{1}{n+1} \sum_{j=1}^{n} f(\tau_j) , \ \tau = e^{i\theta}, \ \tau_j = \exp(2\pi i j/(2n+1)) ,$$

has the highest degree of accuracy for all trigonometric polynomials of order n i.e. it is exact for every trigonometric polynomial of order n, but there exists a trigonometric polynomial of order $(n+1)$ for which it is not exact. This implies that the formula

$$S\phi = \frac{1}{\pi i}\int_\gamma \frac{\phi(\tau)-\phi(t)}{\tau - t}\,d\tau + \phi(t) \approx \frac{2}{n+1}\sum_{j=1}^{n+1}\frac{\phi(\tau_j)-\phi(t)}{\tau_j - t}\,\tau_j + \phi(t)\ ,$$

(11.36)

$$\frac{\phi(\tau_j)-\phi(t)}{\tau_j - t} = \phi'(\tau_j)\ ,\ t = \tau_j\ ,$$

is an identity if ϕ is any polynomial of the form $\sum_{k=-n}^{n+1} \alpha_k\tau^k$. Starting from this, it is not difficult to obtain estimates of the errors in (11.36) for different classes of functions ϕ (see [61, 82]).

We shall now consider the use of these approximate formulae in equations (8.37). If a, b and f possess derivatives of order r which are Hölder continuous on Γ, it is more appropriate to rewrite $x^\pm(t)$ in the form

$$x^\pm(t) = \prod_{k=1}^{2s}(d_k-t)^{\lambda_k}\prod_{j=1}^{s}\left(\frac{d_{2j}-t}{d_{2j-1}-t}\right)^{R(t)/2\pi i}\times$$

(11.37)

$$\times\exp\left\{\pm\frac{1}{2}\psi(t) + \frac{1}{2\pi i}\int_\Gamma \frac{\psi(\tau)-R(t)}{\tau - t}\,d\tau\right\}\ ,$$

where R is a Hermite polynomial,(3) which satisfies the conditions

(11.38) $$R^{(\nu)}(d_k) = \psi^{(\nu)}(d_k),\ k = 1,2,\dots,2s,\ \nu = 0,1,2,\dots,r\ .$$

The evaluation of

(11.39) $$\omega(t) = \exp\left\{\pm\frac{1}{2}\psi(t) + \frac{1}{2\pi i}\int_\Gamma \frac{\psi(\tau) - R(t)}{\tau - t}\,dt\right\}$$

presents no great difficulties, and can be done by the methods given above. We obtain a different problem when we evaluate

(11.40) $$\tilde{g}^\pm(t) = \pm\frac{g(t)}{x^+(t)} + \frac{1}{2\pi i}\int_\Gamma \frac{g(\tau)d\tau}{x^+(\tau)(\tau-t)}\ .$$

To evaluate the integral in (11.40), we assume $f(t) = g(t)/\omega(t)$ and write

$$y(t) = \prod_{j=1}^{s}\left(\frac{d_{2j}-t}{d_{2j-1}-t}\right)^{-R(t)/2\pi i}\prod_{k=1}^{2s}(d_k-t)^{-\lambda_k}\ ,$$

$$I(t) = \frac{1}{\pi i}\int_\Gamma \frac{g(\tau)d\tau}{x^+(\tau)(\tau-t)} = I_1(t) + f(t)\,I_2(t)\ ,$$

(3) Instead of R we could also use the piecewise-polynomial function

$$w(t) = R_j(t)\ (t \in \Gamma_j,\ R_j^{(\nu)}(d_i) = \psi^{(\nu)}(d_i),\ i = 1,2,\dots,s;\ \nu = 0,1,2,\dots,r)$$

where

(11.41) $$I_1(t) = \frac{1}{\pi i} \int_\Gamma \frac{y(\tau)[f(\tau)-f(t)]}{\tau - t}\, d\tau,\quad I_2(t) = \frac{1}{\pi i} \int_\Gamma \frac{y(\tau)d\tau}{\tau - t} .$$

The function

(11.42) $$Y(z) = \prod_{k=1}^{2s} (d_k - z)^{-\lambda_k} \prod_{j=1}^{s} \left(\frac{d_{2j}-z}{d_{2j-1}-z} \right)^{-R(z)/2\pi i}$$

is analytic in the plane cut along Γ, and takes the value $y^+(t) = y(t)$ on the left-hand side of Γ, and $y^-(t) = y(t)\exp[R(t)]$ on the right-hand side of Γ. Hence from the Sokhotski-Plemelj formulae we have that

(11.43) $$\Psi(z) = \frac{1}{2\pi i} \int_\Gamma \frac{y(\tau)d\tau}{\tau - z} - \frac{Y(z)}{1-\exp R(z)} ,$$

satisfies the equation $\psi^+(t) = \psi^-(t)$ (for all $t \in \Gamma$ except perhaps at the zeros of R and the points $d_1, d_2, \ldots, d_{2s}$). Consequently

(11.44) $$I_2(t) = y(t) \frac{1+\exp R(t)}{1-\exp R(t)} + 2\psi(t) ,$$

where Ψ is analytic in the entire plane except at the finite number of singular points at which $\Psi(z)$ coincides with $Y(z)/[\exp R(z)-1]$. Hence

(11.45) $$\Psi(t) = \sum_{\nu=1}^{N} \frac{A_\nu}{t-z_\nu} + \sum_{\nu=0}^{\infty} B_\nu t^\nu ,$$

where $z_1, z_2, \ldots, z_N$ are the zeros[4] of the polynomial R,

(11.46) $$A_\nu = \lim_{z \to z_\nu} \frac{Y(z)(z-z_\nu)}{\exp R(z)-1} ,$$

and B_ν are the coefficients in the Taylor series expansion of $\left\{ Y(z)/[\exp R(z)-1] - \sum_{j=1}^{N} A_\nu/(z-z_\nu) \right\}$ about $z = 0$.

For small values of r,s the above methods for the evaluation of I_2 will obviously be reasonably effective[5]. We can calculate $I_1(t)$ by using a piecewise-polynomial approximation for $f(t)$ and evaluating integrals of the form

(4) The case when R has multiple zeros can be analyzed similarly.

(5) If we define R by the condition $R^{(\nu)}(d_k) = G^{(\nu)}(d_k)$ for $k = 1,2,\ldots,2s$, $\nu = 0,1,\ldots,r$, then the corresponding function ψ will be rational, but it can happen (if a zero of R is on Γ) that $\omega(t) = \exp\left\{ \pm \psi(t)/2 + \frac{1}{2\pi i} \int_\Gamma \frac{\psi(\tau)-\ln R(\tau)}{\tau - t}\, dt \right\}$ will not be differentiable on Γ.

$$\int_{\Gamma_j} (d_{2j-1}-t)^{\left(\frac{\psi(d_{2j-1})}{2\pi i} - \lambda_{2j-1}\right)} (d_{2j}-t)^{\left(-\frac{\psi(d_{2j})}{2\pi i} - \lambda_{2j}\right)} \times S(t) \exp[Q(t) \ln P(t)]dt \,, \tag{11.47}$$

where S, Q and P are polynomials with $Q(d_{2j-1}) = Q(d_{2j}) = 0$. These are determined by splitting up the integrals along Γ_j,(6) into integrals along γ_j and $\Gamma- \gamma_j$, where γ_j is that part of Γ_j lying inside the circle with centre at d_{2j-1} and radius $|d_{2j}-d_{2j-1}|/2$, and expanding the integrands in series.

Not infrequently in applications (see N.I. Muskhelishvili [107]) $\psi(t) \equiv c$ = constant $\neq 0$. In this case

$$x^{\pm}(t) = \prod_{j=1}^{s} \left(\frac{d_{2j}-t}{d_{2j-1}-t} \right)^{c/2\pi i} \prod_{k=1}^{2s} (d_k-t)^{\lambda_k} \exp(\pm c/2) \,, \tag{11.48}$$

so that from (11.44) and (11.45), we obtain

$$I_2(t) = \frac{1}{\pi i} \int_{\Gamma} \frac{\exp(\pm c/2)}{x^{\pm}(\tau)(\tau-t)} d\tau = \left(\frac{1+e^c}{1-e^c}\right) \frac{\exp(\pm c/2)}{x^{\pm}(t)} + 2 \sum_{\nu=0}^{\infty} B_\nu t^\nu \,, \tag{11.49}$$

where B_ν are the coefficients of z^ν in the Taylor series expansion about $z = 0$ of the function

$$\prod_{j=1}^{s} \left(\frac{d_{2j}-z}{d_{2j-1}-z}\right)^{-c/2\pi i} \prod_{k=1}^{2s} (d_k-z)^{-\lambda_r} \frac{1}{\exp(c)-1} \,.$$

Approximating to g on Γ by a polynomial, we can similarly calculate the integral

$$I(t) = \frac{1}{\pi i} \int_{\Gamma} \frac{g(\tau)d\tau}{x^{+}(\tau)(\tau-t)} \,,$$

which is of interest to us. (For more details, see [107, §110]).

Finally we note that an approximate method for the evaluation of

$$\int_{-\infty}^{\infty} \frac{\phi(y)dy}{y - x}$$

in the case when ϕ decreases sufficiently rapidly at infinity has been described in [78].

(6) Using Cauchy's theorem, we can easily reduce the problem to an integration along the interval $[d_{2j-1}, d_{2j}]$ instead of along Γ_j.

§12 THE CASE OF A NORMALLY SOLVABLE EQUATION WITH CLOSED CONTOUR AND SUFFICIENTLY SMOOTH INPUT DATA

We shall consider the equation

$$K\phi \equiv a(t)\phi(t) + \frac{b(t)}{\pi i}\int_\gamma \frac{\phi(\tau)d\tau}{\tau - t} + \lambda\int_\gamma k(\tau,t)\phi(\tau)d\tau = f(t) \quad , \tag{12.1}$$

$t \in \gamma$, where γ is the unit circle. We shall assume that a, b, f and k (uniformly with respect to τ) either (i) have derivatives of order r ($r \geq 0$) which belong to $H(\alpha,\gamma)$ or (ii) are analytic in the annulus $R_1 \leq |t| \leq R_2$, $R_1 < 1$, $R_2 > 1$. Furthermore, we shall assume that λ is a complex parameter, and $a^2(t) - b^2(t) \neq 0$ on γ.

From equations (9.19) and (9.20) it follows that the more general equation in which the contour of integration is a sufficiently smooth Lyapunov curve, can also be reduced to the above form.

§12.1 Application of the Method of Least Squares

In §3, we described the method of least squares in a general functional setting. When considering a particular equation, the following details must be resolved: the choice of a suitable Hilbert space; the determination of $D(A)$, $L(A)$ and $R(A)$; the choice of the coordinate system of elements $\{e_k\}$; methods for the approximate calculation of the elements $y_k = Ae_k$ and the given scalar product; and an effective analysis of the errors.

We shall consider equation (12.1) as an equation in the Hilbert space $L_2(\gamma)$ with a scalar product defined by

$$(f_1,f_2)_{L_2(\gamma)} = \frac{1}{2\pi}\int_\gamma [f_1(t)\overline{f_2(t)}/(it)]dt = \frac{1}{2\pi}\int_0^{2\pi} f_1(t)\overline{f_2(t)}d\theta \quad , \tag{12.2}$$

$t = e^{i\theta}$. We shall assume that the domain $D(K) \equiv L_2(\gamma)$.

It is not difficult to verify that

$$\|K\|_{L_2(\gamma)} \leq \|a\|_{C(\gamma)} + \|b\|_{C(\gamma)} + |\lambda| \left(\int_\gamma\int_\gamma |k(\tau,t)|^2 |d\tau||dt|\right)^{\frac{1}{2}} \tag{12.3}$$

Together with (12.1), we shall introduce the equation

$$\tilde{K}\tilde{\phi} \equiv (a(t) + b(t))\tilde{\phi}^+(t) - (a(t) - b(t))t^{-\varkappa}\tilde{\phi}^-(t) + \\ + \lambda\int_\gamma k(\tau,t)[\tilde{\phi}^+(\tau) - \tau^{-\varkappa}\tilde{\phi}^-(\tau)]d\tau = f(t) \quad , \tag{12.4}$$

$t \in \gamma$, where $\varkappa$ is the index of equation (12.1) and of the operator K (i.e. $\varkappa = (2\pi)^{-1}[\arg (a-b)/(a+b)]_\gamma$) . The functions $\tilde{\phi}^+$ and $\tilde{\phi}^-$ are the boundary values on γ of the Cauchy integral

$$\frac{1}{2\pi i} \int_\gamma \frac{\tilde{\phi}(\tau)d\tau}{\tau - z} ,$$

from inside and outside of γ, respectively.

Theorem 12.1. *If λ is not a characteristic value, then equation* (12.4) *has a unique solution in $L_2(\gamma)$ for any right-hand side $f \in L_2(\gamma)$. If* (12.1) *is solvable, then its solution is given by $\phi = \tilde{\phi}^+ - t^{-\varkappa}\tilde{\phi}^-$, where ϕ^- has a zero of the highest possible order at infinity.*

Consequently when $\varkappa \leq 0$, equation (12.1) has a unique solution. When $\varkappa > 0$, all the linearly independent solutions of the homogeneous equation $K\phi = 0$ are given by the functions $\phi_\nu, (\nu = 1,2,..,\varkappa)$ defined by

$$\varphi_\nu(t) = \tilde{\phi}_\nu^+(t) - t^{-\varkappa}\tilde{\phi}_\nu^-(t) - t^{-\nu} \tag{12.5}$$

where $\tilde{\phi}_\nu = \tilde{\phi}_\nu^+ - \tilde{\phi}_\nu^-$ and

$$\tilde{K}\tilde{\phi}_\nu = Kt^{-\nu} , \text{ for } \nu = 1,2,\dots,\varkappa . \tag{12.6}$$

This implies that the null space $L(K)$ of the operator K is empty when $\varkappa \leq 0$, and has dimension $\varkappa$ for $\varkappa > 0$. In the latter case, all the linearly independent elements of $L(K)$ can be determined by solving equations (12.6) since the operator $\tilde{K}$ has an inverse in $L_2(\gamma)$. Recalling that the index of the conjugate operator K^* is equal to $-\varkappa$ (see §9), then for $\varkappa < 0$, the range $R(K)$ of the operator K will be a subspace which is orthogonal to the subspace $L(K^*)$. For $\varkappa \geq 0$, $R(K) \equiv L_2(\gamma)$.

Let us choose the coordinate system $\{e_k\}$ to be the functions t^k where $k = 0, \pm 1, \pm 2,.., \pm n,...$.

Then

$$y_k = Kt^k = \begin{cases} [(a(t)+b(t))]t^k + \lambda \int_\gamma k(\tau,t)\tau^k d\tau , & k \geq 0 , \\ [(a(t)-b(t))]t^k + \lambda \int_\gamma k(\tau,t)\tau^k d\tau , & k < 0 , \end{cases} \tag{12.7}$$

The integrals $\int_\gamma k(\tau,t)\tau^k d\tau$ are calculated either by the use of a quadrature formula, or by approximating $k(\tau,t)$ (considered as a function of τ) in such a way that the resulting integrals may be evaluated exactly. In particular, if we write

$$k(\tau,t) \approx u_n(k,\tau) = \sum_{s=-n}^{n} r_s(t)\tau^s \ , \quad u_n(k,\tau_j) = k(\tau_j,t),$$

where $\tau_j = \exp\ (\ 2\pi ij/(2n+1)\ )$, we obtain

$$(12.8) \qquad \int_\gamma k(\tau,t)\tau^k d\tau \ \approx \ \int_\gamma u_n(k,\tau)\tau^k d\tau = \begin{cases} 2\pi i\ r_{-k-1}(t), & -n \le k+1 \le n, \\ 0, & |k+1| > n\ . \end{cases}$$

From (10.26), it follows that

$$(12.9) \qquad \begin{aligned} \left|\int_\gamma k(\tau,t)\tau^k d\tau - \int_\gamma u_n(k,\tau)\tau^k d\tau\right| &\le \int_\gamma |k(\tau,t) - u_n(k,\tau)|\,|d\tau| \\ &\le (2\pi)\,\|\,k(\tau,t) - u_n(k,\tau)\|_{L_2(\gamma)} \le 2(2\pi)E_n(k,\gamma)\ . \end{aligned}$$

For equation (12.1) under assumption (i), $E_n(k,\gamma)$ may be estimated using (10.7) with $p = \infty$; under assumption (ii), we can use (10.35).

The scalar product, defined by equation (12.2), can be calculated either by use of a quadrature formula or by using a method similar to that just described. In particular if we take

$$f_1(t) \approx u_n(f_1,t) = \sum_{k=-n}^{n} a_k^{(1)} t^k \ , \quad f_2(t) \approx u_n(f_2,t) = \sum_{k=-n}^{n} a_k^{(2)} t^k \ ,$$

then

$$(12.10) \qquad (f_1,f_2) \approx \frac{1}{2\pi i}\int_\gamma [u_n(f_1,t)\overline{u_n(f_2,t)}/t]dt = \sum_{k=-n}^{n} a_k^{(1)}\ \overline{a_k^{(2)}}$$

and

$$(12.11) \qquad \begin{aligned} \left|(f_1,f_2) - \sum_{k=-n}^{n} a_k^{(1)}\ \overline{a_k^{(2)}}\right| &\le \|\, f_1(t) - u_n(f_1,t)\|\ \|f_2\| \ + \\ &+ \max_j\ |f_1(\tau_j)|\ \|f_2(t)-u_n(f_2,t)\| \\ &\le 2\left(\max_t\ |f_2(t)|\right)E_n(f_1) + 2\left(\max_t\ |f_1(t)|\right)E_n(f_2)\ . \end{aligned}$$

Assuming that λ is not a characteristic value, we shall obtain an approximate solution to equation (12.1) in the form

$$(12.12) \qquad \phi_n(t) = \sum_{k=0}^{n} \alpha_k t^k \ + t^{-\varkappa} \sum_{k=-n}^{-1} \alpha_k t^k \ ,$$

by writing $\tilde{\phi}_n(t) = \sum_{k=-n}^{n} \alpha_k t^k$ as the approximate solution of equation (12.4) and using the method of least squares. The coefficients α_k are determined by any algorithm which minimizes

$$(12.13) \qquad \tilde{\Delta}_n(P) = \|f - \sum_{k=-n}^{n} \alpha_k\ \tilde{y}_k\|^2 \quad ,$$

$$P = P(\alpha_{-n}, \alpha_{-n+1}, \ldots, \alpha_n) \ ,$$

where

$$(12.14) \quad \tilde{y}_k = \tilde{K} t^k = \begin{cases} (a(t) + b(t))t^k + \lambda \int\limits_\gamma k(\tau,t)\tau^k d\tau \ , & k \geq 0, \\ (a(t) - b(t))t^{k-\varkappa} + \lambda \int\limits_\gamma k(\tau,t)t^{k-\varkappa} d\tau \ , & k < 0. \end{cases}$$

A detailed examination of many such algorithms has been given in §3.

Since the operator $\tilde{K}$ has a bounded inverse, the error $\|\tilde{\phi} - \tilde{\phi}_n\|$ can be estimated by using (2.28), (2.30), (3.8) and (3.12). As is obvious, the first three estimates are applicable in general only as a proof of convergence and not as an estimation of the rate of convergence. In (3.12), an effective estimate of the error may be determined if we can calculate m_n, where

$$(12.15) \qquad m_n = \min_{\alpha_k} \| \sum_{k=-n}^{n} \alpha_k \tilde{y}_k \|^2 \ , \text{ with } \sum_{k=-n}^{n} |\alpha_k|^2 = 1 \ ,$$

and find ϵ_n for which

$$(12.16) \qquad I_n(\tilde{\phi}) \leq \epsilon_n \|\tilde{\phi}\| \ ,$$

where $I_n(\tilde{\phi})$ denotes the measure of the deviation of the best approximation to $\tilde{\phi}$ in $L_2(\gamma)$ by functions of the form $\sum\limits_{k=-n}^{n} \alpha_k t^k$. The estimation of m_n can be obtained (see §3.1) if we can calculate an approximate value of the norm of the inverse matrix of the normal system of equations corresponding to (12.13). The inequality (12.16) can be determined with the help of (9.17). It is obvious however that the estimates obtained in this complicated way will usually be rather conservative and of doubtful practical value.

This is not surprising since although we have assumed that $a^2 - b^2 \neq 0$ on γ and λ is not a characteristic value, we have used no information concerning the closeness of $a^2 - b^2$ to zero, and of λ to the set of characteristic values. This latter information is usually not known *a priori*. We are therefore forced to make a judgement concerning the closeness of ϕ_n to ϕ with respect to either the size of the minimum value of $\tilde{\Delta}_n(P)$, when a β-approximation is required to the solution of (12.1), or to the tendency of the coefficients α_k to minimize $\tilde{\Delta}_n(P)$ as n increases, when an α- approximation is required.

§12.2 Application of the Method of Collocation

We shall now examine equations (12.1) and (12.4) as linear equations in the Banach space $W(\gamma)$ (see §7, Theorem 7.4), with domain $D(K) = D(\tilde{K}) = W(\gamma)$.

It is not difficult to verify that

$$\|K\|_{W(\gamma)} \leq \max\left[\max_{t\varepsilon\gamma}\left(|c(t)| + \frac{1}{\pi}\int_\gamma \left|\frac{c(\tau) - c(t)}{\tau - t}\right| |d\tau|\right),\right.$$

(12.17)

$$\left.\max_{t\varepsilon\gamma}\left(|d(t)| + \frac{1}{\pi}\int_\gamma \left|\frac{d(\tau) - d(t)}{\tau - t}\right| |d\tau|\right)\right] + 2\pi|\lambda|\max_{t\varepsilon\gamma}\|k(\tau,t)\|_{W(\gamma)},$$

where $c(t) = a(t) + b(t)$ and $d(t) = a(t) - b(t)$.

We note that for any function $g \varepsilon H(\alpha,\gamma)$, $0 < \alpha \leq 1$,

$$\frac{1}{\pi}\int_\gamma \left|\frac{g(\tau) - g(t)}{\tau - t}\right| |d\tau| \leq \frac{H_g^\alpha}{\pi}\int_\gamma |\tau - t|^{\alpha-1}|d\tau| = \frac{2H_g^\alpha}{\pi}\int_0^\pi \left(2\sin\frac{\theta}{2}\right)^{\alpha-1} d\theta$$

(12.18)

$$\leq \frac{2^\alpha H_g^\alpha}{\pi}\int_0^\pi \left(\frac{\theta}{\pi}\right)^{\alpha-1} d\theta = 2^\alpha H_g^\alpha/\alpha .$$

Assuming for simplicity that $a+b \equiv 1$, we again seek an approximate solution of (12.1) in the form (12.12), with the coefficients α_k now being defined by the system of equations

(12.19) $$(\tilde{K}\tilde{\phi}_n)(t_j) = f(t_j), \quad (j = -n, -n+1, \cdot\cdot, n) .$$

On writing out these equations in more detail, we have

(12.20)
$$\sum_{k=0}^{n}\alpha_k t_j^k + (a-b)_j t_j^{-\varkappa}\sum_{k=-n}^{-1}\alpha_k t_j^k +$$

$$+ \lambda\int_\gamma k(\tau,t_j)\left(\sum_{k=0}^{n}\alpha_k\tau^k + \sum_{k=-n}^{-1}\alpha_k\tau^{k-\varkappa}\right)d\tau = f_j, j=-n,-n+1,\ldots,n.$$

The points t_j divide the circle γ into $(2n + 1)$ equal arcs with $t_j = \exp(2\pi ij/(2n + 1))$, and $(a - b)_j = a(t_j) - b(t_j)$, $f_j = f(t_j)$.

First, we shall analyze the convergence of the method in the case when $k(\tau,t) \equiv 0$, $\tilde{K} \equiv \tilde{K}^0$, $K \equiv K^0$. The following lemma is of some importance.

Lemma. *Assume that the polynomials* $\chi_n^+(t) = \sum_{k=0}^{n} c_k t^k$ *and* $\chi_n^-(t) = \sum_{k=-n}^{0} d_k t^k$ *do not have a common zero and that* $\chi_n^{\pm} \neq 0$ *on* γ . *Then the algebraic system*

$$(12.21) \qquad \sum_{k=0}^{n} \alpha_k t_j^k + \left(\frac{\chi_n^+}{\chi_n^-}\right)_j \sum_{k=-n}^{-1} \alpha_k t_j^k = f_j \ , \ j = -n,-n+1,\dots,n \ ,$$

always has a unique solution.

Proof. We construct the polynomial $R_n(t) = \sum_{k=-n}^{n} r_k t^k$ such that $R_n(t_j) = f_j \chi_n^-(t_j)$

From the theorem concerning the expansion of the rational function $R_n t^n / [\chi_n^+ \chi_n^- t^n]$ as a sum of partial fractions, we find that the α_k are defined by the identity

$$(12.22) \qquad \chi_n^-(t) \sum_{k=0}^{n} \alpha_k t^k + \chi_n^+(t) \sum_{k=-n}^{-1} \alpha_k t^k \equiv R_n(t) \ .$$

The lemma now follows at once. #

Theorem 12.2. *For sufficiently large n, the system of equations* (12.20), (*with* $k(\tau,t) \equiv 0$), *has a unique solution* $\{\alpha_k^0\}$. *In addition, if* $\tilde{\phi}_n^{(0)}(t) = \sum_{k=-n}^{n} \alpha_k^0 t^k$, *then*

$$(12.23) \qquad \|\tilde{\phi}_n^0\|_{W(\gamma)} < C_1 \max_{-n \le j \le n} |f_j| \ \ln n \ ,$$

where C_1 *does not depend on n.*

Proof. Following the method given by L.V. Kantorovich [74], we introduce operators ψ and ψ^{-1} which define a one-to-one correspondence between the elements of $\bar{\alpha}\ (\alpha_{-n},\dots,\alpha_n)$ and $\bar{\phi}(\phi_{-n},\dots,\phi_n)$ where

$$\phi_j = \sum_{k=-n}^{n} \alpha_k t_j^k \quad \text{for } j = -n,\dots,n. \text{ i.e. we have } \bar{\alpha} = \psi^{-1}\bar{\phi}, \quad \bar{\phi} = \psi\bar{\alpha} \ .$$

We rewrite equations (12.20) as

$$(12.24) \qquad N\bar{\phi} = \bar{K}^0\bar{\alpha} = \bar{f} \ ,$$

and consider this as a linear equation in a Banach space $\bar{W}$, the elements of which comprise all the quantities $\bar{\phi}$, and with norm defined by

$$(12.25) \qquad \|\bar{\phi}\|_{\bar{W}} = \Big\| \sum_{k=-n}^{n} \alpha_k t^k \Big\|_{W(\gamma)} \ ,$$

where $\bar{\alpha} = \psi^{-1}\bar{\phi}$. Similarly, equations (12.21) are written as

(12.26) $$N_1\bar{\phi}_1 = \bar{f}\,.$$

We take for $\chi_n^{\pm}$ the polynomials of best approximation in $C(\gamma)$ to the functions $\chi^{\pm}$ respectively, where

$$[\,a(t) - b(t)]\, t^{-\varkappa} = \chi^{+}(t)/\chi^{-}(t)\quad ,$$

$$\chi^{\pm}(t) = \exp\left\{ \pm\frac{1}{2}\,\ln\,[(a(t) - b(t))\, t^{-\varkappa}\,] + \frac{1}{2\pi i}\int_{\gamma} \frac{\ln[(a(\tau)-b(\tau))\tau^{-\varkappa}]d\tau}{\tau - t}\right\}.$$

We find an n such that the conditions of the lemma are satisfied; then the operator N_1 has an inverse N_1^{-1}. We show that

(12.27) $$\|N_1\|_{\bar{W}} \leq C_2\,\ln n\ ,\quad \|N_1^{-1}\|_{\bar{W}} \leq C_3\,\ln n\ ,$$

where C_2, C_3 are independent of n. In fact, from Theorem 10.1, we have

$$\|N_1\bar{\phi}_1\|_{\bar{W}} = \|u_n(f,t)\|_W \leq C_4(\max_j |f_j|)\ln n \leq C_2\|\bar{\phi}_1\|_{\bar{W}}\ \ln n\quad .$$

Then the solution $\{\alpha_k^{(1)}\}$ of (12.21) is given by the coefficients of the solution of the Riemann boundary value problem (8.3), which can be represented in the form (see (8.8)),

$$\sum_{k=0}^{n} \alpha_k^{(1)} t^k = \frac{R_n(t)}{2\chi_n^{+}(t)} + \frac{\chi_n^{+}(t)}{2\pi i}\int_{\gamma} \frac{R_n(\tau)d\tau}{\chi_n^{+}(\tau)\chi_n^{-}(\tau)(\tau-t)}\quad ,$$

(12.28)

$$\sum_{k=-n}^{-1} \alpha_k^{(1)} t^k = -\frac{R_n(t)}{2\chi_n^{-}(t)} + \frac{\chi_n^{-}(t)}{2\pi i}\int_{\gamma} \frac{R_n(\tau)d\tau}{\chi_n^{+}(\tau)\chi_n^{-}(\tau)(t-t)}\quad .$$

Hence

$$\|N_1^{-1}\bar{f}\|_{\bar{W}} = \max_t \Big|\sum_{k=0}^{n} \alpha_k^{(1)} t^k\Big| + \max_t \Big|\sum_{k=-n}^{-1} \alpha_k^{(1)} t^k\Big|$$

$$\leq C_5\|R_n\|_{W(\gamma)} \leq C_6(\max_j |f_j\chi_n^{-}(t_j)|)\ln n$$

$$\leq C_3\|\bar{f}\|_{\bar{W}}\ \ln n\ ,$$

which is what we required.

Let us now estimate the norm of $E = N - N_1$.

We have

$$\|E\bar{\Phi}\|_{\bar{W}} = C_7 \ln n \left(\max_j \left| \left[(a-b)_j t_j^{-\varkappa} - \left(\frac{\chi_n^+}{\chi_n^-}\right)\right]_j \sum_{k=-n}^{-1} \alpha_k t_j^k \right| \right)$$

$$\leq C_7 \ln n \left(\max_t \left| ((a(t)-b(t))t^{-\varkappa} - \chi_n^+(t)/\chi_n^-(t) \right| \right) \|\bar{\Phi}\|_{\bar{W}} ,$$

so that

$$\|E\|_{\bar{W}} \leq C_7 \ln n \left(\max_t \left| (a(t)-b(t))t^{-\varkappa} - \chi_n^+(t)/\chi_n^-(t) \right| \right) \tag{12.29}$$

$$\leq C_8 n^{-\alpha} \ln n ,$$

where α is the Hölder index of the function $a-b$, and C_8 is independent of n.

We choose n so that

$$\|E\| \, \|N_1^{-1}\| < C_8 C_3 n^{-\alpha} \ln^2 n < 1 . \tag{12.30}$$

Then

$$N^{-1} = N_1^{-1} - N_1^{-1} E N_1^{-1} + N_1^{-1}(EN_1^{-1})^2 - N_1^{-1}(EN_1^{-1})^3 + \ldots . \tag{12.31}$$

and

$$\|\phi_n^0\|_{W(\gamma)} = \|N^{-1}\bar{f}\|_{\bar{W}} \leq \|N_1^{-1}\bar{f}\| (1 - \|EN_1^{-1}\|)^{-1}$$

$$\leq C_1 (\max_j |f_j|) \ln n ,$$

which proves the theorem. #

From (12.23) it is not difficult to estimate the rate of convergence of $\tilde{\phi}_n^0$ to $\tilde{\phi}^0$. For a given n let $T_n^\pm$ be the polynomial of best approximation to $\tilde{\phi}^\pm$ respectively, and let $\tilde{\phi} = \tilde{\phi}^+ - \tilde{\phi}^-$ be the solution of equation (12.4) with $k(\tau,t) \equiv 0$. Then it is obvious that

$$T_n^+ - (a-b)t^{-\varkappa} T_n^- = f - [\tilde{\phi}^+ - T_n^+ - (a-b)t^{-\varkappa} (\tilde{\phi}^- - T_n^-)] .$$

Hence $T_n^{\pm}$ is obtained as the solution of (12.20) when, on the right-hand side, we substitute $f_j - \epsilon_j$ instead of f_j where
$\epsilon_j = [\tilde{\phi}^{+} - T_n^{+} - (a-b)t^{-\varkappa}(\tilde{\phi}^{-} - T_n^{-})]_j$. From (12.23), we have

$$\|\tilde{\phi}_n^0 - T_n^{+} + T_n^{-}\|_{W(\gamma)} \leq C_1 (\max_j |\epsilon_j|) \ln n. \tag{12.32}$$

We now immediately obtain the final estimate,

$$\begin{aligned} \|\tilde{\phi} - \tilde{\phi}_n^0\|_{W(\gamma)} &\leq \max_t |\tilde{\phi}^{+} - T_n^{+}| + \max_t |\tilde{\phi}^{-} - T_n^{-}| + \|\tilde{\phi}_n^0 - T_n^{+} + T_n^{-}\|_{W(\gamma)} \\ &\leq C_9\, \rho_n(\tilde{\phi}) \ln n . \end{aligned} \tag{12.33}$$

Theorem 12.2 is a generalization of Theorem 10.1. The essential difference when compared with the interpolation polynomial $u_n(\phi,t)$, is that in general $\tilde{\phi}_n^0$ can only be obtained for sufficiently large n. For example, for $-\pi \leq \theta \leq \pi(1 - 1/(2m+1))$, let $a-b = t, t = e^{i\theta}$, and define $a-b$ on the remaining points of γ so that the index is equal to zero. Then equations (12.20), with $k(\tau,t) \equiv 0$, take the form

$$\sum_{k=0}^{n} \alpha_k t_j^k + \sum_{k=-n}^{-1} \alpha_k t_j^{k+1} = f_j,\ j = -n,-n+1,\ldots,n, \tag{12.34}$$

for all $n \leq m$. For these values of n, the determinant of the system of equations is obviously equal to zero so that (12.34) will have no solution for any f_j.

We now proceed to an analysis of convergence in the general case.

<u>Theorem 12.3</u>. *Let the operator* $\tilde{K}$ *have an inverse, and equation* (12.1) *have the solution* ϕ_* *which is related to the solution of* $\tilde{K}\tilde{\phi} = f$ *by the equations* $\phi_*^{+} = \tilde{\phi}^{+}, \phi_*^{-} = t^{-\varkappa}\tilde{\phi}^{-}$. *Then there exists an* n_0 *such that for all* $n \geq n_0$, *equations* (12.20) *have the unique solution* $\{\alpha_k^0\}$ *with*

$$\|\phi_* - \phi_n^0\|_{C(\gamma)} \leq \|\tilde{\phi} - \tilde{\phi}_n^0\|_{W(\gamma)} \leq C_{10} \|\tilde{\phi} - T_n\|_{W(\gamma)} \ln n \quad , \tag{12.35}$$

where

$$\phi_n^0 = \sum_{k=0}^{n} \alpha_k^0 t^k + \sum_{k=-n}^{-1} \alpha_k^0 t^{k-\varkappa}, \quad \tilde{\phi}_n^0 = \sum_{k=-n}^{n} \alpha_k^0 t^k \quad ,$$

and $T_n = \sum_{k=-n}^{n} d_k t^k$ *is the polynomial of best approximation to* $\tilde{\phi}$ *with respect to the metric of* $W(\gamma)$.

Proof. We write the equation $\tilde{K}\tilde{\phi} = f$, (recall that $\tilde{K}$ is defined by (12.4)), in the form

(12.36) $$\tilde{K}\tilde{\phi} \equiv \tilde{K}^0\tilde{\phi} + \lambda\tilde{k}\tilde{\phi} = f,$$

where

$$\tilde{K}^0\tilde{\phi} = \tilde{\phi}^+ - (a-b)t^{-\varkappa}\tilde{\phi}^-, \quad \tilde{k}\tilde{\phi} = \int_\gamma k(\tau,t)(\tilde{\phi}^+(\tau) - \tau^{-\varkappa}\tilde{\phi}^-(\tau))d\tau .$$

We choose $n \geq n_1$ where n_1 satisfies (12.30), and following the theory of approximate methods of L.V. Kantorovich [74], we introduce the following spaces and operators :

$$X = Y = W(\gamma) \quad ;$$

X_n is the set of all polynomials $\phi_n = \sum_{k=-n}^{n} \alpha_k t^k$, and $X_n \subset X$; $\overline{X}_n$ is the set of all vectors $\overline{\alpha}_n$ $(\alpha_{-n},\alpha_{-n+1},\ldots,\alpha_n)$; and x_0 is the linear operator which defines a one-to-one correspondence between X_n and $\overline{X}_n$ such that

$$x_0(\sum_{k=-n}^{n} \alpha_k t^k) = \overline{\alpha}_n(\alpha_{-n},\alpha_{-n+1},\ldots,\alpha_n) \quad .$$

We assume that $\|\overline{\alpha}_n\|_{\overline{X}_n} = \|x_0^{-1}\overline{\alpha}_n\|_X$. It is obvious that $\|x_0\| = \|x_0^{-1}\| = 1$. Y_n is defined to be the set of all elements $\psi_n = \sum_{k=-n}^{n} \alpha_k\tilde{K}^0 t^k$, $Y_n \subset Y$; $\overline{Y}_n$ is the set of all elements $\overline{\psi}_n(\psi_{-n},\psi_{-n+1},\ldots,\psi_n)$, where $\|\overline{\psi}_n\|_{\overline{Y}_n} = \max_j|\psi_j|$. The linear operator $\overline{y}_0$ is defined by the one-to-one correspondence between Y_n and $\overline{Y}_n$, such that

$$y_0\psi_n = \overline{\psi}_n \{ \psi_n(t_{-n}),\psi_n(t_{-n+1}),\ldots,\psi_n(t_n) \}, \; t_j = \exp(2\pi ij/(2n+1)) \; .$$

From (12.33), we easily obtain

$$\|y_0^{-1}\overline{\psi}_n\| \leq \|\tilde{K}^0\| C_1 \ln n \max_j|\psi_j|, \qquad \text{i.e. } \|y_0^{-1}\| \leq C_1\|\tilde{K}_0\| ;$$

where y is the linear operator which is the continuation of y_0 on the space Y ,

$$y\psi = \overline{\psi}_n \{ \psi(t_{-n}),\psi(t_{-n+1}),..,\psi(t_n) \},\ \|y\| \le 1.$$

The system of equations (12.20) is written in the form

$$\overline{K}\overline{\alpha} \equiv \overline{K}^0\overline{\alpha} + \lambda\overline{k}\overline{\alpha} = \overline{f} \equiv yf \quad . \tag{12.37}$$

It is readily seen that $\overline{K}^0 = y_0\tilde{K}^0 x_0^{-1}$, and $\overline{k} = y\tilde{k}x_0^{-1}$. Then we obtain the element $\psi_n' \in Y$ such that

$$\|\tilde{k}\tilde{\phi} - \psi_n'\| \le \epsilon_n\|\tilde{\phi}\| \quad , \qquad \text{where } \epsilon_n \to 0 \text{ as } n \to \infty \quad . \tag{12.38}$$

For the determination of this last estimate, it is sufficient to write $k(\tau,t)$ in the form

$$k(\tau,t) = \mu^+(t) - (a(t) - b(t))t^{-\varkappa}\mu^-(t) \ , \tag{12.39}$$

and then to approximate to μ^+ and μ^- by polynomials of best approximation. It is clear that the rate of convergence of ϵ_n to zero will be governed by the analytic properties of $k(\tau,t)$, a and b. (If $a,b,k \in H(\alpha,\gamma)$, the rate of convergence will be less than a constant $\times n^{-\alpha}$). Since $\psi_n' \equiv k_n'\tilde{\phi} \in Y_n$ then $Y\psi_n' = y_0\psi_n'$, and

$$\overline{K} = y_0(\tilde{K}^0 + k_n')x_0^{-1} + \lambda y(\tilde{k} - k_n')x_0^{-1} \ . \tag{12.40}$$

From (12.38), the operator $K_n = \tilde{K}^0 + \lambda k_n' = \tilde{K} - \lambda(\tilde{k} - k_n')$ for $n \ge n_2 \ge n_1$, will have an inverse K_n^{-1}. But then the operator $\overline{K}$ (see (12.40)), for $n \ge n_0 \ge n_2 \ge n_1$, has an inverse $\overline{K}^{-1}$ where $\|\overline{K}^{-1}\| \le C_{11}\ \ell n\ n$.

We choose $n \ge n_0$, and denote by $T_n = \sum_{k=-n}^{n} d_k t^k$, the polynomial of best approximation to the solution $\tilde{\phi}$ of equation (12.4), with respect to the metric of $W(\gamma)$. It is obvious that the coefficients $\{ d_k \}$ are the solution of equation (12.37) with the right-hand side yf replaced by $y\ [f - \lambda\tilde{k}(\tilde{\phi} - T_n)]$. In other words, we have

(12.41) $$\|\sum_{k=-n}^{n} \alpha_k^0 t^k - \sum_{k=-n}^{n} d_k t^k\|_{W(\gamma)} \leq \|\bar{K}^{-1}\|_{\bar{X}_n} \|\tilde{K}\|_{W(\gamma)} \|\tilde{\phi} - T_n\|_{W(\gamma)} ,$$

from which

(12.42) $$\|\phi - \sum_{k=-n}^{n} \alpha_k^0 t^k\|_{W(\gamma)} \leq (1 + \|\bar{K}^{-1}\|_{\bar{X}_n} \|\tilde{K}\|_{W(\gamma)}) \|\tilde{\phi} - T_n\|_{W(\gamma)} .$$

But $$\|\phi_* - \sum_{k=0}^{n} \alpha_k^0 t^k - \sum_{k=-n}^{-1} \alpha_k^0 t^{k-\varkappa}\|_{C(\gamma)} \leq \|\tilde{\phi} - \sum_{k=-n}^{n} \alpha_k^0 t^k\|_{W(\gamma)} ,$$

and the theorem is proved. #

The inequality (12.42) can be used for the determination of a realistic estimate of the error. By analogy with (12.17) and (12.18), we can estimate $\|\tilde{K}\|_{W(\gamma)}$. On transforming the matrix of the system of equations (12.20), we can estimate $\|\bar{K}^{-1}\|_{\bar{X}_n}$. Under the condition of assumptions (i) for equation (12.1), we obtain (see §10.1),

(12.43) $$\|\tilde{\phi} - T_n\|_{W(\gamma)} \leq \frac{\pi \|\tilde{\phi}_\theta^{(r)}(t)\|_{W(\gamma)}}{2(n+1)^r} , \quad t = e^{i\theta} .$$

Regularizing equation (12.4), (see §9.2), it is possible (as in [76] where estimates of the type $\|\phi^{(r)}\|_C \leq \text{const } \|\phi\|_C$ are obtained, ϕ being the solution of a Fredholm integral equation of the second kind), to find the estimate

(12.44) $$\|\tilde{\phi}_\theta^{(r)}\|_{W(\gamma)} \leq C_{12} \|\tilde{\phi}\|_{W(\gamma)} .$$

From (12.42) - (12.44) we obtain

$$\|\tilde{\phi} - \sum_{k=-n}^{n} \alpha_k^0 t^k\|_{W(\gamma)} \leq \frac{C_n \|\tilde{\phi}\|_{W(\gamma)} \ln n}{(n+1)^r}$$

$$\leq \frac{C_n \|\tilde{\phi} - \sum_{k=-n}^{n} \alpha_k^0 t^k\|_{W(\gamma)} \ln n + C_n \|\sum_{k=-n}^{n} \alpha_k^0 t^k\|_{W(\gamma)} \ln n}{(n+1)^r} ,$$

where C_n is a known constant which depends on n. Hence

(12.45) $$\|\tilde{\phi} - \sum_{k=-n}^{n} \overset{0}{\alpha}_k t^k\|_{W(\gamma)} \leq \frac{C_n \ln n}{(n+1)^r} \frac{1}{\left(1 - \frac{C_n \ln n}{(n+1)^r}\right)} \|\sum_{k=-n}^{n} \overset{0}{\alpha}_k t^k\|_{W(\gamma)} ,$$

provided that

(12.46) $$(n+1)^{-r} C_n \ln n < 1 .$$

In conclusion, we should note that although the method of collocation has approximately the same rate of convergence as the method of least squares, it has the advantage when compared with equations of the form (3.5) (with $\alpha = 0$) that the system (12.20) is simpler to construct. On the other hand the order of the system at which it first becomes solvable, is not known. This shortcoming may be overcome by simultaneously determining those values of λ for which the determinant of the system (12.20) is equal to zero, (see §12.4).

§12.3 Other Methods

It is not difficult to see that the Ritz-Galerkin method of §4 and the iterative methods of §5 could be used for the approximate solution of equation (12.1).

The Ritz-Galerkin method is examined in [54, 58, 64, 67]. It is similar to the collocation method and will not be discussed further.

The application of iterative methods requires computation of equations (5.3), (5.70), (5.71), (5.87) and (5.86). These require repeated evaluation of ordinary and singular integrals involving complicated functions. For one step of the iterative process, these methods may require considerably more arithmetic operations than for one step of some of the steep descent algorithms (§3). The approximation of a, b, f and k in (12.1) by polynomials may enable all the integrals to be evaluated exactly (assuming x_0 is also a polynomial), but the degree of the polynomial x_{n+1} grows quickly and this can be unsatisfactory due to limitations of computer memory.

Thus, in the case of one singular integral equation with sufficiently smooth data, the method of least squares is to be preferred to iterative methods. However, in the case of a system of singular integral equations or if the data is less smooth, it may be more effective to use iterative methods (see §14, §15).

In theory, singular integral equations may be solved by the same methods as used for Fredholm integral equations of the second kind [76, 169]. However, the actual application of these methods to singular integral equations without regularization is non-trivial, and is based on a study of the structure of the solution of the corresponding dominant equation.

We shall now consider the application of a method, which is often used for the approximate solution of integral equations of the second kind, based on the use of a piecewise-linear approximation to the solution.

We look for a solution of (12.4) in the form of a piecewise-linear polynomial

$$(12.47)\qquad \begin{gathered} {}'\tilde{\phi} = \tilde{\phi}_k \frac{t_{k+1} - t}{t_{k+1} - t_k} + \tilde{\phi}_{k+1} \frac{t - t_k}{t_{k+1} - t_k}\ ; \quad t \in \widetilde{t_k t_{k+1}}\ ; \\ t_k = \exp(2\pi i k/n),\ k = 0,1,\dots,n-1,\ \tilde{\phi}_{n\pm s} = \tilde{\phi}_{\pm s}\ ; \end{gathered}$$

and using the discretized equation

$$(12.48)\qquad (\tilde{K}'\tilde{\phi})\ (t_j) = f(t_j),\ j = 0,1,\dots,n-1 \quad .$$

It is clear that (12.48) is a system of n linear algebraic equations in the n unknowns $\tilde{\phi}_k$.

By the use of an example, we can show that in general we are unable to prove theorems about the uniqueness of (12.48) and the convergence of $'\tilde{\phi}$ to $\tilde{\phi}$, which are analogous to corresponding theorems for Fredholm integral equations of the second kind (see [156]).

In particular we consider the equation

$$E\phi \equiv a\phi(t) + \frac{1}{\pi i}\int_\gamma \frac{\phi(\tau)d\tau}{\tau - t} = f(t),\ t \in \gamma, \tag{12.49}$$

where a is a constant. We seek an approximate solution of (12.49) in the form of a piecewise-linear polynomial

$$'\phi(t) = \phi_k \frac{t_{k+1} - t}{t_{k+1} - t_k} + \phi_{k+1} \frac{t - t_k}{t_{k+1} - t_k},\ t \in \widetilde{t_k t_{k+1}},$$

using the discretized equation

$$(E\ '\phi)(t_j) = f(t_j),\ j = 1,2,..,n\ . \tag{12.50}$$

Straightforward calculations give

$$(E\ '\phi)_{t=t_j} = \sum_{k=0}^{n-1} a_{k-j}\,\phi_k,\ j = 0,1,..,n-1,$$

$$a_0 = a + 2/n,$$

$$a_1 = \frac{1}{\pi i}(1+d_1)\ \ln\ (1+d_1^{-1}),\quad a_{n-1} = -\frac{1}{\pi i} d_{n-2}\ \ln\ (1+d_{n-2}^{-1}),$$

$$a_\nu = \frac{1}{\pi i}\left[(1+d_\nu)\ \ln\ (1+d_\nu^{-1}) - d_{\nu-1}\ \ln\ (1+d_{\nu-1}^{-1})\right],\ \nu = 2,3,..,n-2, \tag{12.51}$$

$$a_{n\pm s} = a_{\pm s},\ d_s = \frac{\sin(\pi s/n)}{\sin(\pi/n)}\exp\ [-i\pi(s+1)/n],$$

$$s = 1,2,..n.$$

We must now evaluate the determinant Δ_n of equation (12.50), where

(12.52) $$\Delta_n = \det \begin{pmatrix} a_0 & a_1 & a_2 & \dots & a_{n-1} \\ a_{n-1} & a_0 & a_1 & \dots & a_{n-2} \\ \cdot & \cdot & \cdot & & \cdot \\ \cdot & \cdot & \cdot & & \cdot \\ \cdot & \cdot & \cdot & & \cdot \\ a_1 & a_2 & a_3 & \dots & a_0 \end{pmatrix} .$$

To do this we introduce the polynomial P where

(12.53) $$P(z) = a_0 + a_1 z + \dots + a_{n-1} z^{n-1} ,$$

and multiply Δ_n by the Vandermonde determinant W_n defined by

(12.54) $$W_n = \det \begin{pmatrix} 1 & 1 & \dots & 1 \\ t_0 & t_1 & \dots & t_{n-1} \\ t_0^2 & t_1^2 & \dots & t_{n-1}^2 \\ \cdot & \cdot & & \cdot \\ \cdot & \cdot & & \cdot \\ \cdot & \cdot & & \cdot \\ t_0^{n-1} & t_1^{n-1} & \dots & t_{n-1}^{n-1} \end{pmatrix} .$$

From the theorem that the determinant of the product of two matrices is equal to the product of the determinants of each of the matrices (see [70]), we find

$$\Delta_n W_n = \det \begin{pmatrix} P(t_0) & P(t_1) & \dots & P(t_{n-1}) \\ P(t_0)t_0 & P(t_1)t_1 & \dots & P(t_{n-1})t_{n-1} \\ \cdot & \cdot & & \cdot \\ \cdot & \cdot & & \cdot \\ \cdot & \cdot & & \cdot \\ P(t_0)t_0^{n-1} & P(t_1)t_1^{n-1} & \dots & P(t_{n-1})t_{n-1}^{n-1} \end{pmatrix}$$

$$= P(t_0)\, P(t_1) \dots P(t_{n-1}) W_n .$$

Hence

(12.55) $$\Delta_n = P(t_0)P(t_1) \dots P(t_{n-1}) .$$

It is easy to see that

(12.56) $$P(t_s) = [\,E\,'(t^s)\,]_{t=1} = a + [\,S\,'(t^s)\,]_{t=1}$$

From (11.2) it follows that

$$\lim_{n\to\infty}[\,S\,'(t^s)\,]_{t=1} = 1 \quad\text{and}\quad \lim_{n\to\infty}[\,S\,'(t^{n-s})\,]_{t=1} = -1$$

for all fixed integers $s > 0$.

We shall now show that if we put $n = 2m$ then

(12.57) $$\lim\,[\,S\,'(t^m)]_{t=1} = 0.$$

In fact $t_k^{n/2} = (-1)^k$, so choosing an arbitrary fixed integer $p > 0$ and $n > 4(p+1)$, we obtain from (12.51) that

$$\begin{aligned}
[\,S\,'(t^{n/2})\,]_{t=1} &= \frac{2}{n} - \frac{1}{\pi i}\ln(1+d_1^{-1}) + \frac{1}{\pi i}\ln(1+d_2^{-1}) - \dots + \frac{1}{\pi i}\ln(1+d_{n-2}^{-1}) - \\
&\quad - \frac{2}{\pi i}d_1\ln(1+d_1^{-})+\frac{2}{\pi i}d_2\ln(1+d_2^{-1}) - \dots + \frac{2}{\pi i}d_{n-2}\ln(1+d_{n-2}^{-1}) \\
&= \frac{2}{n} + \frac{1}{\pi i}\sum_{r=1}^{m-1}[(1+2d_{2r})\ln(1+d_{2r}^{-1}) - (1+2d_{2r-1})\ln(1+d_{2r-1}^{-1})] \\
&= I_{n,p} + \frac{1}{\pi i}\sum_{r=p+1}^{m-p-1}\Big\{(1+2d_{2r})(d_{2r}^{-1} - \frac{d_{2r}^{-2}}{2} + \frac{d_{2r}^{-3}}{3} - \dots) - \\
&\qquad - (1+2d_{2r})(d_{2r-1}^{-1} - \frac{d_{2r-1}^{-2}}{2} + \frac{d_{2r-1}^{-3}}{3} - \dots)\Big\} \\
&= I_{n,p} + \frac{1}{\pi i}\sum_{\nu=2}^{\infty}\frac{(\nu-1)}{\nu(\nu+1)}(-1)^{\nu-1}\sum_{r=p+1}^{m-p-1}(d_{2r}^{-\nu} - d_{2r-1}^{-\nu}) \\
&= I_{n,p} + G_{n,p}\,.
\end{aligned}$$

We can estimate $G_{n,p}$ and $I_{n,p}$ separately. We have

$$|d_{2r}^{-\nu} - d_{2r-1}^{-\nu}| = \left| \left(\frac{\sin(\pi/n)\exp[i\pi(2r+1)/n]}{\sin(2r\pi/n)}\right)^{\nu} - \left(\frac{\sin(\pi/n)\exp[i\pi\, 2r/n]}{\sin[(2r-1)\pi/n]}\right)^{\nu} \right|$$

$$\leq \begin{cases} 2\left[\pi/[2(2r-1)]\right]^{\nu}, & 2r-1 < m, \\ 2\left[\pi/[2(n-2r)]\right]^{\nu}, & 2r-1 \geq m. \end{cases}$$

Therefore

$$|G_{n,p}| \leq \frac{4}{\pi} \sum_{\nu=2}^{\infty} \frac{\nu-1}{\nu(\nu+1)} \sum_{r=p+1}^{[\frac{m}{2}]+1} \left[\frac{\pi}{2(2r-1)}\right]^{\nu}$$

$$\leq 2 \sum_{\nu=2}^{\infty} \frac{\nu-1}{\nu(\nu+1)} \left(\frac{\pi}{2}\right)^{\nu-1} \sum_{r=2p+1}^{\infty} \left(\frac{1}{r}\right)^{\nu}$$

$$\leq 2 \sum_{\nu=2}^{\infty} \frac{1}{\nu(\nu+1)} \left(\frac{\pi}{4p}\right)^{\nu-1} \leq \frac{1}{p}.$$

Further, for arbitrary fixed r,

$$\lim_{n\to\infty} \left\{(1+2d_{2r})\,\ln(1+d_{2r}^{-1}) - (1+2d_{2r-1})\,\ln(1+d_{2r-1}^{-1})\right\}$$

$$= (1+2.2r)\,\ln\left(1+\frac{1}{2r}\right) - [1+2(2r-1)]\,\ln\left(1+\frac{1}{2r-1}\right)$$

$$= -\lim_{n\to\infty} \left\{(1+2d_{n-2r})\,\ln(1+d_{n-2r}^{-1}) - (1+2d_{n-2r-1})\,\ln(1+d_{n-2r+1}^{-1})\right\}.$$

Therefore

$$\lim_{n\to\infty} I_{n,p} = 0.$$

It now follows that for sufficiently large n, $[S'(t^{n/2})]_{t=1}$ can be made smaller than any prescribed positive number. This establishes (12.57).

This result implies that on a set of values of a ($a \neq \pm 1$) for which E has an inverse, the operator E_n, where $E_n{}'\phi \equiv {}'(E'\phi)$, does not necessarily have an inverse for all sufficiently large n. For example, choosing $a = -[S'(t^{n/2})]_{t=1}$, the determinant Δ_n of the system of equations (12.50) is always equal to zero.

The convergence of $'\tilde{\phi}$ to $\tilde{\phi}$ in the space $H(\beta,\gamma)$ ($\beta < \alpha$) is not difficult to prove if we minimize $\| '(\tilde{K}'\tilde{\phi}) - 'f\|_H^\beta$ under the additional constraint that $\|'\tilde{\phi}\|_H^\alpha$ < constant. It follows that every bounded set in $H(\alpha,\gamma)$ is compact in $H(\beta,\gamma)$, so that the theory of quasi-solutions (see §2.2) is valid. This method has been used by Lavrentiev [89].

Finally we shall examine briefly a method proposed by Multhopp, [162]. An approximate solution of

$$K\phi \equiv a(t)\phi(t) + \frac{1}{\pi i}\int_\gamma \frac{k(\tau,t)\phi(\tau)d\tau}{\tau - t} = f(t),\ t \in \gamma, \tag{12.58}$$

is sought in the form $\phi_n = \sum_{k=-n}^{n-1} \alpha_k t^k$. The values

$\phi_j = \sum_{k=-n}^{n-1} \alpha_k t_j^k$, $t_j = \exp(i\pi j/n)$ for $j = 0,1,2,..,2n-1$, are found from the system

$$a(t_j)\phi_j + \frac{1}{\pi i}\left[\int_\gamma \frac{[k(\tau,t)\phi(\tau)]_n d\tau}{\tau - t}\right]_{t=t_j} = f(t_j) \tag{12.59}$$

with $j = 0,1,2,..,2n-1$, where

$$[k(\tau,t)\phi(\tau)]_n = \sum_{k=-n}^{n-1} \beta_k(t)\tau^k,\ \beta_k(t) = \frac{1}{2n}\sum_{j=0}^{2n-1} k(t_j,t)\phi_j t_j^{-k}$$

The equation

$$\phi(t) + \frac{1}{2\pi i} \int_\gamma \frac{\phi(\tau)}{\tau} d\tau = f(t) \quad , \tag{12.60}$$

has a unique solution for all $f \in L_2(\gamma)$. The corresponding system of equations (12.59) has zero determinant for arbitrary n, thus it is not possible to prove theorems similar to Theorem 12.3 for Multhopp's method. In all situations where we assume that the operator K of (12.58) has an inverse, in order to prove results on Multhopp's method, it is also necessary to assume that the operator K_1 has an inverse, where

$$K_1\phi \equiv a(t)\phi(t) - \frac{1}{\pi i} \int_\gamma \frac{k(\tau,t)\phi(\tau)d\tau}{\tau - t} \quad , \quad t \in \gamma \quad . \tag{12.61}$$

(For a proof of Multhopp's method under special assumptions, see [71]).

In conclusion we note that in [102, 137], a method is proposed for the approximate solution of singular integral equations which is analogous to the degenerate kernel methods for Fredholm integral integral equations of the second kind.

§12.4 The Calculation of Characteristic Values

Consider the solution of equation (12.1) by the method of collocation.[7] Let us denote the determinant of the system of equations (12.20) by $\Delta(\lambda)$. We shall call the roots of the equation $\Delta(\lambda) = 0$ the *approximate characteristic values* of the nth approximation.

Theorem 12.4. *All the characteristic values of equation* (12.1) *(and only these values) can be obtained as the limits of all possible sequences of the approximate characteristic values.*

---

(7) The results stated below will also be valid for the Ritz-Galerkin method.

<u>Proof</u>. Let U be an arbitrary open set which contains all the characteristic values of the operator $K(\lambda)$. If $\lambda \notin U$, then the operator $\tilde{K}(\lambda)$ has an inverse $\tilde{K}^{-1}(\lambda)$. Consequently from Theorem 12.3, the system of equations (12.20) is uniquely solvable for sufficiently large n; that is $\Delta(\lambda) \neq 0$. Thus all the zeros of $\Delta(\lambda)$, for n large enough, are contained in U. It must now be proved that for sufficiently large n, an arbitrarily small neighbourhood V of any characteristic value λ_0 of the operator $K(\lambda)$, contains at least one root of the equation $\Delta(\lambda) = 0$. Since characteristic values do not have limit points in the finite complex plane (they are the poles of a meromorphic function), we can assume that V does not contain any characteristic values of K, other than λ_0. Inside V, choose a circle C with centre at λ_0, such that there is an annulus inside V which contains C. It is clear that $\phi(\lambda,t) = \tilde{K}^{-1}(\lambda)f$, considered as a function of λ, is uniformly bounded on C.

Consequently, for sufficiently large n, the functions $\tilde{\phi}_n(\lambda,t) = \sum_{k=-n}^{n} \alpha_k^0 t^k$, (where $\{\alpha_k^0\}$ is the solution of (12.20)), are uniformly bounded on C and converge to $\tilde{\phi}(\lambda,t)$ uniformly with respect to $\lambda \in C$. Let us now assume that there is a set of integers n_k with $n_k \to \infty$, such that $\Delta_{n_k}(\lambda) \neq 0$ on V.

In this case, $\|\tilde{\phi}_{n_k}(\lambda,t)\|_{W(\gamma)}$ are analytic functions of λ inside C which converge uniformly to $\|\tilde{\phi}(\lambda,t)\|_{W(\gamma)}$. From a well known theorem of Weierstrass, we conclude that $\|\tilde{\phi}(\lambda,t)\|_{W(\gamma)}$ is also an analytic function of λ inside C. This is not possible, and the theorem is proved. #

The determination of the zeros of $\Delta(\lambda)$ is complicated. To calculate the characteristic values, it is expedient to regularize equation (12.4), so that we can use standard algorithms for the calculation of eigenvalues[8] of matrices (see [8]).

---

(8) Remember that eigenvalues are the inverse of characteristic values.

§ 12.5 The Determination of All Solutions

In [116], an example is constructed which shows that, in general, all the solutions of (12.1) corresponding to the characteristic value $\lambda = \lambda_0$, cannot be obtained as the limits of all possible sequences of approximate solutions $\tilde{\phi}_n(\lambda_0, t) = \sum_{k=-n}^{n} \alpha_k^0 t^k$. Here $\{ \alpha_k^0 \}$ denotes all possible solutions of the system of equations (12.20) with $\lambda = \lambda_0$. We shall now show how this difficulty can be overcome.

Let $\lambda = \lambda_0$ be a characteristic value of the operator $\tilde{K}(\lambda)$. It is known (see [18]) that a solution of equation (12.4) in the neighbourhood of λ_0 can be written as

$$\tilde{\phi}(\lambda) = \tilde{K}^{-1} f = \frac{P_r f}{(\lambda-\lambda_0)^r} + \frac{P_{r-1} f}{(\lambda-\lambda_0)^{r-1}} + \ldots + \frac{P_1 f}{\lambda-\lambda_0} + \Gamma(\lambda) f, \tag{12.62}$$

where the operators P_j are independent of $\lambda (P_r \neq 0)$, and Γ is an analytic function of λ in the neighbourhood of λ_0. In addition, the range of the operator P_r is the set of all solutions of the homogeneous equation corresponding to equation (12.4) with $\lambda = \lambda_0$. This follows [9] because

$$P_r = \frac{1}{2\pi i} \int_C \tilde{K}^{-1}(\lambda) (\lambda-\lambda_0)^{r-1} d\lambda , \tag{12.63}$$

and

$$\tilde{K}(\lambda_0) P_r = \frac{1}{2\pi i} \int_C (\lambda-\lambda_0)^{r-1} d\lambda - \frac{\tilde{k}}{2\pi i} \int_C \tilde{K}^{-1}(\lambda)(\lambda-\lambda_0)^r d\lambda = 0, \tag{12.64}$$

where $\tilde{k}\tilde{\phi} = \int_\gamma k(\tau,t)[\tilde{\phi}^+(\tau) - \tau^{-\varkappa} \tilde{\phi}^-(\tau)] d\tau$, and C is a circle with centre at λ_0 and radius equal to, for example, $|\lambda_1 - \lambda_0|/2$, λ_1 being the characteristic value nearest to, but distinct from, λ_0.

Let $\tilde{\phi}_n(\lambda)$ be an approximate solution of (12.4) obtained by any method, but such that it is uniformly close to $\tilde{\phi}(\lambda)$ on C. Since

---

(9) The results can be proved on the basis of those given in [75, Ch. 13].

(12.65) $$\max_{\lambda \in C} \|\tilde{K}^{-1}(\lambda)\| \leq \text{constant},$$

a $\tilde{\phi}_n(\lambda)$ with this property may be obtained by any of the methods described above. Recalling that

(12.66) $$P_r f = \frac{1}{2\pi i} \int_C \tilde{\phi}(\lambda)(\lambda-\lambda_0)^{r-1} d\lambda,$$

we find that

(12.67) $$P_r f = \frac{1}{2\pi i} \int_C \tilde{\phi}_n(\lambda)(\lambda-\lambda_0)^{r-1} d\lambda,$$

where r is chosen as the smallest positive integer such that

(12.68) $$\int_C \tilde{\phi}(\lambda)(\lambda-\lambda_0)^r \, d\lambda = 0,$$

or

(12.69) $$\int_C \tilde{\phi}_n(\lambda)(\lambda-\lambda_0)^r \, d\lambda \approx 0.$$

It is clear that the determination of $\tilde{\phi}_n(\lambda)$ for any $f \in W(\gamma)$ and $\lambda \in C$ gives rise to many computational difficulties. If we replace the integral in (12.67) by any quadrature formula, for example,

(12.70) $$P_r f \approx \frac{1}{m} \sum_{j=1}^{m} \tilde{\phi}_n(\lambda_j)(\lambda_j-\lambda_0)^r, \quad \lambda_j = \lambda_0 + \delta \exp(2\pi ij/m),$$

where δ is the radius of the circle C, it is then necessary to calculate $\tilde{\phi}_n(\lambda)$ for the set of values λ_j. Choosing a specific f, we can in this way obtain an approximate value for the non-trivial solution of the homogeneous equation $\tilde{K}(\lambda_0)\,\tilde{\phi} = 0$.

A very difficult problem is that of choosing those functions f so that we obtain linearly independent solutions in the above procedure. To determine $\tilde{\phi}_n(\lambda)$ by direct methods, it is appropriate to solve the corresponding system of algebraic equations for arbitrary right-hand sides. This reduces the problem to that of the inversion of matrices and later to the application of an approximate formula for (12.67) in the form

$$P_r f \approx Bf_n \quad , \tag{12.71}$$

where B is a known finite-dimensional operator. Suppose $e_1, e_2, \dots, e_n$ is an arbitrary orthonormalized system, and write $f_n = \sum_{k=1}^{n} a_k e_k$. Then

$$P_r f \approx \sum_{k=1}^{n} a_k \, Be_k \quad . \tag{12.72}$$

We find those indices k for which $Be_k \approx 0$. The number of remaining indices will be approximately equal to the number of non-trivial linearly independent solutions of the equation $\tilde{K}(\lambda_0)\tilde{\phi} = 0$. The required non-trivial linearly independent solutions will then be given approximately by Be_k, with k taken over the appropriate indices.

Let us assume that these solutions are given by $\tilde{\psi}_1, \tilde{\psi}_2, \dots, \tilde{\psi}_\ell$. For $j = 1,2,\dots,\ell$, we construct the functions $\psi_j = \tilde{\psi}_j^+ - t^{-\varkappa}\tilde{\psi}_j^-$. All the ψ_j for which $t^{-\varkappa}\tilde{\psi}_j^-$ become zero at infinity will obviously make up the complete system of linearly independent solutions of the equation $K(\lambda_0)\psi = 0$, for $\varkappa \leq 0$. If $\varkappa > 0$ then the functions $\tilde{\phi}_\nu^+ - t^{-\varkappa}\tilde{\phi}_\nu^- - t^{-\nu}$, $(\nu = 1,2,\dots,\varkappa)$ where $\tilde{\phi}_\nu$ is the solution of the equation $\tilde{K}(\lambda_0)\tilde{\phi}_\nu = K(\lambda_0)t^{-\nu}$, should be added to each ψ_j, $j = 1,2,\dots,\ell$.

If the equation $\tilde{K}(\lambda_0)\tilde{\phi} = f$ is solvable, then obviously $P_j f = 0$ $(j = 1,2,\dots,r)$, and the corresponding partial solution $\tilde{\phi}_*$ can be determined from $\lim \tilde{\phi}_n(\lambda,t)$ as $\lambda \to \lambda_0$ and $n \to \infty$.

§13 THE NORMALLY SOLVABLE EQUATION WITH AN OPEN CONTOUR OF INTEGRATION

§13.1 Reduction to the Case of a Closed Contour

Let us consider the equation†

$$(13.1)\qquad K\phi \equiv a(t)\phi(t) + \frac{b(t)}{\pi i}\int_{\ell}\frac{\phi(\tau)d\tau}{\tau - t} + \int_{\ell} k(\tau,t)\,\phi(\tau)d\tau = f(t) \quad ,$$

$t \in \ell$, in which $\ell \subset \gamma$ and is the union of s non-overlapping arcs $\overset{\frown}{d_1 d_2}$, $\overset{\frown}{d_3 d_4}, \ldots, \overset{\frown}{d_{2s-1} d_{2s}}$. We assume that the functions a, b, f, k possess derivatives of order r, with respect to t, which are Hölder continuous on ℓ, and that $a-b \neq 0$, $a+b = 1$ on ℓ.

We introduce the function $\psi = \ln(a-b)$, and construct the Hermite polynomial(10) R such that

$$(13.2)\qquad R^{(\nu)}(d_k) = \psi^{(\nu)}(d_k) \ , \ \nu = 0,1,\ldots,r \ ; \quad k = 1,2,\ldots,2s.$$

We now solve the homogeneous Riemann problem

$$(13.3)\qquad u^{+} = \exp(R)\ u^{-} \quad ,$$

assuming that the functions $u^{\pm}$ do not become zero outside ℓ, and that $(u^{\pm})^{\pm 1}$ are integrable on ℓ. Applying known methods (see §8.3), we find that

$$(13.4)\qquad u^{\pm}(t) = \prod_{k=1}^{2s} (d_k - t)^{\lambda_k} \exp\left\{\pm \frac{1}{2} R(t) + \frac{1}{2\pi i}\int_{\ell}\frac{R(\tau)d\tau}{\tau - t}\right\} ,$$

where λ_k are integers satisfying

$$(13.5)\qquad -1 < \alpha_k + \lambda_k < 1 \ , \ \alpha_k + i\beta_k = \frac{(-1)^k \psi(d_k)}{2\pi i} \quad , \ k = 1,2,\ldots,2s \ .$$

Let us make the substitutions, (compare with (8.37)),

$$(13.6)\qquad \phi = \begin{cases} (\phi_1/u^+)^+ u^+ - (\phi_1/u^+)^- u^- \ , & \varkappa \leq 0 \ , \\ (\phi_1/u^+)^+ u^+ - (\phi_1/u^+)^- u^- + P_{\varkappa-1}(u^+ - u^-) \ , & \varkappa > 0 \ , \end{cases}$$

where $\varkappa$ is the index of equation (13.1) (and the boundary value problem (13.3)) defined by

---

(10) Instead of a Hermite polynomial we could have used a piecewise polynomial defined by $R(t) = R_j(t)$, $t \in \overset{\frown}{d_{2j-1} d_{2j}}$, $R_j^{(\nu)}(d_i) = \psi^{(\nu)}(d_i)$ for $i = 2j-1, 2j$; $j = 1,2,\ldots,s$; $\nu = 0,1,\ldots,r$.

† Throughout this section, the author consistently omits the complex parameter λ from the regular part of the complete equation, cf. §9, §12. (Eds.)

(13.7)
$$\varkappa = -\sum_{k=1}^{2s} \lambda_k \quad ,$$

and $P_{\varkappa-1}$ is an arbitrary polynomial of degree $\varkappa$-1. We obtain the equation

(13.8)
$$K_1\phi_1 \equiv \frac{1}{2}[1+\exp(\psi(t)-R(t))]\phi_1(t) + \frac{1}{2}[1-\exp(\psi(t)-R(t))]\frac{1}{\pi i}\int_\ell \frac{\phi_1(\tau)d\tau}{\tau - t} +$$
$$+ \int_\ell k_1(\tau,t)\ \phi_1(\tau)d\tau$$
$$= f_1(t) \equiv \begin{cases} f(t) \ , \varkappa \le 0 \ , \\ f(t) - K[P_{\varkappa-1}(u^+ - u^-)] \ , \varkappa > 0, \end{cases}$$

for $t \in \ell$, where

(13.9)
$$k_1(\tau,t) = \frac{1}{2\pi i}\left[\exp[\psi(t) - R(t)] - 1\right]\left[\frac{u^+(\tau) - u^+(t)}{u^+(\tau)(\tau - t)}\right] +$$
$$+ \frac{1}{2}k(\tau,t)\ [1+\exp(-R(\tau))] + \frac{1}{2\pi i\ u^+(\tau)}\int_\ell \frac{k(\xi,t)[u^+(\xi)-u^-(\xi)]}{\tau - \xi}d\xi$$

On continuing the definitions of f_1 and k_1 (with respect to t) to the whole of γ in such a way that we preserve the Hölder continuity of their r^{th} order derivatives, and assuming that

$$a_2(t) = \begin{cases} [1+\exp(\psi(t)-R(t))]/2 \ , & t \in \ell \ , \\ 1 \ , & t \in \gamma-\ell \ ; \end{cases}$$

$$b_2(t) = \begin{cases} [1-\exp(\psi(t)-R(t))]/2 \ , & t \in \ell \ , \\ 0 \ , & t \in \gamma-\ell \ ; \end{cases}$$

and

(13.10)
$$k_2(\tau,t) = \begin{cases} k_1(\tau,t) + \dfrac{b_2(t)}{\pi i}\displaystyle\int_{\gamma-\ell}\frac{k_1(\tau,u)du}{u - t} \ , \tau \in \ell \ , \ t \in \gamma, \\ 0 \ , \ \tau \in \gamma-\ell \ , \ t \in \gamma \ ; \end{cases}$$

$$f_2(t) = f_1(t) + \frac{b_2(t)}{\pi i}\int_{\gamma-\ell}\frac{f_1(\tau)d\tau}{\tau - t} \quad ,$$

we obtain the singular integral equation over the closed contour γ given by

$$(13.11) \qquad K_2\phi_2 \equiv a_2(t)\,\phi_2(t) + \frac{b_2(t)}{\pi i}\int_\gamma \frac{\phi_2(\tau)d\tau}{\tau - t} + \int_\gamma k_2(\tau,t)\phi_2(\tau)d\tau = f_2(t) \;,$$

$t \in \gamma$. From the above construction, the following theorem easily follows.

Theorem 13.1 (i) *Any solution of equation* (13.11) *is, on* ℓ*, a solution of equation* (13.8).

(ii) *If* ϕ_1 *is the solution of equation* (13.8) *then* $\phi_2 = \phi_1$, $t \in \ell$ *and* $\phi_2 = f_1 - \int_\ell k_1(\tau,t)\,\phi_1(\tau)d\tau$ *for* $t \in \gamma - \ell$*, is the solution of equation* (13.11).

(iii) *Every integrable solution of* (13.1) *can be represented in the form given by equation* (13.6).

(iv) *Let* ϕ_1 *be the solution of equation* (13.8). *If* $\mathcal{H} \geq 0$, *then* (13.6) *gives the solution of equation* (13.1). *If* $\mathcal{H} < 0$*, then* (13.6) *gives the solution of equation* (13.1) *provided that the functions* $(\phi_1/u^+)^\pm$ *have zeros of order greater than* $-\mathcal{H}+1$ *at infinity.*

(v) *Every integrable solution of equation* (13.11) *possesses an integrable derivative of order* r *on* γ.

Thus the problem of finding an approximate solution of a singular integral equation taken over open contours can be reduced to a similar problem for a closed contour. The solution of the latter problem will be sufficiently smooth if the original coefficients are sufficiently smooth.

§13.2 . Questions on the Improvement of Convergence

In order to improve the convergence of different methods for the approximate solution of singular integral equations, an analysis of the structure of the solutions of appropriate equations is of considerable importance. We can do this by regularizing the equations since, for Fredholm integral equations of the second kind given by

$$(13.12) \qquad \phi(t) = f(t) - \int_\Gamma T(\tau,t)\,\phi(\tau)d\tau \;,$$

it is not difficult to deduce the basic structure of ϕ from the structure of f and T.

Let us examine this problem from a different point of view, by taking as the canonical equation

$$(13.13) \qquad K\phi \equiv a(t)\,\phi(t) + \frac{b(t)}{\pi i}\int\limits_{|\tau|=1}\frac{\phi(\tau)d\tau}{\tau - t} + \int\limits_{|\tau|=1} k(\tau,t)\,\phi(\tau)d\tau = f(t)\ ,$$

$|t| = 1$, with sufficiently smooth functions a, b, k and f. The original equation is first reduced to this canonical form by various transformations of the variables. This approach corresponds to the situation where, having prepared a standard program for the solution of equations of the form (13.13), we wish to solve other equations with this program. Obviously they must be transformed to the form of equation (13.13). It should be noted that although regularization allows the structure of the required solution to be more readily exhibited, the determination of a specific structure can be very difficult when programming a particular problem, and even more difficult when programming a class of problems. For the construction of a standard program for the solution of a wide class of singular integral equations, it is desirable to represent the answer in a simple analytic form (for example in the form of a polynomial, $\sum\limits_{k=-n}^{n} \alpha_k t^k$). Then the basic estimate for the rate of convergence depends on the degree of smoothness of the solution (which for equation (13.13) can be easily determined from the degree of smoothness of a, b, k and f). A detailed analysis of the structure of the solution is not then absolutely necessary.

Suppose the original equation is given by

$$(13.14) \qquad K\phi \equiv a(t)\,\phi(t) + \frac{b(t)}{\pi i}\int\limits_{\Gamma}\frac{\phi(\tau)d\tau}{\tau - t} + \int\limits_{\Gamma} k(\tau,t)\,\phi(\tau)d\tau = f(t)\ ,$$

$t \in \Gamma$, where Γ is the union of a finite number of non-overlapping smooth arcs, and a, b, f and k satisfy the same conditions on Γ as we had in equation (13.1). To transform equation (13.14) to the form of (13.13), we first reduce it to an equation of the form (13.1) by a change of independent variables. The equation taken over the contour ℓ is then reduced to the required form by the methods described in the previous section.

Let us now assume that in equation (13.13), the functions a, b, f and k (considered as a function of t), possess derivatives of order r which are Hölder continuous everywhere on γ ($|t| = 1$), except at the points $d_1, d_2, \ldots, d_s \in \gamma$, where a, b, f and k may have finite discontinuities. We assume that the points $d_1, d_2, \ldots, d_s$ are arranged on γ in the positive (counter-clockwise) direction. The contour γ is then considered as comprising the arcs $\breve{d_1 d_2}$, $\breve{d_2 d_3}, \ldots, \breve{d_{s-1} d_s}$, $\breve{d_s d_1}$. In this way, equation (13.13) is converted into into an equation over a set of open arcs where the assumptions on a, b, f and k are the same as those used when considering equation (13.1). The only difference is that the arcs comprising the contour of integration now have common points. A careful analysis of the results of the previous section shows that these will remain valid if each of the common points d_j is considered as a double point. One of the double points is taken as belonging to the arc $\widetilde{d_{j-1}d_j}$ with the function values at this point coinciding with the limiting values of the corresponding functions along $\widetilde{d_{j-1}d_j}$; the other belongs to the arc $\widetilde{d_jd_{j+1}}$ $(d_{s+1} = d_1)$ with the function values coinciding with their limiting values along the arc $\widetilde{d_jd_{j+1}}$. In addition, for each arc $\widetilde{d_{j-1}d_j}$ we construct its Hermite polynomial.

Thus considering equation (13.14) as an equation over the arcs $\widetilde{d_1d_2}, \widetilde{d_2d_3}, \ldots, \widetilde{d_sd_1}$, we can, by the methods of the previous section, reduce it to the form (13.13) with coefficients possessing derivatives of order r on γ.

A similar analysis can be done when a, b, f and k possess lower order derivatives at a finite number of points than they have at the remaining points.

Let a, b and k in equation (13.13) possess derivatives of order r which are Hölder continuous on γ, and let f have the same degree of smoothness except at a finite number of points $d_1, d_2, \ldots, d_s$ of γ, where it is integrable. Assuming, for simplicity, that $a+b \equiv 1$, we construct the Hermite polynomial R where

$$R^{(\nu)}(c_k) = \psi^{(\nu)}(c_k) \quad , \quad k = 1,2,\ldots,s \; ; \; \nu = 0,1,\ldots,r \; , \tag{13.15}$$
$$\psi = \ln(a-b) \quad ,$$

and make the substitution

$$\phi = \phi_2 + (f/u^+)^+u^+ - (f/u^+)^-u^- \quad , \tag{13.16}$$

where $u^{\pm}$ are solutions of the Riemann boundary value problem

$$u^+(t) = \exp[R(t)] \, . \, u^-(t) \; , \; t \in \gamma \; , \tag{13.17}$$

which do not vanish on γ. We obtain the equation

$$\begin{aligned} K\phi_2 &\equiv a(t)\,\phi_2(t) + \frac{b(t)}{\pi i}\int_\gamma \frac{\phi_2(\tau)d\tau}{\tau - t} + \int_\gamma k(\tau,t)\,\phi_2(\tau)d\tau \\ &= (\exp[\psi(t) - R(t)] - 1)\left(\frac{f}{u^+}\right)^-(t)\,u^+(t) \\ &\quad - \int_\gamma k(\tau,t)\left[\left(\frac{f}{u^+}\right)^+(\tau)\,u^+(\tau) - \left(\frac{f}{u^+}\right)^-(\tau)\,u^-(\tau)\right]d\tau \\ &= f_2(t) \; . \end{aligned} \tag{13.18}$$

It can be verified that f_2 possesses a Hölder continuous derivative of order $(r-1)$ on γ.

In the case when f, a, b and k (considered as a function of t) have infinite (but integrable) discontinuities at a finite number of points, then multiplying both sides of (13.13) by a suitably chosen function (where it is necessary to ensure that the new coefficients a,b do not become zero simultaneously), we again reduce the equation to the above case.

So far we have been considering improvement of the convergence of direct methods (i.e. least squares, Ritz-Galerkin, collocation etc.) with respect to n, where n is the number of parameters in the representation of the approximate solution. Another problem is that of improving the convergence of the iterative algorithms (i.e. coordinate-wise descent, gradient descent etc.) where the parameters are determined for a fixed n. Also we are interested in the improvement of the rate of convergence of general iterative processes of steepest descent type, and of simple iteration. We shall merely note some literature on this topic. L.A. Lyusternik [95] discusses the improvement of convergence of the Gauss-Seidel algorithm. The improvement of convergence of iterative processes of steepest descent type, by the construction of multistep processes, is discussed by L.V. Kantorovich and G.P. Akilov [75]. A discussion of the improvement of convergence of the method of simple iteration is given by Householder [141], (see also §5.1).

§13.3 Application of the Method of Least Squares

Consider equation (13.14) in which Γ is the union of a finite set of non-overlapping arcs Γ_j with end-points d_{2j-1} and d_{2j}, for $j = 1,2,\ldots,s$.

We calculate

$$(13.19)\qquad \alpha_k + i\beta_k = (-1)^k\psi(d_k)/2\pi i \quad , \quad \psi(t) = \ln[a(t) - b(t)] \ ,$$

$k = 1,2,\ldots,2s$, and select λ_k according to the conditions

$$(13.20)\qquad -1 < \alpha_k + \lambda_k < 1 \quad , \quad k = 1,2,\ldots,2s \ .$$

Let $d_1, d_2,\ldots,d_m$ be the non-special ends, and let the solution of equation (13.14) be required which belongs to the class $h(d_{m_1+1},\ldots,d_m)$, (see §8.3). Putting

$$(13.21)\qquad \begin{aligned} &0 < \delta_k = -\alpha_k - \lambda_k < 1 \quad , \quad k = 1,2,\ldots,m_1 \ , \\ &0 < \delta_k = \alpha_k + \lambda_k < 1 \ , \quad k = m_1+1,\ldots,m \ , \end{aligned}$$

we consider equation (13.14) to be a linear equation in the Hilbert space $L_2(\Gamma,\rho)$, where ρ is defined by equation (7.7) with $p = 2$.

As in §12.1, we examine only those aspects of the application of the method of least squares which were not considered in §3, and which are concerned with the specific character of equation (13.14) in the space $L_2(\Gamma,\rho)$.

We construct the function ω where

$$(13.22) \qquad \omega(t) = R_j(t) = \psi(d_{2j-1}) + \frac{\psi(d_{2j}) - \psi(d_{2j-1})}{d_{2j} - d_{2j-1}} (t - d_{2j-1}) \quad ,$$

$t \in \Gamma_j$, $j = 1,2,\ldots,s$, and the functions $u^{\pm}$ (integrable on Γ) given by

$$(13.23) \qquad u^{\pm}(t) = \prod_{k=1}^{2s} (d_k - t)^{\lambda_k} \exp\left[\pm \frac{1}{2} \omega(t) + \frac{1}{2\pi i} \int_\Gamma \frac{\omega(\tau) d\tau}{\tau - t} \right] \quad ,$$

which together comprise a solution of the Riemann boundary value problem

$$(13.24) \qquad u^+(t) = \exp[\omega(t)] \, . \, u^-(t) \quad , \quad t \in \Gamma \, .$$

Together with equation (13.14), we introduce the equation

$$\tilde{K}\tilde{\phi} \equiv u^+(t)\, \tilde{\phi}^+(t) - u^+(t) \exp(\psi(t) - \omega(t))\, \tilde{\phi}^-(t) +$$

$$(13.25) \qquad + \int_\Gamma k(\tau,t) \, [\tilde{\phi}^+(\tau)\, u^+(\tau) - \tilde{\phi}^-(\tau)\, u^-(\tau)] d\tau$$

$$= f(t) \quad , \quad t \in \Gamma \quad ,$$

which we shall consider as a linear mapping from $L_2(\Gamma, \rho^{-1})$ into $L_2(\Gamma,\rho)$.

We apply the method of least squares to equation (13.25), assuming that the operator $\tilde{K}$ has an inverse. This last condition implies that if equation (13.14) is solvable then the equations

$$(13.26) \qquad \phi^+ = \tilde{\phi}^+ u^+ \quad , \quad \phi^- = \tilde{\phi}^- u^- \quad , \quad \phi = \phi^+ - \phi^- \quad ,$$

give that solution of equation (13.14) for which the functions $\phi^{\pm}$ have zeros of the highest order at infinity with respect to all the remaining solutions of the chosen class.

From the structure of equations (8.37), we choose for the coordinate system $e_1, e_2, \ldots, e_n, \ldots$, the functions $t^k/u^+(t)$, where $k = 0,1,\ldots$. For the approximate solution of equation (13.25), we put

$$(13.27) \qquad \tilde{\phi}_n(t) = \sum_{k=0}^{n} \alpha_k^0 \frac{t^k}{u^+(t)} \quad , \qquad ^{(11)}$$

(11) MacCamy [161] has proposed to seek the approximate solution of a singular integral equation in the form of a set of linear manifolds with respect to α_k, each of which has the structure of the required solution in a neighbourhood of one of the ends, and only in this neighbourhood does it approximate the solution.

and choose the coefficients α_k^0 so as to minimize

$$\Delta_n(P) = \|f(t) - \sum_{k=0}^{n} \alpha_k y_k(t)\|_{L_2(\Gamma,\rho)}$$

(13.28)

$$P(\alpha_0,\alpha_1,\ldots,\alpha_n) \ , \ y_k = \tilde{K}(t^k/u^+) \ .$$

Let us write out y_k in more detail. We have

$$y_k(t) = t^k + (1 - \exp(\psi(t) - w(t)))\, u^+(t) \left(\frac{t^k}{u^+}\right)^-(t)$$

(13.29)

$$+ \int_\Gamma k(\tau,t) \left[\left(\frac{\tau^k}{u^+}\right)^+(\tau)\, u^+(\tau) - \left(\frac{\tau^k}{u^+}\right)^-(\tau)\, u^-(\tau) \right] d\tau \ ,$$

$$\left(\frac{t^k}{u^+}\right)^{\pm}(t) = \pm \frac{t^k}{2u^+(t)} + \frac{1}{2\pi i} \int_\Gamma \frac{\tau^k d\tau}{u^+(\tau)(\tau - t)} \ .$$

With this cumbersome representation, the use of the simplest quadrature formula with mesh refinement in the neighbourhood of the ends $d_1,d_2,\ldots,d_{2s}$, will give an appropriate method for the calculation of the functions y_k, and also for the calculation of the corresponding scalar products in $L_2(\Gamma,\rho)$. In particular cases, methods incorporating the singularities may also be useful. (see §11.2, §11.3 and [73]).

Instead of the piecewise-linear function (13.22), we can use a piecewise-polynomial function w which, together with its derivatives up to a certain order, coincides with the function ψ at the end points of the contour Γ. Then the function ϕ_1 where

$$\phi_1(t) = u^+(t)\, \tilde{\phi}(t) = f(t) - \int_\Gamma k(\tau,t)[\tilde{\phi}^+(\tau)u^+(\tau) - \tilde{\phi}^-(\tau)u^-(\tau)]d\tau$$

(13.30)

$$- (1 - \exp(\psi(t) - w(t)))u^+(t)\, \tilde{\phi}^-(t) \ ,$$

and the functions y_k (see (13.29)) will possess derivatives on Γ provided that f and k are sufficiently smooth.

Often, it is more advantageous to seek an approximate solution of equation (13.25) in the form

$$\tilde{\phi}_n(t) = \sum_{k=0}^{n} \alpha_k^0 \frac{T_k(t)}{u^+(t)} \ , \tag{13.31}$$

where T_k are the polynomials which are orthogonal on Γ with respect to the weight function $|u^+|^{-1}$. Instead of the weight function ρ of equation (7.7), it is also more convenient to take $|u^+|^{-1}$. Let us examine in more detail equations which frequently arise (see [107]) in which $\alpha_k + \lambda_k = \pm 1/2$, and ψ is a constant. We shall assume for simplicity, that the contour of integration Γ is the line segment $[a,b]$ of the real axis. The weight function ρ is given by one of the four functions

$$\rho(x) = [(b-x)(x-a)]^{\frac{1}{2}} (b-x)^{\epsilon_1} (x-a)^{\epsilon_2} , \quad \epsilon_1,\epsilon_2 = 0,-1 ,$$

which differs from the corresponding function $|u^+|^{-1}$, only by a constant.

Let us consider the behaviour at infinity of the integrals

$$\frac{1}{2\pi i} \int_a^b \frac{\delta(x)\ T_k^*(x)dx}{x - z} ,$$

where $\delta(x) = \rho^{-1}(x)$, and T_k^* denotes the polynomial of degree k which is orthogonal on $[a,b]$ with respect to the weight function δ. It is known (see [102]) that

$$\frac{1}{2\pi i} \int_a^b \frac{\delta(x)dx}{x - z} = - \frac{P_k(z)}{T_k^*(z)} + o\left(\frac{1}{z^{2k+1}}\right) ,$$

(13.32)

$$\text{where} \qquad P_k(z) = \frac{1}{2\pi i} \int_a^b \delta(x) \frac{T_k^*(x) - T_k^*(z)}{x - z} dx .$$

Hence

$$(13.33) \qquad \frac{1}{2\pi i} \int_a^b \frac{\delta(x)\ T_k^*(x)dx}{x - z} = P_k(z) + \frac{T_k^*(z)}{2\pi i} \int_a^b \frac{\delta(x)dx}{x - z} = o\left(\frac{1}{z^{k+1}}\right) .$$

We assume that equation (13.14), considered as a linear mapping from $L_2([a,b],\rho)$ into $L_2([a,b],\rho)$, has a unique solution ϕ_* for which the integral $\int_a^b \phi_*(x)dx/(x-z)$ has a zero of highest order at infinity with respect to all the remaining solutions. We shall look for an approximate solution of equation (13.14) in the form

$$(13.34) \qquad \phi_n(x) = \sum_{k=\varkappa}^{n} a_k\ \delta(x)\ T_k^*(x) , \quad n \geq \varkappa,$$

where $\varkappa = - \sum_{k=1}^{2s} \lambda_k$ is the index of (13.14) which, from the above assumption, must be greater than or equal to zero. The values of a_k are determined from the condition that they minimize

$$\Delta_n = \|f(x) - \sum_{k=\varkappa}^{n} a_k\, y_k(x)\|_{L_2([a,b],\rho)} \quad , \tag{13.35}$$

where $y_k = K(\delta T_k^*)$.

Suppose we consider the operator K as a linear mapping from the space $\tilde{L}_2([a,b],\rho)$, with elements ϕ which can be written as $\phi(x) = \sum_{k=\varkappa}^{\infty} a_k\, \delta(x)\, T_k^*(x)$, into the space $L_2([a,b],\rho)$. Then the best conditions for the application of the method of least squares are obtained when the operator K has an inverse.[12]

Using equations (7.12), we obtain

$$y_k(x) = a(x)\,\delta(x)\,T_k^*(x) + ib(x)\,T_k^{\epsilon_1,\epsilon_2}(x) + \int_a^b k(x_0, x)\,\delta(x_0)\,T_k^*(x_0)dx_0 \quad ,$$

where

$$T_k^{\epsilon_1,\epsilon_2}(x) = \begin{cases} -\,i\,T_k(x) \;, & (\epsilon_1 = 0,\ \epsilon_2 = -1) \;, \\ i\,T_k(x) \;, & (\epsilon_1 = -1,\ \epsilon_2 = 0) \;, \\ -\,i\,T_{k-1}(x) \;, & (\epsilon_1 = \epsilon_2 = 0) \;, \\ i\,T_{k+1}(x) \;, & (\epsilon_1 = \epsilon_2 = -1) \;, \end{cases} \tag{13.36}$$

and T_k are polynomials which are orthogonal on $[a,b]$ with respect to the weight function ρ, and have the same normalization as T_k^*.

To use the various least squares algorithms (see §3) it is necessary to be able to calculate the scalar products (y_k, y_j) and (f, y_j) in $L_2([a,b],\rho)$. It is obvious from equations (13.36) that these can be calculated exactly when f, a, b and k are polynomials.

As an illustrative example, let us calculate the upward force P_y due to a steady stream of incompressible, inviscid fluid flowing past a circular arc of length $2a$, whose chord lies along the x-axis. The required force is found (see [89]) from the solution of the equation

$$KT \equiv \frac{1}{2\pi i}\int_{-a}^{a} \cot\left(\frac{x_0 - x}{2}\right) T(x_0)dx_0 = 2\sin x \;, \tag{13.37}$$

which is bounded in neighbourhoods of the points $\pm a$; and the equation

$$P_y = d\ \left|\int_{-a}^{a} T(x)dx\right| \;, \tag{13.38}$$

---

(12) If $\varkappa < 0$, we look for an approximate solution in the form used when $\varkappa = 0$. The operator K will not now have an inverse.

the coefficient d being given by the product of the density of the fluid and its speed at infinity. If we rewrite equation (13.37) in the form

$$(13.39)\quad KT \equiv \frac{i}{\pi i}\int_{-a}^{a}\frac{T(x_0)dx_0}{(x_0-x)} + \frac{1}{2\pi}\int_{-a}^{a}\left[-\frac{(x_0-x)}{6} - \frac{(x_0-x)^3}{360} + \ldots\right] T(x_0)dx_0$$

$$= 2\sin x\ ,$$

we see that in this example, $a = 0$, $b = i$ and consequently

$$(13.40)\quad \begin{aligned} &G = \frac{a-b}{a+b} = -1\ , \quad \alpha_1 + i\beta_1 = -\ln(-1)/2\pi i = -1/2\ , \\ &\alpha_2 + i\beta_2 = \ln(-1)/2\pi i = 1/2\ , \quad \lambda_1 = 1\ , \lambda_2 = 0\ , \\ &\varkappa = -1\ , \quad \rho(x) = (1-x^2)^{-\frac{1}{2}} \\ &T_n(x) = (a^2 - x^2)^{\frac{1}{2}}\,[\alpha_0 T_0^*(x) + \alpha_1 T_1^*(x) + \ldots + \alpha_n T_n^*(x)]\ . \end{aligned}$$

If we normalize T_k^* by the condition that

$$\frac{2}{\pi a^2}\int_{-a}^{a}(a^2 - x^2)^{\frac{1}{2}}\,[T_k^*(x)]^2 dx = 1\ ,$$

then

$$(13.41)\quad \begin{aligned} &T_k^*(x) = a\sin[(k+1)\arccos(x/a)]\,.\,(a^2 - x^2)^{-\frac{1}{2}}\ , \\ &T_k(x) = a\cos[k\arccos(x/a)]\ . \end{aligned}$$

From the form of the approximate solution, it is sufficient to know only α_0 for the evaluation of P_y, since

$$(13.42)\quad P_y \approx d\ \left|\int_{-a}^{a} T_n(x)dx\right| = d\pi a^2|\alpha_0|/2\ .$$

Let us approximate to equation (13.39) by the equation

$$(13.43)\quad K_1T_1 = \frac{1}{\pi}\int_{-a}^{a}\frac{T_1(x_0)dx_0}{x_0 - x} - \frac{1}{12\pi}\int_{-a}^{a}(x_0 - x)T_1(x_0)dx_0$$

$$= 2(x - x^3/6)\ ,$$

and calculate T_1 from the first four terms

$$(13.44)\quad T_{1,4}(x) = (a^2-x^2)^{\frac{1}{2}}\,(\alpha_0 T_0^*(x) + \alpha_1 T_1^*(x) + \alpha_2 T_2^*(x) + \alpha_3 T_3^*(x))\ .$$

We easily find that

$$T_0^*(x) = 1 \ , \ T_1^*(x) = 2x/a \ ; \ \ T_2^*(x) = 4(x/a)^2 - 1 \ ,$$

$$T_0(x) = a \ , \ T_1(x) = x \ , \ T_2(x) = (2/a)x^2 - a \ , \ T_3(x) = (4/a^2)x^2 - 3x \ ,$$

$$K\delta T_0^* = -(1 - a^2/24)T_1 \ , \ K\delta T_1^* = -(a^2/48)T_0 - T_2 \ ,$$

$$K\delta T_k^* = - T_{k+1} \ , \ k \geq 2 \ ;$$

$$2(x - x^3/6) = (2 - a^2/4)T_1(x) - (a^3/12)T_3(x) \ .$$

With respect to α_0, we obtain the equation

$$(1 - a^2/24)^2\alpha_0 = -(1 - a^2/24)(2 - a^2/4) \ ,$$

from which

$$\alpha_0 = -(2 - a^2/4)(1 - (a^2/24) + \ldots) \ , \ |a| < \sqrt{24} \ .$$

Thus

$$P_y \simeq d\pi(a^2 - a^4/12) \ , \quad |a| < \sqrt{24} \ . \tag{13.45}$$

The exact value of P_y is given by

$$P_y = 4d\pi \sin^2(a/2) = d\pi(a^2 - a^4/12 + \ldots) \ . \tag{13.46}$$

§14 OTHER CASES OF A SINGLE EQUATION

§14.1 Equations in the Exceptional Case

We shall now consider equation (12.1) assuming that $a^2 - b^2$ has zeros of integer order at isolated points of γ, (see §8.2). In this case, equation (12.1) will not be normally solvable, (see §9.5). Consequently for its approximate solution it is necessary to apply either the method of regularization (see §§2.1, 3 and 5), or the method of quasi-solutions (§2.2).

In both these methods, a basic weakness is the absence of a general method for finding an effective estimate of the error $\delta_R = \|\overset{*}{x} - A^* \omega_R\|$. Here $\overset{*}{x}$ denotes the unique solution of $Ax = y$ which is orthogonal to all the zeros of the operator A, A^* is the operator conjugate to A, and $\|\omega_R\| = R$.

We shall now show that in the case of equation (12.1), such estimates may be obtained. Let us introduce the equations

$$K^0\phi \equiv a(t)\phi(t) + \frac{b(t)}{\pi i}\int_\gamma \frac{\phi(\tau)d\tau}{\tau - t} = f(t) \tag{14.1}$$

and (see [33, §24.4])

$$K^{0*} \equiv \overline{a(t)}\psi(t) + \frac{1}{\pi i}\int_\gamma \frac{\overline{b(\tau)}\psi(\tau)d\tau}{\tau - t} = \phi(t) \quad . \tag{14.2}$$

Let

$$a(t) - b(t) = \left[\prod_{k=1}^{\mu} (t-\alpha_k)^{m_k}\right] r(t) \;, \quad a(t) + b(t) = \left[\prod_{j=1}^{\nu} (t-\beta_j)^{p_j}\right] s(t), \tag{14.3}$$

where r and s do not vanish on γ, and suppose that the function $G_1 = r/s$ has index $\varkappa_1$ on γ so that

$$\varkappa_1 = \frac{1}{2\pi}\left[\arg \frac{r(t)}{s(t)}\right]_\gamma \quad . \tag{14.4}$$

On using the notation of §8.2, since $g(t) = f(t)\Big/\left[s(t)\prod_{j=1}^{\nu}(t-\beta_j)^{p_j}\right]$, then from equation (8.14) the general solution of equation (14.1) is given by

$$\begin{aligned} \phi(t) &= \phi^+(t) - \phi^-(t) \;, \\ \phi^+(t) &= \frac{\exp[\psi_1^+(t)]}{\prod_{j=1}^{\nu}(t-\beta_j)^{p_j}}\,[g_1^+(t) + P^{(1)}_{\varkappa_1+m-1}(t)] \;, \\ \phi^-(t) &= \frac{\exp[\psi_1^-(t)]}{t^{\varkappa_1}\prod_{k=1}^{\mu}(t-\alpha_k)^{m_k}}\,[g_1^-(t) + P^{(1)}_{\varkappa_1+m-1}(t)] \;, \end{aligned} \tag{14.5}$$

where

$$g_1^{\pm}(t) = \pm \frac{f(t)}{2s(t)\exp[\psi_1^{+}(t)]} + \frac{1}{2\pi i}\int_\gamma \frac{f(\tau)d\tau}{s(\tau)\exp[\psi_1^{+}(\tau)].(\tau-t)} ,$$

(14.6)

$$\psi_1^{\pm}(t) = \pm\frac{1}{2}\ln[t^{-\varkappa_1} G_1(t)] + \frac{1}{2\pi i}\int_\gamma \frac{\ln[\tau^{-\varkappa_1} G_1(\tau)]d\tau}{\tau - t} .$$

In (14.5), $P^{(1)}_{\varkappa_1+m-1}$ denotes an arbitrary polynomial of degree $\varkappa_1+m-1$, such that

$$P^{(1)}_{\varkappa_1+m-1} \equiv 0 \quad \text{for } \varkappa_1+m-1 < 0 .$$

To solve equation (14.2) we introduce the Cauchy integral

$$\chi(z) = \frac{1}{2\pi i}\int_\gamma \frac{\overline{b(\tau)}\psi(\tau)d\tau}{\tau - z} ,$$

and note that

$$\overline{b(t)}\psi(t) = \chi^{+}(t) - \chi^{-}(t) , \quad \frac{1}{\pi i}\int_\gamma \frac{\overline{b(\tau)}\psi(\tau)d\tau}{\tau - t} = \chi^{+}(t) + \chi^{-}(t) . \tag{14.7}$$

From equations (14.2) and (14.7), it follows that

$$\overline{a(t)}\psi(t) = \phi(t) - \chi^{+}(t) - \chi^{-}(t) . \tag{14.8}$$

From (14.7) and (14.8) we obtain

$$(\overline{a(t)} + \overline{b(t)})\psi(t) = \phi(t) - 2\chi^{-}(t) \text{ and } (\overline{a(t)} - \overline{b(t)})\psi(t) = \phi(t) - 2\chi^{+}(t), \tag{14.9}$$

from which

$$\chi^{+}(t) = \left(\frac{\overline{a(t)} - \overline{b(t)}}{\overline{a(t)} + \overline{b(t)}}\right)\chi^{-}(t) + \frac{\overline{b(t)}\phi(t)}{\overline{a(t)} + \overline{b(t)}} . \tag{14.10}$$

We solve this Riemann boundary value problem in the same way as in §8.2, taking into account that

$$\overline{a(t)} - \overline{b(t)} = \left[\prod_{k=1}^{\mu} (1-t\bar{\alpha}_k)^{m_k}\right] \overline{r(t)} . t^{-m} ,$$

(14.11)

$$\overline{a(t)} + \overline{b(t)} = \left[\prod_{j=1}^{\nu} (1-t\bar{\beta}_j)^{p_j}\right] \overline{s(t)} . t^{-p} ,$$

and that the function $G_2 = [\overline{r/s}]t^{p-m}$ has index

$$\varkappa_2 = -\varkappa_1 + p - m . \tag{14.12}$$

We obtain

$$\chi^+(t) = \frac{\exp[\psi_2^+(t)]}{\prod_{j=1}^{\nu}(1-t\bar{\beta}_j)^{p_j}}\,[g_2^+(t) + P^{(2)}_{-\varkappa_1+p-1}(t)\,] ,$$

(14.13)

$$\chi^-(t) = \frac{\exp[\psi_2^-(t)]}{t^{-\varkappa_1+p-m}\prod_{k=1}^{\mu}(1-t\bar{\alpha}_k)^{m_k}}\,[g_2^-(t) + P^{(2)}_{-\varkappa_1+p-1}(t)\,] ,$$

where

$$g_2^{\pm}(t) = \pm\frac{\overline{b(t)}t^p\,\phi(t)}{s(t)\exp[\psi_2^+(t)]} + \frac{1}{2\pi i}\int_\gamma \frac{\overline{b(\tau)}\tau^p\,\phi(\tau)d\tau}{\overline{s(\tau)}(\tau-t)\exp[\psi_2^+(\tau)]} ,$$

(14.14)

$$\psi_2^{\pm}(t) = \pm\frac{1}{2}\,\ell n\,[t^{-\varkappa_2}G_2(t)] + \frac{1}{2\pi i}\int_\gamma \frac{\ell n[\tau^{-\varkappa_2}G_2(\tau)]d\tau}{\tau - t} .$$

From equation (14.9), we finally obtain

$$\psi(t) = \frac{\phi(t) - 2\chi^+(t)}{\overline{a(t)} - \overline{b(t)}}$$

(14.15)

$$= \left\{\frac{t^m\exp[\psi_2^+(t)]}{\overline{r(t)}}\right\}\frac{\left\{\phi(t)\exp[-\psi_2^+(t)]\prod_{j=1}^{\nu}(1-t\bar{\beta}_j)^{p_j} - 2g_2^+(t) - 2P_{-\varkappa_1+p-1}(t)\right\}}{\prod_{j=1}^{\nu}(1-t\bar{\beta}_j)^{p_j}\prod_{k=1}^{\mu}(1-t\bar{\alpha}_k)^{m_k}} .$$

Let us assume that equation (14.1) is solvable in a class of bounded and sufficiently smooth functions. This implies that in equation (14.5) we can select a polynomial $P^{(1)}_{\varkappa_1+m-1}$ so that ϕ^+ and ϕ^- are bounded, and also such that ϕ^- becomes zero at infinity. If ϕ is to possess a derivative of order r at all points of γ, it is sufficient that both G_1 and g_1 have derivatives of order r on γ and in addition that g_1 has derivatives of order $r+p_j$ and $r+m_k$ at the points β_j and α_k respectively.

It is easily seen that the solvability of equation (14.1) does not necessarily imply the solvability of equation (14.2) (in the class of bounded functions) where ϕ is chosen as the solution of equation (14.1). If $-\varkappa_1 < m$, then $\chi^{\pm}$ and ψ will not, in general, be bounded. Nevertheless as we have shown in §2.1, it is always possible to find a function ψ_R, with $\|\psi_R\|_{L_2(\gamma)} = R$, such that $\delta_R = \|K^{0*}\psi_R - \overset{*}{\phi}\|_{L_2(\gamma)} \to 0$ as $R \to \infty$, where $\overset{*}{\phi}$ is the unique solution of equation (14.1) which is orthogonal to all the zeros of the operator K^0.

Let us construct the function ψ_R. To do this we note that the general solution of the equation $K^0 u = 0$ has the form

$$u^+(t) = \frac{\exp[\psi_1^+(t)]\, P^{(1)}_{\varkappa_1+m-1}(t)}{\prod_{j=1}^{\nu}(t-\beta_j)^{p_j}}, \quad u^-(t) = \frac{\exp[\psi_1^-(t)]\, P^{(1)}_{\varkappa_1+m-1}(t)}{t^{\varkappa_1}\prod_{k=1}^{\mu}(t-\alpha_k)^{m_k}}. \tag{14.16}$$

Consequently a non-trivial solution in the class of bounded functions only exists if $\varkappa_1 > p$. In addition

$$P^{(1)}_{\varkappa_1+m-1}(t) = \prod_{j=1}^{\nu}(t-\beta_j)^{p_j}\prod_{k=1}^{\mu}(t-\alpha_k)^{m_k}\, P_{\varkappa_1-p-1}(t),$$

where $P_{\varkappa_1-p-1}$ is an arbitrary polynomial of degree $\varkappa_1-p-1$. Hence the system of functions $\{u_\ell\}$ defined by

$$u_\ell(t) = u_\ell^+(t) - u_\ell^-(t),$$

$$u_\ell^+(t) = \exp[\psi_1^+(t)]\, t^\ell \prod_{k=1}^{\mu}(t-\alpha_k)^{m_k}, \quad u_\ell^-(t) = \exp[\psi_1^-(t)]\, t^{\ell-\varkappa_1}\prod_{j=1}^{\nu}(t-\beta_j)^{p_j} \tag{14.17}$$

for $\ell = 0,1,\ldots, \varkappa_1-p-1$,

will be a complete system for the non-trivial solutions of the equation $K^0u = 0$. From the solutions (14.5) it is not difficult to select a function $\overset{*}{\phi}$ which is orthogonal to all the functions u_ℓ. Consequently, the function $\overset{*}{\phi}$ will be of the form given in (14.5), but all the coefficients of the polynomial $P^{(1)}_{\varkappa_1+m-1}$ will now be defined. Let us put

$$\chi_R^+(t) = \frac{\overset{*}{g}{}_2^+(t)\exp[\psi_2^+(t)]}{P_p^+(t)}, \quad \chi_R^-(t) = \frac{\overset{*}{g}{}_2^-(t)\exp[\psi_2^-(t)]}{t^{-\varkappa_1+p-m}\,Q_m^-(t)},$$

$$\begin{aligned}\psi_R(t) &= \frac{\overset{*}{\phi}(t)[\overline{b(t)}/\overline{b_R(t)}] - 2\chi_R^-(t)}{\overline{a_R(t)} - \overline{b_R(t)}}\\ &= \frac{\overset{*}{\phi}(t)[\overline{b(t)}/\overline{b_R(t)}] - 2\chi_R^+(t)}{\overline{a_R(t)} + \overline{b_R(t)}},\end{aligned} \tag{14.18}$$

$$\|\psi_R\|_{L_2(\gamma)} = R,$$

where P_p^+ and Q_m^- are polynomials of degree p,m respectively, having no zeros on γ. In addition we have

$$\overline{a_R(t)} - \overline{b_R(t)} = Q_m^-(t)\ t^{-m}\ \overline{r(t)} \ , \tag{14.19}$$
$$\overline{a_R(t)} + \overline{b_R(t)} = P_p^+(t)\ t^{-p}\ \overline{s(t)} \ ,$$

and $\overset{*}{g}{}_2^{\pm}$ are determined by the equations used to define $g_2^{\pm}$, but with $\overset{*}{\phi}$ substituted in place of ϕ.

Theorem 14.1. *Let γ_s^+ $(s = 1,2,\dots,p)$ and δ_r^- $(r = 1,2,\dots,m)$ be the zeros of the polynomials P_p^+ and Q_m^- respectively.*(13) *If, as $R \to \infty$, the following conditions are satisfied*

$$\|(1-b/\bar{b}_R)\overset{*}{\phi}\| \to 0 \ , \ \|(\bar{a}-\bar{b})/(\bar{a}_R-\bar{b}_R) - 1\| \to 0 \ ,$$
$$\|(\bar{a}+\bar{b})/(\bar{a}_R+\bar{b}_R) - 1\| \to 0 \ , \tag{14.20}$$
$$\max_{1\le r\le m} \prod_{k=1}^{\mu} |1 - \delta_r^- \bar{\alpha}_k|^{m_k} \|1/Q_m^-\| \to 0 \ ,$$

and

$$\max_{1\le s\le p} \prod_{j=1}^{\nu} |1 - \gamma_s^+ \bar{\beta}_j|^{p_j} \|1/P_p^+\| \to 0 \ ,$$

then

$$\|K^{0*}\ \psi_R - \overset{*}{\phi}\| \to 0 \quad as \quad R \to \infty \ . \tag{14.21}$$

Proof. We shall first show that χ_R^- becomes zero at infinity. If $\mathcal{H}_1 \le p$, this is obvious. Suppose $\mathcal{H}_1 > p$, then

$$(\overset{*}{\phi},\ u_\ell) = 0 \ , \ \ell = 0,1,\dots,\mathcal{H}_1 - p - 1 \ , \tag{14.22}$$

where the functions u_ℓ are defined by equations (14.17). We introduce the function

$$\overset{*}{g}_2(z) = \frac{1}{2\pi i} \int_\gamma \frac{\overline{b(\tau)}\ \tau^p\ \overset{*}{\phi}(\tau)d\tau}{\overline{s(\tau)}\ \exp[\psi_2^+(\tau)](\tau-z)} \ .$$

If we represent $1/(\tau-z)$ in the neighbourhood of infinity as

$$\frac{1}{\tau-z} = -\frac{1}{z} - \frac{\tau}{z^2} - \frac{\tau^2}{z^3} - \dots\dots \ , \tag{14.23}$$

and write down the conditions for $\overset{*}{g}_2$ to have a zero of order $\mathcal{H}_1+1-p$ at infinity (which are the same as for the function $\overset{*}{g}{}_2^-$) then it follows from

(13) We shall assume that these zeros are simple.

(14.23) that these conditions are given by

$$\frac{1}{2\pi i}\int_\gamma \overset{*}{\phi}(\tau)\,\frac{\overline{b(\tau)}\tau^{p+\ell}d\tau}{\overline{s(\tau)}\,\exp[\psi_2^+(\tau)]} = 0\ ,\ \ell = 0,1,\dots,\varkappa_1-p-1\ ,$$

or equivalently

$$(\overset{*}{\phi},\ \frac{b}{t^{p+\ell+1}\ s\ \overline{\exp[\psi_2^+]}}) = 0\ ,\ \ell = 0,1,\dots,\varkappa_1-p-1\ . \tag{14.24}$$

It is not difficult to see that (14.24) follows from (14.22), and conversely. In fact, if we observe that for any function f, $\ell n\,\bar f = \overline{\ell n\,f}$, $\exp\bar f = \overline{\exp f}$, and (see [33, §18.1])

$$\overline{\frac{1}{\pi i}\int_\gamma \frac{f(\tau)d\tau}{\tau - t}} = -\frac{1}{\pi i}\int_\gamma \frac{\overline{f(\tau)}d\tau}{\tau - t} + \frac{1}{\pi i}\int_\gamma \frac{\overline{f(\tau)}d\tau}{\tau}\ , \tag{14.25}$$

we readily find that

$$\psi_2^\pm(t) = \pm\frac{1}{2}\,\ell n\,[\overline{t^{-\varkappa_1}G_1(t)}] + \frac{1}{2\pi i}\int_\gamma \frac{\overline{\ell n\,[\tau^{-\varkappa_1}G_1(\tau)]}}{\tau - t}d\tau$$

$$= -\overline{\psi_1^\mp(t)} + \frac{1}{2\pi i}\int_\gamma \frac{\overline{\ell n\,[\tau^{-\varkappa_1}G_1(\tau)]}}{\tau}d\tau\ ,$$

$$u_\ell(t) = \exp[\psi_1^-(t)]\,t^{\ell-\varkappa_1}\left(\frac{-2b(t)}{s(t)}\right)$$

$$= \frac{\text{const.}\ b(t)\ t^{\ell-\varkappa_1}}{s(t)\ \overline{\exp[\psi_2^+(t)]}}\ .$$

It remains to note that for $\ell = 0,1,2,\dots,\varkappa_1-p-1$ the sequences $\{t^{\ell-\varkappa_1}\}$ and $\{t^{-p-\ell-1}\}$ are the same.

From the result just established, and equation (14.18), it follows that ψ_R is the solution of the equation

$$K_R^{0*}\,\psi \equiv \overline{a_R(t)}\psi(t) + \frac{1}{\pi i}\int_\gamma \frac{\overline{b_R(\tau)}\psi(\tau)d\tau}{\tau - t} = \overset{*}{\phi}(t)\overline{b(t)}/\overline{b_R(t)}\ . \tag{14.26}$$

Hence

$$\|K^{0*}\psi_R - \overset{*}{\phi}\| \le \|(K^{0*} - K_R^{0*})\psi_R\| + \|(1 - \bar{b}/\bar{b}_R)\phi^*\|$$

(14.27)
$$= \|(1 - \bar{b}/\bar{b}_R)\phi^*\| + \|[((P_p/P_p^+) - 1)\phi^*(\bar{b}/\bar{b}_R)]^+ - [((Q_m/Q_m^-) - 1)\phi^*(\bar{b}/\bar{b}_R)]^- - 2[((P_p/P_p^+) - 1)\chi_R^-]^+ - 2[((Q_m/Q_m^-) - 1)\chi_R^+]^-\| ,$$

where it is assumed that $P_p = \prod_{j=1}^{\nu} (1-t\bar{\beta}_j)^{p_j}$ and $Q_m = \prod_{k=1}^{\mu} (1-t\bar{\alpha}_k)^{m_k}$. From equation (14.18) and the Sokhotski-Plemelj formulae,

(14.28)
$$\left[\left(\frac{Q_m}{Q_m^-} - 1\right)\chi_R^+\right]^- = \left[\frac{(Q_m - Q_m^-)\exp[\psi_2^+]g_2^{*+}}{Q_m^- P_p^+}\right]^- = \frac{T_{m-1}}{Q_m^-} ,$$

where T_{m-1} is the polynomial of degree $(m-1)$ defined by

(14.29)
$$T_{m-1}(\delta_r^-) = Q_m(\delta_r^-)\ \chi_R^+(\delta_r^-) , \quad r = 1,2,\ldots,m .$$

Similarly,

(14.30)
$$\left[\left(\frac{P_p}{P_p^+} - 1\right)\chi_R^-\right]^+ = \frac{R_{p-1}}{P_p^+} ,$$

where R_{p-1} is the polynomial of degree $p-1$ which is defined by

(14.31)
$$R_{p-1}(\gamma_s^+) = P_p(\gamma_s^+)\ \chi_R^-(\gamma_s^+) , \quad s = 1,2,\ldots,p .$$

It is not difficult to show that $|P_p/P_p^+|$ and $|Q_m/Q_m^-|$ cannot increase indefinitely as $R \to \infty$. Furthermore $|T_{m-1}| \le C_1 \max_{1<r<m} |Q_m(\delta_r^-)|$ and $|R_{p-1}| \le C_2 \max_{1\le s\le p} |P_p(\gamma_s^+)|$, where C_1, C_2 are independent of R. Using this together with (14.20) and (14.27), we complete the proof of the theorem. #

Conditions (14.20) will be partially satisfied in the case when b becomes zero on γ if $|P_p^+ - P_p|$ and $|Q_m^- - Q_m|$ tend to zero as $R \to \infty$, where γ_s^+ and δ_r^- are chosen so that the distances from them to the nearest points of γ and to the nearest of the points $\bar{\beta}_j$ and $\bar{\alpha}_k$ have the same order of smallness as $R \to \infty$. If we start with equation (14.2), then its conjugate is given by equation (14.1) and similar results will be valid without additional restrictions on b.

In the case of equation (12.1), it is possible to take for ψ_R the function

$$\psi_R = K^* \left(\frac{f - K\overset{*}{\phi}{}^{(\alpha)}}{\alpha} \right), \quad \alpha > 0 , \tag{14.32}$$

where K^* is the operator conjugate to K, (see equation (9.8)), and $\overset{*}{\phi}{}^{(\alpha)}$ is the solution of

$$\alpha\phi + K^* K\phi = K^* f \ . \tag{14.33}$$

This is a normally solvable singular integral equation, and so an approximate solution can be found without having to use the methods of regularization or quasi-solutions.

§14.2 Minimum Restrictions on the Smoothness of the Input Data, and on the Condition for Inversion when $a \pm b$ is zero on γ

Equation (9.1) is a linear equation in $L_2(\Gamma,\rho)$ and under the conditions that a and b are bounded almost everywhere and measurable on γ (see [47]), and that k and f satisfy conditions 1) of §9.1, the operator is bounded. Under these general assumptions, the most appropriate algorithms to use will be the iteration algorithmsof §5, since they require no conditions on the smoothness of the input data. In order to calculate the integrals appearing in equations (5.3), (5.70), (5.71) and (5.87), it is necessary to apply quadrature formulae (see §11.2 and [113]) with a refined mesh in the neighbourhoods of the singular points of the integrands.

For the approximate solution of equation (14.1), we shall discuss a particular method which allows the method of simple iteration to be applied under minimal restrictions on the smoothness of the coefficients, although additional conditions are necessary.

We make the substitution(14)

$$\phi = \lambda\phi_1 + \mu S\phi_1 \quad , \tag{14.34}$$

where λ,μ are arbitrary complex constants with $\lambda^2 \neq \mu^2$. Then it follows that

$$S\phi = \mu\phi_1 + \lambda S\phi_1$$

and instead of equation (14.1), we have

---

(14) This substitution is due to I.B. Simonenko.

$$\phi_1 = -\frac{a\mu + b\lambda}{a\lambda + b\mu} S\phi_1 + \frac{f}{a\lambda + b\mu} . \tag{14.35}$$

Let $A\phi_1 = -[(a\mu+b\lambda)/(a\lambda+b\mu)]S\phi_1$, and find the conditions on a and b such that $\|A\|$ does not exceed $q < 1$. If $\mu = 0$, then we require $\|(b/a)S\| \leq q$, and if $\lambda = 0$ then we must have $\|(a/b)S\| \leq q$. We shall assume that neither λ nor μ is zero. Then

$$A\phi_1 = -\frac{z+\delta}{z\delta + 1} S\phi_1 , \text{ where } z = \frac{a}{b} , \ \delta = \frac{\lambda}{\mu} ,$$

and we have

$$\|A\| = \left\| \frac{z+\delta}{z\delta+1} S \right\| .$$

In $L_2(\gamma)$,

$$\left\| \frac{z+\delta}{z\delta+1} S \right\| \leq \operatorname*{ess\,sup}_{t\epsilon\gamma} \left| \frac{z+\delta}{z\delta+1} \right| ,$$

and so for

$$\operatorname*{ess\,sup}_{t\epsilon\gamma} \left| \frac{z+\delta}{z\delta+1} \right| \leq q , \quad z = \frac{a(t)}{b(t)} , \quad \|A\| \leq q. \tag{14.36}$$

Putting $(z+\delta)/(z\delta+1) = \zeta$, we have $z = (\zeta-\delta)/(1-\delta\zeta)$. It follows that $\|A\|_{L_2(\gamma)} \leq q$ if the values of the function $z = a/b$ lie in any domain into which the disc $|\zeta| \leq q$ is mapped by the linear fractional substitution $(\zeta-\delta)/(1-\delta\zeta)$, for any $\delta \neq \pm 1$. Rewriting $z = (\zeta-\delta)/(1-\delta\zeta)$ as a sequence of transformations

$$z_3 = \zeta - \frac{1}{\delta} , \quad z_2 = \frac{1}{z_3} , \quad z_1 = \left(1 - \frac{1}{\delta^2}\right) z_2 , \quad z = z_1 - \frac{1}{\delta} ,$$

it is not difficult to obtain the parameters of the circle into which the circle $|\zeta| = q$ is mapped. In the z-plane, the centre z_0 and radius R_z of the circle are given by

$$z_0 = \frac{\bar{\delta}q^2 - \delta}{1 - |\delta q|^2} , \quad R_z = \frac{|\delta^2 - 1|q}{|1 - |\delta q|^2|} . \tag{14.37}$$

The disc $|\zeta| \leq q$ is mapped into the interior of this circle when $|\delta q| < 1$, and into the exterior of the circle when $|\delta q| > 1$. When $|\delta q| = 1$, the disc $|\zeta| \leq q$ is mapped into a half-plane, by the transformation $z = (\zeta-\delta)/(1-\delta\zeta)$.

Thus, for arbitrary $z = a/b$, only if $\|A\|_{L_2(\gamma)} \leq q < 1$ and $f/(a\lambda+b\mu) \in L_2(\gamma)$, can equation (14.35) (and also (14.1)) have a unique solution in $L_2(\gamma)$ which can be obtained by the method of simple iteration, (§5.1).

If the values of z do not satisfy the required property, then in many cases the substitution $\phi^{\pm} = \phi_2^{\pm} \psi^{\pm}$, where $\psi^{\pm}$ are known functions, can be useful.

In conclusion, we describe a method which has been proposed by I.B. Simonenko [122] and G.F. Mandzhavidze [99]. Suppose $\varkappa$ is the index of equation (14.1). If we approximate to $\chi = \ln [t^{-\varkappa} (a-b)/(a+b)]$ by the polynomial $u_n(\chi,t)$ (in the case of sufficiently smooth χ), or by the piecewise-linear function

$$'\chi(t) = \chi(t_{j+1}) \left(\frac{t - t_j}{t_{j+1} - t_j}\right) + \chi(t_j) \left(\frac{t_{j+1} - t}{t_{j+1} - t_j}\right) ,$$

$j = 1,2,\ldots,n$, we find that

$$\left| \frac{a(t) - b(t)}{a(t) + b(t)} t^{-\varkappa} \frac{\chi_n^-(t)}{\chi_n^+(t)} - 1 \right| \le q < 1 , \tag{14.38}$$

where $\chi_n^{\pm}(t)$ denotes either $\exp[u_n^{\pm}(\chi,t)]$ or $\exp['\chi^{\pm}(t)]$. We now introduce the equation

$$\psi_n = B_n\psi_n + g_n ,$$

$$B_n\psi_n \equiv \left[\frac{a - b}{a + b} t^{-\varkappa} \frac{\chi_n^-}{\chi_n^+} - 1\right] \psi_n^+ , \tag{14.39}$$

$$g_n \equiv \frac{f}{(a + b)\chi_n^+} .$$

This can be solved by the method of simple iteration since $\|B_n\|_{L_2(\gamma)} \le q < 1$, and the solution of equation (14.1) is given by

$$\phi_* = \psi_n^+ \chi_n^+ - t^{-\varkappa} \psi_n^- \chi_n^- .$$

In the case of the complete singular integral equation (9.1), if we write this as

$$\begin{aligned} \psi_n &= D_n\psi_n + g_n , \\ D_n\psi_n &\equiv B_n\psi_n + \frac{\lambda}{(a+b)\chi_n^+} \int_\Gamma k(\tau,t)(\psi_n^+\chi_n^+ - \tau^{-\varkappa} \psi_n^-\chi_n^-)d\tau , \end{aligned} \tag{14.40}$$

it is possible to find ψ_n by simple iteration provided q and $|\lambda|$ are sufficiently small. We can then apply the method of analytic continuation with respect to the parameter λ, for arbitrarily large values of $|\lambda|$, (see §5.1).

§15 SYSTEMS OF SINGULAR INTEGRAL EQUATIONS

§15.1 Reduction to a System of Fredholm Integral Equations of the Second Kind

Firstly, we shall examine the dominant system of equations

$$K^0\phi \equiv a(t)\phi(t) + \frac{b(t)}{\pi i}\int_\Gamma \frac{\phi(\tau)d\tau}{\tau - t} = f(t) \ , \ t \in \Gamma \ , \tag{15.1}$$

where Γ is a closed Lyapunov contour containing the origin of the complex plane; a and b are functional matrices of order m satisfying the condition that $\det(a\pm b) \neq 0$ for $t \in \Gamma$; $f = (f_1, f_2, \ldots, f_m)$ is a given vector, and $\phi = (\varphi_1, \varphi_2, \ldots, \varphi_m)$ is the unknown vector. Let us assume that $a,b \in H(\alpha,\Gamma)$ i.e. their elements are Hölder continuous on Γ with index α, and $f \in L_2(\Gamma)$ i.e. its components are square integrable on Γ. Unlike the case of a single equation, the system of equations (15.1) in general does not possess a solution in closed form. (For the detailed theory, see N.P. Vekua, [21]).

By applying the Sokhotski-Plemelj formulae

$$S\phi = \frac{1}{\pi i}\int_\Gamma \frac{\phi(\tau)d\tau}{\tau - t} = \phi^+(t) + \phi^-(t) \ ; \ \phi(t) = \phi^+(t) - \phi^-(t) \ , \tag{15.2}$$

which maps any vector $\phi \in L_2(\Gamma)$ into vectors ϕ^+ and ϕ^- which also belong to $L_2(\Gamma)$ and are the boundary values from within and without Γ respectively of the piecewise analytic vector function

$$\Phi(z) = \frac{1}{2\pi i}\int_\Gamma \frac{\phi(\tau)d\tau}{\tau - z} \ ,$$

we obtain

$$K^0\phi \equiv (a+b)\phi^+ - (a-b)\phi^- = f \ . \tag{15.3}$$

Hence

$$\begin{gathered} \phi^+ = G\phi^- + g, \\ G = (a+b)^{-1}(a-b) \ , \ g = (a+b)^{-1}f \ . \end{gathered} \tag{15.4}$$

Recalling that

$$S\phi^+ = \phi^+ \ , \ S\phi^- = -\phi^- \ , \tag{15.5}$$

we obtain

$$(15.6)\qquad \begin{aligned} S(G\phi^- + g) &= G\phi^- + g, \\ GS\phi^- &= - G\phi^- , \end{aligned}$$

from which

$$(15.7)\qquad \begin{aligned} G\phi^- &= \frac{1}{2\pi i}\int_\Gamma \frac{G(\tau) - G(t)}{\tau - t}\,\phi^-(\tau)d\tau + g^- , \\ \phi^+ &= G\phi^- + g. \end{aligned}$$

Thus each solution of equation (15.1) gives rise (by means of equations (15.2)) to a solution of the Riemann boundary value problem (15.4) and the system (15.7). Under certain conditions the converse is possible.

Theorem 15.1 (see [164]). *If the equation*

$$(15.8)\qquad a\phi - bS\phi = 0$$

has only the trivial solution, then any solution of the Fredholm integral equation of the second kind

$$(15.9)\qquad \begin{aligned} G\phi_1 &= \frac{1}{2\pi i}\int_\Gamma \frac{G(\tau) - G(t)}{\tau - t}\,\phi_1(\tau)d\tau + g^-(t) , \\ \phi_2 &= G\phi_1 + g , \end{aligned}$$

corresponding to equation (15.1), *is both the solution of equation* (15.7) *and the boundary value problem,* (15.4). *In addition* $\phi = \phi_2 - \phi_1$ *is the solution of equation* (15.1).

Proof. It is sufficient to show that if ϕ_1 and ϕ_2 satisfy (15.9), then $\phi_1^+ = \phi_2^- = 0$. The first of equations (15.9) can be rewritten in the form

$$G(\phi_1/2 + S\phi_1/2) = - G\phi_1/2 + S(G\phi_1)/2 + g^-$$

or

$$G\phi_1^+ = (G\phi_1 + g)^- = \phi_2^- .$$

If we assume that ϕ_1^+ and ϕ_2^- are not zero vectors, then they will be a non-trivial solution of the so-called *accompanying Riemann boundary value problem of* (15.4)

$$(15.10)\qquad \psi^+(t) = G^{-1}(t)\psi^-(t) .$$

In this case $\phi = \phi_1^+ - \phi_2^-$ will be a non-trivial solution of equation (15.8), which is not possible. #

As a consequence of this theorem, it follows that if equations (15.8) and

$$(15.11)\qquad a\phi + bS\phi = 0 ,$$

have only trivial solutions, then equation (15.1) and the corresponding system (15.9) will have a unique solution for any right-hand side. Alternatively, if the system (15.9) has a unique solution for any right-hand side and equation (15.1) is solvable, then equations (15.8) and (15.11) have only trivial solutions. In this favourable situation, an approximate solution of equation (15.1) may be obtained by using any method for finding an approximate solution of the system (15.9), (see [76]).

We note that the system of equations (15.9) possesses a kernel of the special form

(15.12) $$T(\tau,t) = \frac{G(\tau) - G(t)}{\tau - t} .$$

If G has a derivative of order $r(\geq 1)$ which is Hölder continuous on Γ, then from the equation

(15.13) $$T(\tau,t) = \int_{\overset{\frown}{01}} G^{(1)}\,[t + u(\tau-t)]du ,$$

where $\overset{\frown}{01}$ denotes the image of the arc $\overset{\frown}{t\tau}$ of Γ under the mapping $u = (v-t)/(\tau-t)$, $v \in \overset{\frown}{t\tau}$, we find

(15.14) $$T_{\tau}^{(r-1)} = \int_{\overset{\frown}{01}} G^{(r)}\,[t + u(\tau-t)]u^{r-1}du ,$$
$$T_{t}^{(r-1)} = \int_{\overset{\frown}{01}} G^{(r)}\,[t + u(\tau-t)](1-u)^{r-1}du .$$

It follows that $T_{\tau}^{(r-1)}$ and $T_{t}^{(r-1)}$ are Hölder continuous on Γ.

We shall now state sufficient conditions for equations (15.8) and (15.11) to have only trivial solutions. If we write $S\phi = \psi$, then equations (15.8) and (15.11) can be written either as $\phi = a^{-1}bS\phi$ and $\phi = -a^{-1}bS\phi$, respectively; or as $\psi = b^{-1}aS\psi$ and $\psi = -b^{-1}aS\psi$, respectively. Suppose we consider $a^{-1}bS$ and $b^{-1}aS$ to be operators in a Hilbert space of vectors $f = (f_1, f_2, \ldots, f_m)$ with a scalar product defined by

(15.15) $$(f^{(1)}, f^{(2)}) = \frac{1}{m}\sum_{k=1}^{m} (f_k^{(1)}, f_k^{(2)})$$
and $$(f_k^{(1)}, f_k^{(2)}) = \frac{1}{L}\int_{\Gamma} f_k^{(1)}(t)\,\overline{f_k^{(2)}(t)}\,|dt| ,$$

where L is the length of the contour Γ. Then the required sufficient conditions will be given by either $\|a^{-1}bS\| < 1$ or $\|b^{-1}aS\| < 1$.(15)

---

(15) With these conditions we note that equation (15.1), or the equation $\psi = -b^{-1}aS\psi + b^{-1}f$, can be solved by the method of simple iteration.

If $\Gamma \equiv \gamma$, the unit circle, then it follows that $\|S\| = 1$. In this case the sufficient condition takes the simpler form of either $\|a^{-1}b\| < 1$ or $\|b^{-1}a\| < 1$. The following theorem gives more general sufficient conditions.

Theorem 15.2. *If the constant matrices* λ *and* μ *can be chosen so that*

$$\|(a\lambda + b\mu)^{-1}(a\mu + b\lambda)S\| < 1 , \tag{15.16}$$

then equations (15.8) *and* (15.11) *will have only trivial solutions.*

The proof is similar to the above discussion if we write $\varphi = \lambda\phi + \mu S\psi$.

An important theorem in the theory and application of systems of singular integral equations is the following.

Theorem 15.3. (J. Plemelj, [164]). *Every solution of the equation*

$$t^r G(t)\tilde{\phi}_1(t) = \frac{1}{2\pi i}\int_\Gamma \frac{\tau^r G(\tau) - t^r G(t)}{\tau - t}\tilde{\phi}_1(t)d\tau + g^-(t) , \tag{15.17}$$

for all sufficiently large positive integers r, *has the property*

$$\tilde{\phi}_1^+(t) = 0 . \tag{15.18}$$

If equations (15.8) and (15.11) do not have trivial solutions, and we require all the solutions of equation (15.1), then the problem is reduced to either the determination of all the linearly independent solutions of equation (15.9) and the selection from them of those satisfying $\phi_1^+ = \phi_2^- = 0$, or the determination of all the solutions of equation (15.17) for sufficiently large r which have a zero of order greater than $(r+1)$ at infinity.[16] The required solutions are then determined by the equation

$$\phi(t) = \tilde{\phi}^+(t) - t^r\tilde{\phi}^-(t) . \tag{15.19}$$

We note that we can find approximate solutions of Fredholm integral equations of the second kind by two different methods. Either

1) the reduction of the solution of the original system to the solution of a system with a degenerate kernel (see [102]); or
2) the introduction of a parameter λ and an approximate solution of the system found by any method so that an approximation is obtained to the resolvent which represents all solutions of the homogeneous system (see §12.5).

---

(16) Good estimates for r may be obtained from known results in the theory of singular integral equations (see [21]).

Let us now examine the system of complete singular integral equations

$$K\phi \equiv a(t)\phi(t) + \frac{b(t)}{\pi i}\int_\Gamma \frac{\phi(\tau)d\tau}{\tau - t} + \int_\Gamma k(\tau,t)\phi(\tau)d\tau = f(t) , \tag{15.20}$$

$t \in \Gamma$, under the following assumptions: Γ is the union of a finite number of mutually disjoint open or closed Lyapunov curves; a,b are functional matrices of order m such that $a,b \in C(\Gamma; d_1,d_2,\ldots,d_s)$ i.e. a and b are continuous everywhere on Γ except at the points d_i where finite discontinuities may occur[17]; $\det(a\pm b) \neq 0$ at all points of Γ; $f = (f_1,f_2,\ldots,f_m) \in L_2(\Gamma,\rho)$ where the weight vector $\rho = (\rho_1,\rho_2,\ldots,\rho_m)$ has each ρ_j of the form given by equation (9.2); the kernel $k(\tau,t)$ is such that $k\phi \equiv \int_\Gamma k(\tau,t)\phi(\tau)d\tau$ is a completely continuous operator in $L_2(\Gamma,\rho)$[18]; $\phi = (\phi_1(t), \phi_2(t),\ldots,\phi_m(t))$ is the unknown vector function in the space $L_2(\Gamma,\rho)$. It should be noted that since the coefficients of equation (15.20) can have discontinuities of the first kind, the case of open contours of integration can be reduced to the case of closed contours (see [21],and also compare with §13.1).

In [143], a method of reduction of equation (15.20) to a Fredholm integral equation of the second kind has been given which, unlike the method described for equation (15.1), does not make use of the Riemann boundary value problem.[19]

<u>Theorem 15.4</u> (B.V. Khvedelidze, [143]). *The operator M defined by*

$$M\psi \equiv a_0\psi - S(b_0\psi) , \tag{15.21}$$

where

$$S\phi = \frac{1}{2\pi i}\int_\Gamma f(\tau)d\tau/(\tau-t) ,$$

$$\begin{aligned} a_0 &= [(a+b)^{-1} + (a-b)^{-1}]/2 , \\ b_0 &= [-(a+b)^{-1} + (a-b)^{-1}]/2 , \end{aligned} \tag{15.22}$$

is both a left and right regularizator of equation (15.20). *In addition*

$$MK\phi \equiv \phi + T\phi , \quad KM\psi \equiv \psi + \tilde{T}\psi \tag{15.23}$$

---

(17) Whenever it is said that a matrix has a certain analytic property, it is to be understood that each of its elements has that property.

(18) A sufficient condition for k to be completely continuous is given by equation (9.3).

(19) B.V. Khvedelidze gave his analysis under more general conditions than we shall assume. In particular he considers the space $L_p(\Gamma,\rho)$, $\rho > 1$, whereas we consider $L_2(\Gamma,\rho)$.

where

$$T = -\frac{1}{\pi i} T_{b_0 a} - \frac{1}{\pi i} T_{b_0 b} S + Mk ,$$

(15.24)

$$\tilde{T} = -\frac{a}{\pi i} T_{b_0} + \frac{b}{\pi i} T_{a_0} + kM .$$

The operator T_ω *is defined by*

(15.25) $$T_\omega \phi = \int_\Gamma \frac{\omega(\tau) - \omega(t)}{\tau - t} \phi(\tau) d\tau ,$$

and is a completely continuous operator in $L_2(\Gamma,\rho)$.

It should be noted that in general, the equations

(15.26) $$MK\phi = Mf , \quad KM\psi = f ,$$

are not equivalent to equation (15.20), since not every solution of the first of equations (15.26) is necessarily a solution of (15.20), and the number of linearly independent solutions of the second of equations (15.26) does not necessarily coincide with the number of linearly independent solutions of (15.20). The problem of the equivalent regularization of systems of singular integral equations has been discussed in detail by F.D. Gakhov, [31].

§15.2 Application of the Method of Least Squares

As in §12.1 we shall examine only those aspects, concerning the application of the method of least squares to systems of singular integral equations, that were not considered in §3.

We shall look upon equation (15.1) as a linear equation in the Hilbert space $L_2(\Gamma)$ with a scalar product as defined by equations (15.15). For the coordinate system of elements $e_1, e_2, \ldots$ we choose the system of vector functions

(15.27)
$$e_{2pm+j} = t^p e_j' , \quad e_{(2p+1)m+j} = t^{-p-1} e_j'$$
$$\text{for } j = 1,2,\ldots,m ; \; p = 0,1,2,\ldots$$

where

$$e_1' = (1,0,\ldots,0), \; e_2' = (0,1,0,\ldots,0), \ldots, e_m' = (0,0,\ldots,0,1) .$$

For this system of elements, we have

(15.28) $$Se_k = \frac{1}{\pi i} \int_\Gamma \frac{e_k(\tau) d\tau}{\tau - t} = \begin{cases} e_k(t) , & k = 2pm+j , \\ -e_k(t) , & k = (2p+1)m+j , \end{cases}$$

and

$$(15.29) \qquad y_k = K^0 e_k = \begin{cases} (a+b)e_k \ , & k = 2pm+j \ , \\ (a-b)e_k \ , & k = (2p+1)m+j \ . \end{cases}$$

We note that certain advantages arise if the method of least squares is applied, not to equation (15.1), but to the first of equations (15.9) which we write as

$$(15.30) \qquad K_1^0\phi_1 \equiv G\phi_1 - \frac{1}{2\pi i}\int_\Gamma \frac{G(\tau) - G(t)}{\tau - t}\phi_1(\tau)d\tau = g^-(t) \quad ,$$

$t \in \Gamma$. Recalling that we are interested only in those solutions of (15.30) such that $\phi_1^+ = 0$, (see §15.1), we choose as the coordinate system of elements $e_1, e_2, \ldots$ the functions

$$(15.31) \qquad e_{pm+j} = t^{-p-1}\, e_j' \quad \text{for} \quad j = 1,2,\ldots,m \quad \text{and} \quad p = 0,1,2,\ldots \ .$$

Then we obtain

$$(15.32) \qquad y_k = K_1^0\, t^{-p-1}\, e_j' = (Gt^{-p-1}\, e_j')^- \ .$$

If we define the approximate solution of equation (15.30) by

$$(15.33) \qquad \phi_{1n} = \sum_{j=1}^{m}\sum_{p=0}^{n} \alpha_{pm+j}\, e_{pm+j} \quad ,$$

we take the approximate solution of (15.1) to be

$$(15.34) \qquad \phi_n = \phi_{2n} - \phi_{1n} \ , \text{ where } \ \phi_{2n} = G\phi_{1n} + g \ .$$

The obvious advantages of this approach are the smaller number of coefficients α_{pm+j} that are required, and the uniform convergence of

$$(15.35) \qquad \tilde{\phi}_{1n} = \frac{G^{-1}(t)}{2\pi i}\int_\Gamma \frac{G(\tau) - G(t)}{\tau - t}\phi_{1n}(\tau)d\tau + G^{-1}(t)g^-(t) \quad ,$$

to ϕ_1, provided that $G \in H(\alpha,\Gamma)$ with $\alpha > 1/2$. The convergence of $\tilde{\phi}_{1n}$ to ϕ_1 is assumed to be in $L_2(\Gamma)$.

Now let us consider equation (15.20) as a linear equation in the Hilbert space $L_2(\Gamma,\rho)$, with a scalar product defined by

$$(15.36) \quad (f^{(1)},f^{(2)}) = \frac{1}{m}\sum_{k=1}^{m}(f_k^{(1)},f_k^{(2)}),\ (f_k^{(1)},f_k^{(2)}) = \frac{1}{L}\int_\Gamma \rho_k(t)f_k^{(1)}(t)\,\overline{f_k^{(2)}(t)}\,|dt|.$$

If it is known that the solution of equation (15.20) can be written as

(15.37) $$\phi_j(t) = \prod_{k=1}^{s} |t - d_k^{(j)}|^{\delta_k^{(j)}} \psi_j(t) + \chi_j(t) \;, \quad j = 1,2,\dots,m \;,$$

where ψ_j, χ_j are at least differentiable on Γ and $-1 < \delta_k^{(j)} < 1$, then it is sensible to seek an approximate solution ϕ_n in the form (20)

(15.38) $$\phi_n = \sum_{j=1}^{m} \sum_{p=-n}^{n} \left(\prod_{k=1}^{s} |t - d_k^{(j)}|^{\delta_k^{(j)}} \alpha_{pm+j} + \beta_{pm+j} \right) e_{pm+j} \;,$$

where $e_{pm+j} = t^p e_j' \;.$

The parameters α_{pm+j} and β_{pm+j} are chosen so that they minimize

(15.39) $$\Delta_n = \|f - K\phi_n\|_{L_2(\Gamma,\rho)} \;,$$

where

(15.40) $$\rho_j(t) = \prod_{k=1}^{s} |t - d_k|^{-\delta_k^{(j)}} \;, \quad \rho = (\rho_1, \rho_2, \dots, \rho_m) \;.$$

If the assumptions on equation (15.20) are satisfied then the quantities $\delta_k^{(j)}$ can be defined (see [21]) by the equations

(15.41) $$\exp\,(2\pi i\; \rho_k^{(j)}) = \lambda_k^{(j)} \;, \quad \delta_k^{(j)} = \mathrm{Re}\; \rho_k^{(j)} \;,$$

where $\lambda_k^{(j)}$ for $j = 1,2,\dots,m$ are the roots of the equation

(15.42) $$\det[G^{-1}(d_k+0)\; G(d_k-0) - \lambda I] = 0 \;,$$
where $$G = (a+b)^{-1}(a-b) \;.$$

Having determined $\delta_k^{(j)}$, we obtain the favourable situation mentioned above where ψ_j, χ_j are sufficiently smooth provided that a, b, f and k are sufficiently smooth everywhere except at the points d_k. Generally the functions ψ_j and χ_j will have logarithmic singularities at the points d_k. If this is not the case, then the derivatives of ψ_j will have singularities of the form $(t-d_k)^{\delta_k^{(j)}}$ or $\ln(t-d_k)$. A construction similar to that of §13.1 for systems of singular integral equations is very difficult to implement.

These difficulties increase if a, b, f and k have minimal smoothness, or if $\det(a\pm b)$ vanishes at a set of points of Γ. In these cases, writing an approximate solution in the form of equation (15.38) is not appropriate.

---

(20) It is assumed that the origin of the complex plane lies inside the domain which is on the left-hand side of Γ with respect to the positive direction.

For most practical problems, we shall find that the solution of equation (15.20) will be continuous, except at a finite number of points of Γ, and integrable everywhere on Γ. This implies that the integral of the solution of equation (15.20), defined by

$$(15.43)\qquad \psi_1(t) = \int_{t_1}^{t} \phi(\tau)d\tau \quad , \quad t_1 \in \Gamma \quad ,$$

where the integration is taken along the arc $\widetilde{t_1 t}$ of Γ, is a piecewise differentiable vector on Γ. If $\phi \in L_p(\Gamma)$, $p > 1$, then by Hölder's inequality (see [33]), it can be shown that ψ is Hölder continuous on Γ with Hölder index $(p-1)/p$. It therefore seems appropriate to evaluate either $\psi_1(t)$ or even $\psi_2(t) = \int_{t_1}^{t} \psi_1(\tau)d\tau$, and then by numerical differentiation to determine

$$\phi(t) = \frac{d\psi_1}{dt} = \frac{d^2\psi_2}{dt^2} \quad .$$

It should however be noted that in general the equations for ϕ, ψ_1 and ψ_2 will not be normally solvable. Consequently in order to obtain approximate solutions it will be necessary to use the method of regularization (see §§2 and 3).

§15.3 Application of Iterative Methods

Difficulties associated with both the calculation of $\delta_k^{(j)}$ from equation (15.41), and the fact that in equation (15.37) the functions ψ_j are in general not differentiable (or even bounded), are the principal reasons preventing the effective use of the method of least squares. For systems of large order, there will also be storage limitations with the computer.

For such equations, iterative methods are perhaps more appropriate (see §5).

Let us consider in more detail the application of the iterative process defined by equation (5.87) to the approximate solution of equation (15.20) in the space $L_2(\Gamma,\rho)$.

In order to avoid the calculation of $\delta_k^{(j)}$, we assume that

$$(15.44)\qquad \rho_j(t) = \prod_{k=1}^{s} |t - d_k|^{1-\epsilon} \quad , \; j = 1,2,\ldots,m \quad ,$$

for sufficiently small $\epsilon > 0$. For equation (15.20), we can rewrite equation (5.87) as

$$(15.45)\qquad \phi^{(j+1)} = \phi^{(j)} + \frac{\|f - K\phi^{(j)}\|^2}{\|K^*(f-K\phi^{(j)})\|^2} \; K^*(f - K\phi^{(j)}) \quad ,$$

for $j = 0,1,\ldots$. We shall estimate the storage requirements and the number of arithmetical operations for the implementation of the iterative process given by (15.45).

In this case (see §9.1),

$$K\phi^{(j)} = a\phi^{(j)} + \frac{b}{\pi i}\int_0^L \frac{\phi^{(j)}(s)\tau'(s)ds}{\tau(s) - t(u)} + \int_0^L k(s,u)\phi^{(j)}(s)\tau'(s)ds \,, \tag{15.46}$$

and

$$K^* r^{(j)} = a^* r^{(j)} + \frac{\overline{t'(u)}}{\pi i}\int_0^L \frac{b^*(s)r^{(j)}(s)ds}{\overline{\tau(s)} - \overline{t(u)}} + \int_0^L k^*(u,s)\overline{t'(u)}r^{(j)}(s)ds \,, \tag{15.47}$$

where $r^{(j)} = f - K\phi^{(j)}$, and a^*, b^*, k^* are matrices which are conjugate to a, b, k respectively. Let us assume that the input data for the problem (i.e. the function $t'(u)$, $0 \le u \le L$, the coefficients a and b, the kernel k, and the right-hand side f) are given in tabulated form at the points $t_\nu = t(s_\nu)$, $\nu = 1,2,\ldots,N$, which are distributed on Γ. For any matrix F we shall write $F_\nu = F(t_\nu) = F(s_\nu)$; we shall denote the vector $(F_1,F_2,\ldots,F_N)$ by $\tilde{F}$ for any vector function F. Then instead of equation (15.45) we have

$$\tilde{\phi}^{(j+1)} = \tilde{\phi}^{(j)} + \frac{\|\tilde{f} - \tilde{K}\tilde{\phi}^{(j)}\|_N^2}{\|\tilde{K}^*(\tilde{f} - \tilde{K}\tilde{\phi}^{(j)})\|_N^2}\,\tilde{K}^*(\tilde{f} - \tilde{K}\tilde{\phi}^{(j)}) \quad , \tag{15.48}$$

for $j = 0,1,2,\ldots$. Here the operators $\tilde{K}$ and $\tilde{K}^*$ map the complex linear space of Nm-dimensional vectors E_{Nm} into itself, and are defined by

$$(\tilde{K}\tilde{\phi}^{(j)})_\nu = a_\nu\phi_\nu^{(j)} + \frac{b_\nu}{\pi i}\sum_{\zeta=1}^{N}{}' \frac{\phi_\zeta^{(j)}\tau'_\zeta(s_\zeta - s_{\zeta-1})}{\tau_\zeta - t_\nu} + \sum_{\zeta=1}^{N} k_{\zeta\nu}\phi_\zeta^{(j)}\tau'_\zeta(s_\zeta - s_{\zeta-1}) \tag{15.49}$$

and

$$(\tilde{K}^*\tilde{r}^{(j)})_\nu = a^*_\nu r_\nu^{(j)} + \frac{\bar{t}'_\nu}{\pi i}\sum_{\zeta=1}^{N}{}' \frac{\bar{b}^*_\zeta r_\zeta^{(j)}(s_\zeta - s_{\zeta-1})}{\bar{\tau}_\zeta - \bar{t}_\nu} + \sum_{\zeta=1}^{N} k^*_{\nu\zeta}\bar{t}'_\nu r_\zeta^{(j)}(s_\zeta - s_{\zeta-1}) \,, \tag{15.50}$$

in which $\sum'$ denotes that the summation is taken over all $\zeta \neq \nu$, and $s_0 = 0$. For any vector $\tilde{F} \in E_{Nm}$,

$$\|\tilde{F}\|_N^2 = \frac{1}{Lm}\sum_{k=1}^{m}\sum_{\nu=1}^{N}\sum_{r=1}^{s} |t_\nu - d_k|^{1-\epsilon}\,|F_{k\nu}|^2(s_\nu - s_{\nu-1}) \quad . \tag{15.51}$$

From equations (15.48)-(15.51), it is not difficult to calculate that one iteration using (15.48) requires approximately $(3m+6cs+10)N$ operations of multiplication and division, and $(5m+2)N$ operations of addition. The quantity c defines the number of operations required to calculate $|t_\nu - d_k|^{1-\epsilon}$. If we assume that the computer stores the vectors $\tilde{\phi}^{(j)}$, $\tilde{r}^{(j)} = \tilde{f} - \tilde{K}\tilde{\phi}^{(j)}$ and $\tilde{K}^*\tilde{r}^{(j)}$, and in addition holds the input tables of t', a, b, k and f then the required computer storage is approximately $(3Nm + N + 2Nm^2 + N^2m^2 + Nm)$ locations. The storage required for the program will be negligible compared with this, and is not included here.

§15.4 Other Methods

For the approximate solution of systems of singular integral equations, the same methods which have been used for systems of Fredholm integral equations of the second kind can formally be used. However only a few of them can be justified in the present context. (Reasons for this are considered briefly in §22). Of all the general approximate methods described in Chapter 1, only the Ritz-Galerkin method (see [43]) has a non-trivial basis for its connection with systems of singular integral equations. As shown in §4, this method is worse than the method of least squares although it does have advantages over the latter for the approximate calculation of characteristic values(see §12.4).

Let us now examine in more detail, some methods which are specifically suited to the solution of equation (15.1).

Assume that G can be represented, (either exactly or approximately), in the form

$$G(t) = \sum_{k=-n}^{n} A^{(k)} t^k , \tag{15.52}$$

where $A^{(k)}$ are constant matrices of order m. In equation (15.17), let us write $\tilde{\phi}_1 = t^{-r}\phi^-(t)$ and obtain

$$\phi^-(t) = G^{-1}(t) \sum_{k=-n}^{n} A^{(k)} \frac{1}{2\pi i} \int_\Gamma \frac{\tau^{k+r} - t^{k+r}}{(\tau-t)\tau^r} \phi^-(\tau)d\tau + G^{-1}(t)g^-(t) . \tag{15.53}$$

Since $\phi^-(t)$ is the boundary value of a vector which is analytic outside Γ and becomes zero at infinity, it follows that for $r > n$, every solution of equation (15.53) is given by

$$\phi^-(t) = G^{-1}(t) \sum_{k=1}^{n} A^{(k)} \sum_{s=1}^{k} \alpha^{(k-s)} t^{s-1} + G^{-1}(t)g^-(t) , \tag{15.54}$$

or equivalently by

$$\phi^-(t) = G^{-1}(t) \sum_{s=1}^{n} \left(\sum_{k=s}^{n} A^{(k)} \alpha^{(k-s)} \right) t^{s-1} + G^{-1}(t)g^-(t) \tag{15.55}$$

and

(15.56) $$\phi^-(t) = G^{-1}(t)\sum_{s=0}^{n-1}\left(\sum_{k=s}^{n-1} A^{(k+1)}\, t^{k-s}\right)\alpha^{(s)} + G^{-1}(t)g^-(t)\ .$$

The vectors $\alpha^{(\nu)}$ are defined by

(15.57) $$\alpha^{(\nu)} = \frac{1}{2\pi i}\int_\Gamma \phi^-(\tau)\tau^\nu d\tau \quad , \ \nu = 0,1,2,\dots,n-1\ .$$

Substituting the right-hand side of (15.56) into (15.57), we obtain the following algebraic system:

$$\alpha^{(\nu)} = (\alpha_1^{(\nu)},\ \alpha_2^{(\nu)},\dots,\alpha_m^{(\nu)})\quad ,$$

(15.58) $$\alpha^{(\nu)} = \sum_{s=0}^{n-1}\left(\sum_{k=s}^{n-1} A^{(k+1)}\frac{1}{2\pi i}\int_\Gamma G^{-1}(\tau)\tau^{k-s+\nu}d\tau\right)\alpha^{(s)} + \frac{1}{2\pi i}\int_\Gamma G^{-1}(\tau)g^-(\tau)d\tau$$

for $\nu = 0,1,2,\dots,n-1$. It is not difficult to see that in this case

(15.59) $$G^{-1}(t) = t^n\, Q_{2n(m-1)}(t)/P_N(t)\ ,$$

where $Q_{2n(m-1)}$ is a polynomial matrix of degree $2n(m-1)$, and P_N is a polynomial of degree $N \leq 2nm$ which does not become zero on Γ. Let us write $P_N(t) = P^+_{n_1}(t)P^-_{n_2}(t)$, where $P^+_{n_1}$ has its zeros outside of Γ, and $P^-_{n_2}$ has its zeros inside Γ. Thus we can write

(15.60) $$G^{-1}(t) = \frac{Q_{n_1-1}(t)}{P^+_{n_1}(t)} - \frac{Q_{n_2-1}(t)}{P^-_{n_2}(t)}\ ,$$

where Q_{n_1-1} and Q_{n_2-1} are polynomial matrices of degree n_1-1 and n_2-1, respectively. Assume

(15.61) $$\tilde{P}_{n-1}(t) = \sum_{s=1}^{n}\left(\sum_{k=s}^{n} A^{(k)}\,\alpha^{(k-s)}\right)t^{s-1}\ ,$$

and substitute the right-hand side of (15.59) into equation (15.55). We obtain

(15.62) $$\phi^-(t) = \frac{t^n\, Q_{2n(m-1)}(t)}{P^+_{n_1}(t)P^+_{n_2}(t)}\left\{\tilde{P}_{n-1}(t) + g^-(t)\right\}\ .$$

Let the zeros of the polynomial $P^+_{n_1}$ be given by $\beta_1,\beta_2,\dots,\beta_{n_1}$. (We restrict attention to the case of simple zeros). Since ϕ^- is analytic outside Γ, is zero at infinity and the matrix $Q_{2n(m-1)}$ has an inverse, it follows that in the case when equation (15.1) is solvable and G is given by (15.52) with det $A^{(n)} \neq 0$ (i.e. $N = n_1+n_2 = 2nm$), we must have

$$\tilde{P}_{n-1}(\beta_k) = -g^-(\beta_k) , \quad k = 1,2,\dots,n_1 . \tag{15.63}$$

It therefore follows that for $n < n_1$, equation (15.1) will not, in general, be solvable. If $n \geq n_1$, then the polynomial vector $\tilde{P}_{n-1}$ which satisfies equations (15.63) can always be defined.

Let

$$\tilde{P}_{n-1}(t) = \sum_{s=1}^{n} P^{(s)} t^{s-1} . \tag{15.64}$$

On comparing (15.64) with (15.61) we obtain

$$\sum_{k=s}^{n} A^{(k)} \alpha^{(k-s)} = P^{(s)} , \quad s = 1,2,\dots,n . \tag{15.65}$$

If the matrix $A^{(n)}$ has an inverse, then this last system of equations is always solvable for $\alpha^{(\nu)}$.

We now show that the $\alpha^{(\nu)}$ as defined by equations (15.65) give the solution of the system of equations (15.58), or alternatively that ϕ^- as defined by equation (15.62) together with equations (15.63), is the solution of equation (15.53). In fact

$$\sum_{k=1}^{n} A^{(k)} \sum_{s=1}^{k} \alpha^{(k-s)} t^{s-1} = \sum_{k=1}^{n} A^{(k)} \sum_{s=1}^{k} \frac{1}{2\pi i} \int_\Gamma G^{-1}(\tau)[\tilde{P}_{n-1}(\tau) + g^-(\tau)]d\tau$$

$$= \frac{1}{2\pi i} \sum_{k=-n}^{n} A^{(k)} \int_\Gamma \frac{\tau^{k+r} - t^{k+r}}{(\tau-t)\tau^r} G^{-1}(\tau)[\tilde{P}_{n-1}(\tau) + g^-(\tau)]d\tau$$

$$= \frac{1}{2\pi i} \int_\Gamma \frac{\tau^r G(\tau) - t^r G(t)}{(\tau-t)\tau^r} G^{-1}(\tau)[\tilde{P}_{n-1}(\tau) + g^-(\tau)]d\tau$$

$$= \frac{1}{2\pi i} \int_\Gamma \frac{\tilde{P}_{n-1}(\tau) + g^-(\tau)}{\tau - t} d\tau - t^r G(t) \frac{1}{2\pi i} \int_\Gamma \frac{G^-(\tau)[\tilde{P}_{n-1}(\tau) + g^-(\tau)]}{(\tau-t)\tau^r} d\tau$$

$$= \tilde{P}_{n-1}(t) .$$

The required result now follows. Let us summarize the result as a theorem.

Theorem 15.5. *If G is of the form* (15.52) *where* det $A^{(n)} \neq 0$ *and* $n \geq n_1$, *then equation* (15.1) *has* $n-n_1+1$ *linearly independent solutions, each of which can be determined from equation* (15.54), *where* $\alpha^{(\nu)}$ *are the linearly independent solutions of the algebraic system of equations* (15.58).

We note that n_1 can be calculated without knowing $P_{n_1}^{+}$ explicitly. Now $\det[G(t)] = t^{-mn} P_{n_1}^{+}(t)P_{n_2}^{-}(t)$, and the index of this function is given by

$$\mathscr{H} = \frac{1}{2\pi}\,[\arg(\det G(t))]_\Gamma = -\,mn + n_2. \tag{15.66}$$

But $n_1+n_2 = 2nm$ (provided det $A^{(n)} \neq 0$), so that

$$n_1 = nm - \mathscr{H}. \tag{15.67}$$

If det $A^{(n)} = 0$, so that $N < 2nm$, then before we determine the polynomial $\tilde{P}_{n-1}$ from equations (15.63), it is necessary to ensure that the right-hand side of (15.62) becomes zero at infinity. We shall not examine this here. The system of equations (15.58) has a determinant which differs from zero only in rare cases ($n = n_1$, det $A^{(n)} \neq 0$, and the condition det $A^{(n)} = 0$) and such circumstances are of importance. In most cases, if equation (15.1) (with G in the form (15.52)) is solvable, then the system of equations (15.58) will be underdetermined. Thus there arises the typical situation for singular integral equations where one has to solve an underdetermined (or overdetermined) system of algebraic equations. Algorithms for the solution of such systems have been discussed in §§2, 3 and 5.

An algorithm for the solution of the Riemann boundary value problem in the case when $G(t) = \Omega(t)\,U(t)$, where Ω and U are analytic inside and outside of Γ respectively, except at a finite number of points where they can have poles, has been constructed in [31, 21]. But, from a computational point of view, this algorithm, which requires the evaluation of the zeros of the determinants of functional matrices and the solution of homogeneous algebraic systems, is difficult to implement. It should be noted that G.F. Mandzhavidze [99] has proposed a simple iteration process by which any matrix G (det $G \neq 0$), which can be approximated by a rational one, may be reduced to the form ΩU.

In the case of a system of singular integral equations of the form

$$a(y)\phi(y) + b(y)(1-y^2)^{\frac{1}{2}}\frac{1}{\pi i}\int_{-1}^{1}\frac{\phi(u)\,du}{u-y} = f(y)\ , \tag{15.68}$$

$y \in [-1,1]$, where a and b are Hölder continuous on $[-1,1]$, a series of effective approximate methods of solution has been proposed by Kim Ze Pkhen [80]. The basic idea on which these methods are based is as follows. From the Sokhotski-Plemelj formulae, equation (15.68) can be written as

$$(a+b(1-y^2)^{\frac{1}{2}})\phi^+ - (a-b(1-y^2)^{\frac{1}{2}})\phi^- = f^+ - f^- \quad . \tag{15.69}$$

We shall assume that a and b are polynomial matrices (or have been approximated by polynomial matrices). Then $a+b(1-y^2)^{\frac{1}{2}}$ and $a-b(1-y^2)^{\frac{1}{2}}$ can be considered as boundary values,on the right-hand and left-hand sides of $[-1,1]$ respectively, of the piecewise-analytic function $a(z)+b(z)(1-z^2)^{\frac{1}{2}}$. Let Φ and F be the Cauchy integrals of ϕ and f respectively. Then on using Liouville's theorem, it follows from equation (15.69) that

$$[a(z) + b(z)(1-z^2)^{\frac{1}{2}}]\Phi(z) = F(z) + P(z) \quad , \tag{15.70}$$

so that

$$\Phi(z) = [a(z) + (1-z^2)^{\frac{1}{2}}b(z)]^{-1} \quad \{F(z) + P(z)\} \quad , \tag{15.71}$$

where P is a polynomial matrix chosen so that Φ is analytic outside of $[-1,1]$ and zero at infinity.

§16 SYSTEMS OF SINGULAR INTEGRAL EQUATIONS ARISING FROM QUANTUM FIELD THEORY[21]

§16.1 Systems of Equations of the First Type

Systems of equations of the form

$$a(t)\phi(t) + \frac{1}{\pi i}\int_\Gamma \frac{b(\tau)\phi(\tau)d\tau}{\tau - t} + c(t)D(\phi) = f(t),\ t \in \Gamma, \tag{16.1}$$

are said to be of the *first type* (see [163]) when Γ is a closed or open (not necessarily bounded) simple smooth contour; a and b are diagonal matrices of order m; the elements of a, b and f satisfy the usual conditions for singular integral equations; c is a given functional matrix of order m; and $D(\phi)$ is a linear functional vector, $D(\phi) = \{D_1(\phi_1), D_2(\phi_2),\ldots,D_m(\phi_m)\}$.

Theorem 16.1. *The solution of equation* (16.1) *can be obtained in closed form by using singular quadrature and the solution of a certain system of linear algebraic equations.*

Proof. We rewrite equation (16.1) in the form

$$G_j\phi_j \equiv a_j(t)\phi_j(t) + \frac{1}{\pi i}\int_\Gamma \frac{b_j(\tau)\phi_j(\tau)}{\tau - t}d\tau = f_j(t) - \sum_{k=1}^{m} c_{kj}(t)D_k(\phi_k), \tag{16.2}$$

$j = 1,2,\ldots,m$. If we assume that the right hand side of equation (16.2) is known, then we need to solve m linearly independent singular integral equations. Under broad assumptions, these are solvable in closed form by singular quadratures[22] (see [108, 33]). Let us denote the operators which solve these equations by G_j^{-1}, then

$$\phi_j(t) = G_j^{-1}f_j(t) - \sum_{k=1}^{m} D_k(\phi_k)G_j^{-1}(c_{kj}(t)),\ j = 1,2,\ldots,m. \tag{16.3}$$

Applying the operator D_j to both sides of (16.3) we find

$$D_j(\phi_j) = D_j(G_j^{-1}f_j) - \sum_{k=1}^{m} D_k(\phi_k)D_j(G_j^{-1}c_{kj}). \tag{16.4}$$

Determining $D_k(\phi_k)$ from this algebraic system and substituting into (16.3) gives the required solution of $\phi = (\phi_1,\phi_2,\ldots,\phi_n)$. #

---

(21) A brief physical introduction is given in [12].

(22) The method of solution differs only slightly from that described in §8.1 and §8.3. As we see from the subsequent analysis, the non-boundedness of the contour does not introduce any extra difficulties.

Thus for the approximate solution of the system (16.1), the main difficulty will generally be the calculation of the various singular integrals and the functionals $D_k(\phi_k)$.

In an appendix to his paper [163], R. Omnès investigates the behaviour of the integral $SF = \frac{1}{\pi i}\int_0^\infty \frac{F(s)ds}{s-x}$, near 0 and ∞. The results of his analysis are summarised in two tables, but some cases remain unresolved. However these remaining cases can be resolved by making the substitutions $u = 1/(s+1)$, $y = 1/(x+1)$ and applying the results of N.I. Muskhelishvili [108], concerning the behaviour of singular integrals with Cauchy kernel, near the end points of the interval of integration. With the above transformations we can write

$$SF = \frac{1}{\pi i}\int_0^\infty \frac{F(s)ds}{s-x} = -\frac{y}{\pi i}\int_0^1 \frac{F_1(u)du}{u(u-y)} \tag{16.5}$$
$$= -\frac{1}{\pi i}\int_0^1 \frac{F_1(u)du}{u-y} + \frac{1}{\pi i}\int_0^1 \frac{F_1(u)du}{u},$$

where $F_1(u) = F(\frac{1}{u} - 1)$. Thus the system of equations (16.1) with integration over an unbounded contour, may be reduced to a similar system with respect to the bounded contour [0,1], for which a complete analysis can be obtained using well known methods.

Some difficulties arise when, for example, in the case of functions defined on [0,∞), it is assumed that they possess a certain degree of smoothness on any finite interval, and at infinity either decrease or increase in a specified way. A similar situation arises when it is assumed that a function defined on [0,1] has a certain degree of smoothness at all points except at zero where the function has a specified rate of decrease (or growth). Simple illustrative examples are given by the functions $x^{-1}\sin x$ on [0,∞) and $y\sin(1/y)$ on [0,1]. It is readily verified that the function $y\sin(1/y)$ is Hölder continuous on [0,1], is analytic at all points except $y = 0$, and in the neighbourhood of $y = 0$ tends to zero like y. Consider the singular integral

$$d(y) = \frac{1}{\pi i}\int_0^1 \frac{u\sin(1/u)du}{u-y} = \frac{1}{\pi i}\int_0^1 \sin(1/u)du + \frac{y}{\pi i}\int_0^1 \frac{\sin(1/u)du}{u-y}.$$

Since $\sin(1/u)$ is bounded, the function $\frac{1}{\pi i}\int_0^1 \frac{\sin(1/u)du}{u-y}$ is absolutely integrable of any order $p > 1$ on [0,1]. In addition, for sufficiently small $y > 0$

$$\left| \int_0^1 \frac{\sin(1/u) - \sin(1/y)}{u - y} du \right| \leq \int_0^{y-y^2} \frac{2du}{y-u} + \int_{y-y^2}^{y+y^2} \frac{du}{(y-y^2)^2} + \int_{y+y^2}^1 \frac{2du}{u-y}$$

$$\leq \text{const. } |\ln y| \ .$$

Now it is clear that in a neighbourhood of zero, d is continuous and

$$\left| d(y) - \frac{1}{\pi i} \int_0^1 \sin\left(\frac{1}{u}\right) du \right| \leq \text{const. } y|\ln y| \ .$$

Similarly we can investigate functions which are Hölder continuous on $[0,1]$ and decrease like u^α, $\alpha > 0$ in a neighbourhood of $u = 0$.

§16.2 Systems of Equations of the Second Type

Systems of equations of the form

$$\phi(x) + \frac{1}{\pi} \int_1^\infty \frac{b(s)\phi(s)ds}{s - x} + \frac{1}{\pi} \int_1^\infty \frac{c(s)\phi(s)ds}{s + x} = f(x) \ , \tag{16.6}$$

$x \in (-\infty,\infty)$, are said to be of the *second type* (see [144]). The matrices b and c are square matrices of order m which are Hölder continuous on $[1,\infty)$. The vector f is such that $xf = (xf_1, xf_2, \dots, xf_m)$ is Hölder continuous on $(-\infty,\infty)$, i.e. $|f(x_1)-f(x_2)| \leq \text{const.}|x_1^{-1} - x_2^{-1}|^\alpha$ where $\alpha > 0$ and x_1, x_2 are sufficiently large.

If b and c are diagonal matrices and the equation

$$\phi(x) + \frac{1}{\pi} \int_1^\infty \frac{b(s)\phi(s)ds}{s - x} = f_1(x) \equiv f(x) - \frac{1}{\pi} \int_1^\infty \frac{c(s)\phi(s)ds}{s + x} \ , \tag{16.7}$$

$x \in [1,\infty)$ is solvable, then it is not difficult to reduce the system of equations (16.6) to an equivalent system of Fredholm integral equations of the second kind.

If b and c are not diagonal matrices, then in general it is not possible to obtain the solution of (16.7) in closed form. Nevertheless the general theory of systems of singular integral equations turns out to be useful because it enables us to discuss the structure of the solution i.e. its degree of smoothness, presence of singularities etc. Let us introduce the system of equations

$$\phi_1(x) + \frac{b_1(x)}{\pi} \int_{-\infty}^\infty \frac{\phi_1(s)ds}{s - x} + \frac{1}{\pi} \int_{-\infty}^\infty k(s,x)\phi_1(s)ds = f_1(x) \ , \tag{16.8}$$

$x \in (-\infty,\infty)$, where

$$b_1(x) = \begin{cases} b(x), & x \in [1,\infty) , \\ 0 , & x \in (-\infty,1) ; \end{cases}$$

$$c_1(x) = \begin{cases} c(x), & x \in [1,\infty) , \\ 0 , & x \in (-\infty,1) ; \end{cases}$$

$$f_1(x) = \begin{cases} f(x), & x \in [1,\infty) , \\ 0 , & x \in (-\infty,1) ; \end{cases}$$

$$k(s,x) = \frac{c_1(s)}{s+x} + \frac{b_1(s) - b_1(x)}{s-x} , \quad s,x \in (-\infty,\infty) .$$

For $x \in [1,\infty)$ this equation coincides with (16.6). Equation (16.8) is a particular form of singular integral equation with discontinuous coefficients (see [22]).(23)

From Theorem 11.6 and equation (16.6) with $x \in (-\infty,1)$ we obtain the following result.

Theorem 16.2. *Every solution* $\phi = (\phi_1,\phi_2,\ldots,\phi_m)$ *of equation* (16.6) *for which* $\int_1^\infty |\phi_j(x)/x|\,dx < \infty$, *belongs to one of the standard classes i.e. it is a vector which becomes zero at infinity like* $x^{-\mu}$, $\mu > 0$, *and is Hölder continuous at all points of the real axis except perhaps at the points* $x = \pm 1$. *In neighbourhoods of these points we have*

$$(16.9) \qquad |\phi_j(x)| < \frac{\text{const.}}{|x \pm 1|^\alpha} , \quad \alpha < 1 , \; j = 1,2,\ldots,m .$$

In the particular case when the matrices b and c are zero at $x = 1$ and $x = \infty$, the vector ϕ is Hölder continuous everywhere on $(-\infty,\infty)$. In this case equation (16.6) can be rewritten in the form

$$(16.10) \qquad \phi(x) + \frac{1}{\pi}\int_{-\infty}^{\infty} \frac{b_1(s)\phi(s)ds}{s-x} + \frac{1}{\pi}\int_{-\infty}^{\infty} \frac{c_1(s)\phi(s)ds}{s+x} = f(x) ,$$

where b_1, c_1 (see equation (16.8)) and f are Hölder continuous everywhere on the real axis including $\pm\infty$.

The investigation of the cases when the system (16.10) is solvable in closed form is of considerable interest. We shall show first of all that it is possible to reduce the solution of the system of equations (16.10) to the

(23) There is a difference since the contour of integration here is $(-\infty,\infty)$. By using the transformation $x = i(t+1)/(t-1)$, we can transform the problem into one in which the contour is the unit circle.

solution of a "standard" system (see [81]). With this end in view, we shall first consider some other results which are also of interest.

Lemma 1. *Any vector function ϕ which is defined on $(-\infty,\infty)$ can be represented uniquely in the form*

$$\phi = \phi^{r} + \phi^{\mu} \tag{16.11}$$

where $\phi^{r}(x) = [\phi(x) + \phi(-x)]/2$ and $\phi^{\mu}(x) = [\phi(x) - \phi(-x)]/2$.

Uniqueness follows immediately from the fact that only the identically zero function is simultaneously both an even and an odd function on the real axis.

Lemma 2. *For any vector $A\phi$, where A is a matrix,*

$$(A\phi)^{r} = A^{r}\phi^{r} + A^{\mu}\phi^{\mu} \quad \text{and} \quad (A\phi)^{\mu} = A^{\mu}\phi^{r} + A^{r}\phi^{\mu} \quad , \tag{16.12}$$

where A^{r} and A^{μ} are the matrices comprising the even and odd components of A, respectively.

The proof of this lemma is obvious.

Let us introduce the operators

$$\tilde{S}\phi = \frac{1}{\pi}\int_{-\infty}^{\infty}\frac{\phi(s)ds}{s-x} \quad , \quad \tilde{S}_{-}\phi = \frac{1}{\pi}\int_{-\infty}^{\infty}\frac{\phi(s)ds}{s+x} \quad . \tag{16.13}$$

Lemma 3. *For functions ϕ for which the operators* (16.13) *are defined, we have*

$$(\tilde{S}\phi)^{r} = \tilde{S}\phi^{\mu} \quad , \quad (\tilde{S}\phi)^{\mu} = \tilde{S}\phi^{r} \quad . \tag{16.14}$$

The validity of (16.14) is readily shown by making the substitution $s = -v$ in the integral $\tilde{S}\phi$. As a consequence of Lemma 1, we find

$$\tilde{S}_{-}\phi = \frac{1}{\pi}\int_{-\infty}^{\infty}\frac{\phi^{\mu}(s)ds}{s-x} - \frac{1}{\pi}\int_{-\infty}^{\infty}\frac{\phi^{r}(s)ds}{s-x} \quad . \tag{16.15}$$

Now equation (16.10) can be written in the following way:

$$\begin{aligned} \phi^{r}(x) + \frac{1}{\pi}\int_{-\infty}^{\infty}\frac{[b_1^{\mu}(s)+c_1^{\mu}(s)]\phi^{r}(s) + [b_1^{r}(s)+c_1^{r}(s)]\phi^{\mu}(s)}{s-x}\,ds &= f^{r}(x) \; , \\ \phi^{\mu}(x) + \frac{1}{\pi}\int_{-\infty}^{\infty}\frac{[b_1^{r}(s)-c_1^{r}(s)]\phi^{r}(s) + [b_1^{\mu}(s)-c_1^{\mu}(s)]\phi^{\mu}(s)}{s-x}\,ds &= f^{\mu}(x) \; . \end{aligned} \tag{16.16}$$

This pair of equations can be written more succinctly as

(16.17)
$$\tilde{\phi}(x) + \frac{1}{\pi i}\int_{\infty}^{\infty} \frac{B(s)\tilde{\phi}(s)ds}{s - x} = \tilde{f}(x) ,$$

where

(16.18)
$$B = i\begin{pmatrix} b_1^{\mu}+c_1^{\mu} & b_1^{r}+c_1^{r} \\ b_1^{r}-c_1^{r} & b_1^{\mu}-c_1^{\mu} \end{pmatrix} , \tilde{f} = \begin{pmatrix} f^{r} \\ f^{\mu} \end{pmatrix} .$$

Thus we have proved the following theorem.

Theorem 16.3. (Kim Ze Pkhen, [81]). *All solutions of* (16.10) *are determined by*

(16.19)
$$\phi(x) = \phi^{r}(x) + \phi^{\mu}(x) ,$$

where $\tilde{\phi} = (\phi^{r},\phi^{\mu})$ *are those solutions of equation* (16.17) *in which the first* m *components are even functions, and the last* m *components are odd.*

We note that if (16.17) has a unique solution $\tilde{\phi} = (\phi_1,\phi_2)$ then $\phi_1 = \phi^{r}$ and $\phi_2 = \phi^{\mu}$. To show this, assume that $(\phi^{r}+\psi^{\mu}, \phi^{\mu}+\psi^{r})$ is a solution of (16.17). Then we must have

(16.20)
$$\psi^{\mu}(x) + \frac{1}{\pi}\int_{-\infty}^{\infty} \frac{(b_1^{\mu}+c_1^{\mu})\psi^{\mu}(s) + (b_1^{r}+c_1^{r})\psi^{r}(s)}{s - x}\, ds = 0 ,$$
$$\psi^{r}(x) + \frac{1}{\pi}\int_{-\infty}^{\infty} \frac{(b_1^{r}-c_1^{r})\psi^{\mu}(s) + (b_1^{\mu}-c_1^{\mu})\psi^{r}(s)}{s - x}\, ds = 0 .$$

But from the given condition, the homogeneous equation corresponding to (16.17) can have only a trivial solution. This implies that $\psi^{\mu} = \psi^{r} = 0$, and the result follows.

Let us reduce the solution of equation (16.17) to that of a Riemann boundary value problem. In order to do this we introduce the piecewise analytic vector Ψ defined by

(16.21)
$$\Psi(z) = \frac{1}{2\pi i}\int_{-\infty}^{\infty} \frac{B(s)\tilde{\phi}(s)}{s - z}\, ds .$$

Taking into account that

(16.22)
$$B(x)\tilde{\phi}(x) = \psi^{+}(x) - \psi^{-}(x) , \quad \frac{1}{\pi i}\int_{-\infty}^{\infty} \frac{B(s)\tilde{\phi}(s)ds}{s - x} = \psi^{+}(x) + \psi^{-}(x) ,$$

we have that equation (16.17) is equivalent to the problem of determining a vector $\tilde{\phi}$ from the class "μ" (see Theorem 16.2), and a piecewise analytic vector Ψ of the form (16.21) such that

$$\tilde{\phi}(x) = \tilde{f}(x) - \psi^{+}(x) - \psi^{-}(x) ,$$

(16.23)

$$B(x)\tilde{\phi}(x) = \psi^{+}(x) - \psi^{-}(x) .$$

These equations, in turn, are equivalent to

$$(I + B)\tilde{\phi} = \tilde{f} - 2\psi^{-}$$

(16.24)

$$(I - B)\tilde{\phi} = \tilde{f} - 2\psi^{+}$$

or

$$\tilde{\phi} = (I + B)^{-1}(\tilde{f} - 2\psi^{-}) ,$$

(16.25)

$$\tilde{\phi} = (I - B)^{-1}(\tilde{f} - 2\psi^{+}) .$$

On comparing the right-hand sides, we obtain the Riemann boundary value problem

(16.26) $$\psi^{+}(x) = G(x)\psi^{-}(x) + \frac{1}{2}[I - G(x)]\tilde{f}(x) ,$$

where $$G = (I - B)(I + B)^{-1} ,$$

and we look for the solution which vanishes at infinity. Having solved this problem, we obtain the solution of equation (16.17) by using either of equations (16.25).

As mentioned in §8.4, if the problem of the factorisation of G can be solved, then the solution of the inhomogeneous problem (16.26) does not pose any major difficulties, and can be given in closed form. In this section we stated sufficient conditions for the factorisation of G. The same method can also be used in the case when $m = 1$.

§17 THE COMPUTER SOLUTION OF A SYSTEM ARISING IN DISPERSION THEORY

In this section we shall consider in some detail the solution of a complicated system of integral equations, via the use of the coordinate-wise descent algorithm.

We shall consider the approximate solution of a system of equations proposed by V. Tsellner [144]. This system is given by

$$(17.1)\qquad \begin{aligned} D(E) &= \frac{F(E)}{E} + \frac{1}{\pi E}\int_1^\infty \epsilon\,\frac{\alpha(\epsilon)D(\epsilon)+\beta(\epsilon)A(\epsilon)}{\epsilon - E}\,d\epsilon \; + \\ &\quad + \frac{1}{\pi E}\int_1^\infty \epsilon\,\frac{\alpha'(\epsilon)D(\epsilon)-\beta'(\epsilon)A(\epsilon)}{\epsilon + E}\,d\epsilon , \\ A(E) &= \alpha(E)D(E) + \beta(E)A(E) , \quad E \in (-\infty,\infty) \end{aligned}$$

where

$$(17.2)\qquad F(E) = \frac{f^3}{E}\left(\frac{1}{\nu_1}+\frac{1}{\nu_2}\right)\lambda + \frac{f^3}{E}\left(\frac{1}{\nu_1}-\frac{1}{\nu_2}\right)\lambda' .$$

In these equations, f, ν_1 and ν_2 are real parameters with $\nu_1+\nu_2 = 1$; λ and λ' are vectors (given numerically);

$$\alpha(E) = \sum_{j=1}^{3} [R_j(E)A_j + R_j(E\nu_1)B_j + R_j(E\nu_2)C_j] ,$$

$$\beta(E) = \sum_{j=1}^{3} [I_j(E)A_j + I_j(E\nu_1)B_j + I_j(E\nu_2)C_j] ,$$

$$\alpha'(E) = \sum_{j=1}^{3} [R_j(E)A'_j + R_j(E\nu_1)B'_j + R_j(E\nu_2)C'_j] ,$$

$$\beta'(E) = \sum_{j=1}^{3} [I_j(E)A'_j + I_j(E\nu_1)B'_j + I_j(E\nu_2)C'_j] ,$$

where A_j, B_j, C_j, A'_j, B'_j, C'_j are given matrices of order ten. In addition, for $j = 1,2,3$,

$$(17.3)\qquad \left\{ \begin{aligned} & R_j(\epsilon) = \frac{v_j}{1+v_j^2} , \quad I_j(\epsilon) = \frac{1}{1+v_j^2} , \quad v_j = \frac{\epsilon \operatorname{Re}[g_j(\epsilon)]}{\lambda_j p^3(\epsilon)v^2(\epsilon)} , \\ & \text{for } \epsilon \in (1,\infty), \text{ and} \\ & R_j(\epsilon) = I_j(\epsilon) = 0 \text{ for } \epsilon \in (-\infty,1]; \text{ where} \\ & p(\epsilon) = (\epsilon^2-1)^{\frac{1}{2}} , \quad v^2(\epsilon) = \exp[-(\epsilon^2-1)/\eta] , \end{aligned} \right.$$

with λ_j and η real parameters. The values of $\mathrm{Re}[g_j(\epsilon)]$ are given graphically, (see G. Salzman and F. Salzman, [166]). We wish to compute the vector-functions A and D.

Thus the system (17.1) consists of ten singular integral equations and ten algebraic equations, for the determination of twenty unknown functions. If we make the substitution

$$D(E) = [F(E) + \phi(E)]/E \quad ,$$

we obtain

$$A(E) = [I - \beta(E)]^{-1}\,\alpha(E)[F(E) + \phi(E)]/E \quad , \tag{17.4}$$

and the algebraic system of equations will be satisfied. We therefore obtain a system of ten singular integral equations for the vector $\phi = (\phi_1(E),\phi_2(E),\ldots,\phi_{10}(E))$, given by†

$$\phi(E) - \frac{1}{\pi}\int_{-\infty}^{\infty}\frac{b(\epsilon)\phi(\epsilon)d\epsilon}{\epsilon - E} - \frac{1}{\pi}\int_{-\infty}^{\infty}\frac{c(\epsilon)\phi(\epsilon)d\epsilon}{\epsilon + E} = f(E) \quad , \tag{17.5}$$

$E \in (-\infty,\infty)$, where

$$\text{(17.6)}\qquad \begin{aligned} &b(\epsilon) = [I - \beta(\epsilon)]^{-1}\alpha(\epsilon) \;,\quad c(\epsilon) = \alpha'(\epsilon) - \beta'(\epsilon)b(\epsilon) \;, \\ &f(E) = \frac{1}{\pi}\int_{-\infty}^{\infty}\frac{b(\epsilon)F(\epsilon)d\epsilon}{\epsilon - E} + \frac{1}{\pi}\int_{-\infty}^{\infty}\frac{c(\epsilon)F(\epsilon)d\epsilon}{\epsilon + E} \;. \end{aligned}$$

Let us introduce the operators

$$\tilde{S}\phi = \frac{1}{\pi}\int_{-\infty}^{\infty}\frac{\phi(\epsilon)d\epsilon}{\epsilon - E} = iS\phi \;,\quad \text{and}\quad \tilde{S}_-\phi = \frac{1}{\pi}\int_{-\infty}^{\infty}\frac{\phi(\epsilon)d\epsilon}{\epsilon + E} = -\frac{1}{\pi}\int_{-\infty}^{\infty}\frac{\phi(-\epsilon)d\epsilon}{\epsilon - E} = -\tilde{S}\phi_- \;,$$

where ϕ_- denotes $\phi(-x)$. Recalling that for a vector $\phi \in L_2(-\infty,\infty)$,

$$S^2\phi = \phi \;,\quad S\phi^+ = \phi^+ \;,\quad S\phi^- = -\phi^- \;,$$

we find

$$\text{(17.7)}\qquad \begin{aligned} &\tilde{S}^2\phi = -\phi \;,\quad \tilde{S}_-^2\phi = \phi \;,\quad \tilde{S}\tilde{S}_-\phi = \phi_- \;,\quad \tilde{S}_-\tilde{S}\phi = -\phi_- \;, \\ &\tilde{S}\phi^+ = i\phi^+ \;,\quad \tilde{S}\phi^- = -i\phi^- \;. \end{aligned}$$

Applying the operator $\tilde{S}$ to both sides of (17.5), then using (17.6) and (17.7) we obtain

† The author obviously intends that we define $b(E) = c(E) = 0$ for $E \in (-\infty,1)$. (Eds.)

(17.8) $$A\phi \equiv \tilde{S}\phi + b(E)\phi(E) - c(-E)\phi(-E) = [-b(E) - c(-E)]F(E) \equiv \psi \ .$$

We shall consider equation (17.8) as a linear equation in the Hilbert space $L_2(-\infty,\infty)$. Assuming that a solution ϕ exists with $\phi \in L_2(-\infty,\infty)$, we shall investigate its analytic properties. In connection with this we note (see [166]) that each Re g_j satisfies the equation

(17.9) $$\mathrm{Re}[g_j(E)] = 1 - \frac{\lambda_j E}{\pi}\int_1^\infty \frac{p^3(\epsilon)v^2(\epsilon)d\epsilon}{\epsilon^2(\epsilon-E)} - \frac{E}{\pi}\int_1^\infty \frac{p^3(\epsilon)v^2(\epsilon)H_j(\epsilon)d\epsilon}{\epsilon^2(\epsilon+E)} \ ,$$

where

(17.10) $$H_j(\epsilon) = -|g_j(\epsilon)|^2 \sum_{k=1}^{3} \frac{B_{jk}\lambda_k}{|g_k(\epsilon)|^2} \ , \quad \mathrm{Im}[g_j(\epsilon)] = \frac{\lambda_j p^3(\epsilon)v^2(\epsilon)}{\epsilon} \ ,$$

$\epsilon \in [1,\infty)$, B_{jk} being real numbers.

Consider now the integral

(17.11) $$F(z) = \frac{1}{2\pi i}\int_1^\infty \frac{p^3(\epsilon)v^2(\epsilon)d\epsilon}{\epsilon^2(\epsilon-z)} \ , \quad z \notin [1,\infty) \ .$$

From the Sokhotski-Plemelj equations we have

(17.12) $$F^+(E) - F^-(E) = p^3(E)v^2(E)/E^2 \ , \quad F^+(E) + F^-(E) = \frac{1}{\pi i}\int_1^\infty \frac{p^3(\epsilon)v^2(\epsilon)d\epsilon}{\epsilon^2(\epsilon-E)} \ ,$$

$E \in (1,\infty)$. On the other hand, for the function

(17.13) $$\Phi(z) = (z^2-1)^{3/2}\exp[-(z^2-1)/\eta]/2z^2 \ ,$$

we obtain

(17.14) $$\Phi^+(E) - \Phi^-(E) = p^3(E)v^2(E)/E^2 \ , \quad \Phi^+(E) + \Phi^-(E) = 0 \ , \quad E \in (1,\infty) \ .$$

Therefore the function

(17.15) $$\Psi(z) = F(z) - \Phi(z) \ ,$$

has the property

(17.16) $$\Psi^+(E) = \Psi^-(E) = \Psi(E) \ , \quad E \in (1,\infty) \ ,$$

and consequently it is analytic on $[0,\infty)$. Thus

(17.17) $$F^+(E) + F^-(E) = 2\Psi(E) \ , \quad E \in (1,\infty) \ .$$

If we assume that $p^3 v^2 H_j/\epsilon^3$ is integrable on $[1,\infty)$ for each j, then we have that Re g_j is analytic on $[1,\infty)$, and on $[0,1]$ it is given by

$$\text{Re}[g_j(E)] = \lambda_j(1-E^2)^{3/2}\exp[(1-E^2)/\eta]/E + \chi(E) \,, \tag{17.18}$$

where χ is analytic on $[0,1]$. Hence the functions Re g_j have bounded derivatives of all orders from the right at $E = 1$; their second and higher order derivatives are zero at ∞. The values of (Re g_j)$'$ from both the left and the right at $E = 1$ coincide. In addition from equation (17.9), it follows that the values of Re g_j and all their derivatives tend to finite limits as $E \to \infty$.

From equations (17.3), it follows that the matrices b and c are piecewise analytic on the real axis with singularities at the points $\delta = 1,\ 1/\nu_1,\ 1/\nu_2$. In neighbourhoods of these points we have

$$\begin{aligned} b(E) &= \begin{cases} \chi_1(E)(E^2-\delta^2)^{3/2} + \chi_2(E) \,, & E \geq \delta \,, \\ \chi_2(E) \,, & E \leq \delta \,; \end{cases} \\ c(E) &= \begin{cases} \chi_3(E)(E^2-\delta^2)^{3/2} + \chi_4(E) \,, & E \geq \delta \,, \\ \chi_4(E) \,, & E \leq \delta \,; \end{cases} \end{aligned} \tag{17.19}$$

where the functions χ_i are analytic at the point δ. As $E \to \infty$, b, c and their derivatives of any order, tend to zero. Applying arguments similar to those used in §16.2, we have that provided $\det(I \pm b) \neq 0$ or ∞, if the solution ϕ of equation (17.8) belongs to $L_2(-\infty,\infty)$, then ϕ is analytic everywhere except at the points $\delta = \pm 1,\ \pm 1/\nu_1,\ \pm 1/\nu_2$. At these points we can only guarantee that ϕ' is Hölder continuous with index 1/2. As $E \to \infty$, ϕ and all its derivatives tend to zero.

Let

$$E = i\,\frac{1+t}{1-t} \,,\quad \epsilon = i\,\frac{1+\tau}{1-\tau} \,,\quad \phi(E) = \phi(t) = (1-t)\phi_1(t) \,. \tag{17.20}$$

Then equation (17.8) becomes

$$\begin{aligned} A_1\phi_1 &= \frac{1}{\pi}\int_\gamma \frac{\phi_1(\tau)d\tau}{\tau - t} + b(t)\phi_1(t) + \frac{1}{t}\,c(\bar{t})\phi_1(\bar{t}) \\ &= [-b(t) - c(\bar{t})]F(t)/(1-t) \,. \end{aligned} \tag{17.21}$$

It is easily seen that if equation (17.8) has a unique solution, then so does equation (17.21). We shall look for an approximate solution of (17.21) in the form

$$\phi_1' = \sum_{k=1}^{n} \delta_k \ell_k \,, \tag{17.22}$$

under the condition that the unknown parameters δ_k minimize the expression

$$(17.23) \qquad \|\psi_1' - \sum_{k=1}^{n} \delta_k y_k\|_{L_2(\gamma)} , \quad y_k \equiv A_1' \ell_k$$

where $n = 10 \times 2N$; i.e.

$$(17.24) \qquad \begin{aligned} &\ell_{10(2p-2)+s} = t^{p-1}\ell_s' , \quad \ell_{10(2p-1)+s} = t^{-p}\ell_s' , \quad s = 1,2,\dots,10 , \\ &p = 1,2,\dots,N , \\ &\ell_1' = (1,0,\dots,0) , \quad \ell_2' = (0,1,0,\dots,0),\dots, \quad \ell_{10}' = (0,0,\dots,0,1) ; \end{aligned}$$

$$(17.25) \qquad \begin{aligned} y_k \equiv A_1'\ell_k &= \frac{1}{\pi} \int_\gamma \frac{\ell_k(\tau)d\tau}{\tau - t} + b'(t)\ell_k(t) + \frac{1}{t} c'(\bar{t})\ell_k(\bar{t}) \\ &= \begin{cases} [it^{p-1} + b'(t)t^{p-1} + c'(\bar{t})t^{-p}]\ell_s' , & k = 10(2p-2)+s , \\ [-it^{-p} + b'(t)t^{-p} + c'(\bar{t})t^{p-1}]\ell_s' , & k = 10(2p-1)+s , \end{cases} \end{aligned}$$

$s = 1,2,\dots,10$, $p = 1,2,\dots,N$. By ψ_1', b', c' we mean approximate representations of ψ_1, b, c respectively, given by

$$(17.26) \qquad \begin{aligned} &\psi_{1j}' = \sum_{k=-m}^{m-1} a_k^{(j)} t^k \equiv \{a_k^{(j)}\}, \quad a_k^{(j)} = \frac{1}{2m} \sum_{r=-m}^{m-1} \psi_{1j}(\epsilon_r)t_r^k , \\ &\psi_{1j}'(\epsilon_r) = \sum_{s=1}^{10} [-b_{js}(\epsilon_r) - c_{js}(-\epsilon_r)]F_s(\epsilon_r)/(1-t_r) , \\ &t_r = \exp(i\pi r/m) , \quad \epsilon_r = -\cot(\pi r/2m) , \\ &b_{js}' = \sum_{k=-m}^{m-1} b_k^{(j)} t^k \equiv \{b_k^{(js)}\}, \quad b_k^{(js)} = \frac{1}{2m} \sum_{r=-m}^{m-1} b_{js}(\epsilon_r)t_r^k , \\ &c_{js}' = \sum_{k=-m}^{m-1} c_k^{(js)} t^k \equiv \{c_k^{(js)}\}, \quad c_k^{(js)} = \frac{1}{2m} \sum_{r=-m}^{m-1} c_{js}(\epsilon_r)t_r^k , \\ &j = 1,2,\dots,10 . \end{aligned}$$

Recalling that $b = c = 0$ for $E \in (-\infty,1)$ and for $E = +\infty$, the coefficients given in (17.26) may be simplified when m is even to give

$$a_k^{(j)} = \frac{1}{2m} \sum_{r=-(m/2)+1}^{-1} \sum_{s=1}^{10} \left[b_{js}(\epsilon_r) \frac{t_r^k}{t_r - 1} - c_{js}(\epsilon_r) \frac{\bar{t}_r^k}{\bar{t}_r - 1} \right] F_s(\epsilon_r) \ ,$$

(17.27)

$$b_k^{(js)} = \frac{1}{2m} \sum_{r=-(m/2)+1}^{-1} b_{js}(\epsilon_r) t_r^k \ , \quad c_k^{(js)} = \frac{1}{2m} \sum_{r=-(m/2)+1}^{-1} c_{js}(\epsilon_r) t_r^k \ .$$

From equations (17.25) and (17.26), we can write out the vectors y_k as follows: for $k = 10(2p-2)+s$; $s = 1,2,\dots,10$; $p = 1,2,\dots,N$;

$$j \neq s \ , \quad y_{kj} = b'_{js}(t)t^{p-1} + c'_{js}(\bar{t})t^{-p} \equiv \{\alpha_\nu^{(kj)}\} \ ,$$

$$\alpha_\nu^{(kj)} = b_{\nu-p+1}^{(js)} + c_{-\nu-p}^{(js)} \ ;$$

$$j = s \ , \quad y_{ks} = it^{p-1} + b'_{ss}(t)t^{p-1} + c'_{ss}(\bar{t}\ )t^{-p} = \{\alpha_\nu^{(ks)}\} \ ,$$

$$\alpha_\nu^{(ks)} = \begin{cases} b_{\nu-p+1}^{(ss)} + c_{-\nu-p} \ , & \nu \neq p-1 \ , \\ b_0^{(ss)} + c_{-2p+1}^{(ss)} + i \ , & \nu = p-1 \ , \end{cases}$$

$$-m-N \leq \nu \leq m+N \ ; \quad b_{\nu-p+1}^{(js)} = c_{-\nu-p}^{(js)} = 0 \ , \quad \nu-p+1 > m-1 \ ,$$

(17.28)

$$\nu-p+1 < -m \ , \quad -\nu-p > m-1 \ , \quad -\nu-p < -m \ ;$$

$$k = 10(2p-1)+s \ ; \quad s = 1,2,\dots,10 \ ; \quad p = 1,2,\dots,N \ ;$$

$$j \neq s \qquad y_{kj} = b'_{js}(t)t^{-p} + c'_{js}(\bar{t})t^{p-1} \equiv \{\alpha_\nu^{(kj)}\} \ ,$$

$$\alpha_\nu^{(kj)} = b_{\nu+p}^{(js)} + c_{-\nu+p-1}^{(js)} \ ;$$

$$j = s \qquad y_{ks} = -it^{-p} + b'_{ss}(t)t^{-p} + c'_{ss}(\bar{t})t^{p-1} \equiv \{\alpha_\nu^{(ks)}\} \ ,$$

$$\alpha_\nu^{(ks)} = \begin{cases} b_{\nu+p}^{(ss)} + c_{-\nu+p-1}^{(ss)} \ , & \nu \neq 1 \ , \\ b_0^{(ss)} + c_{2p-1-i}^{(ss)} \ , & \nu = -p \ , \end{cases}$$

$$-m-N \leq \nu \leq m+N \ ; \quad b_{\nu+p}^{(js)} = c_{-\nu+p-1}^{(js)} = 0 \ , \quad \nu+p > m-1 \ ,$$

$$\nu+p < -m \ , \quad -\nu+p-1 > m-1 \ , \quad -\nu+p-1 < -m \ .$$

Determining y_k sequentially, we can apply the coordinate-wise descent algorithm (see §3.4) to find the unknown values δ_k. Arbitrarily choosing an initial point $P_n^{(0)}\ (\delta_1^{(0)}, \delta_2^{(0)}, \dots, \delta_n^{(0)})$, the initial pair of vectors is given by

$u_n^{(0)} = \psi_1'\{\psi'_{1,1}, \psi'_{1,2},\dots,\psi'_{1,10}\}$, $\psi'_{1j} = \{a_\nu^{(j)}\}$, $a_\nu^{(j)} = 0$ for $\nu > m-1$ and $\nu < -m$; and

$v_1^{(1)} = y_1\{y_{1,1}, y_{1,2},\dots,y_{1,10}\}$, $y_{1j} = \{\alpha_\nu^{(1j)}\}$, $j = 1,2,\dots,10$.

The initial value of the residual is

$$\Delta_n^{(0)} = \|u_n^{(0)}\|^2_{L_2(\gamma)} = \sum_{j=1}^{10} \sum_{\nu=-m-N}^{m+N} |a_\nu^{(j)}|^2 , \tag{17.29}$$

and we find $\delta_1^{(1)}$ from the condition

$$\|u_n^{(0)} - \delta_1^{(1)} v_1^{(1)}\| = \min_\delta \|u_n^{(0)} - \delta v_1^{(1)}\| . \tag{17.30}$$

We know that

$$\delta_1^{(1)} = \frac{(u_n^{(0)}, v_1^{(1)})}{\|v_1^{(1)}\|^2} = \frac{\sum_{j=1}^{10} \sum_{\nu=-m-N}^{m+N} a_\nu^{(j)} \bar{\alpha}_\nu^{(1j)}}{\sum_{j=1}^{10} \sum_{\nu=-m-N}^{m+N} |\alpha_\nu^{(1j)}|^2} . \tag{17.31}$$

Having found $\delta_1^{(1)}$, we write down $P_1^{(1)}\ (\delta_1^{(0)} + \delta_1^{(1)}, \delta_2^{(0)},\dots,\delta_n^{(0)})$, $u_1^{(1)} = u_n^{(0)} - \delta_1^{(1)} v_1^{(1)} = (\psi'_{1,1} - \delta_1^{(1)} y_{1,1},\dots,\psi'_{1,10} - \delta_1^{(1)} y_{1,10})$, $\psi'_{1j} - \delta_1^{(1)} y_{1j}^{(1)} \equiv \{a_\nu^{(j)} - \delta_1^{(1)}\alpha_\nu^{(1j)}$ and $v_2^{(1)} = y_2\{y_{2,1},\dots,y_{2,10}\}$, $y_{2,j} = \{\alpha_\nu^{(2j)}\}$, $j = 1,2,\dots,10$. We calculate

$$\Delta_1^{(1)} = \|u_1^{(1)}\|^2 = \sum_{j=1}^{10} \sum_{\nu=-m-N}^{m+N} |a_\nu^{(j)} - \delta_1^{(1)}\alpha_\nu^{(1j)}|^2 ,$$

and find $\delta_2^{(1)}$ from the condition

$$\|u_1^{(1)} - \delta_2^{(1)} v_2^{(1)}\| = \min_\delta \|u_1^{(1)} - \delta v_2^{(1)}\| ,$$

using the result

$$\delta_2^{(1)} = \frac{(u_1^{(1)}, v_2^{(1)})}{\|v_2^{(1)}\|^2} = \frac{\sum_{j=1}^{10} \sum_{\nu=-m-N}^{m+N} (a_\nu^{(j)} - \delta_1^{(1)}\alpha_\nu^{(1j)})\bar{\alpha}_\nu^{(2j)}}{\sum_{j=1}^{10} \sum_{\nu=-m-N}^{m+N} |\alpha_\nu^{(2j)}|^2} ,$$

and so on. At the $(n+1)^{\text{st}}$ step, having found $\delta_n^{(1)}$, we write down $P_n^{(1)}(\delta_1^{(0)} + \delta_1^{(1)}, \delta_2^{(0)} + \delta_2^{(1)},\dots,\delta_n^{(0)} + \delta_n^{(1)})$, $u_n^{(1)} = u_{n-1}^{(1)} - \delta_n^{(1)} v_n^{(1)}$, $(v_n^{(1)} = y_n)$, $v_1^{(2)} = y_1$. We calculate $\Delta_n^{(1)} = \|u_n^{(1)}\|^2$ and determine $\delta_1^{(2)}$ from the equation

$$\delta_1^{(2)} = \frac{(u_n^{(1)}, v_1^{(2)})}{\|v_1^{(2)}\|^2} \quad .$$

We write down $P_1^{(2)}(\delta_1^{(0)} + \delta_1^{(1)} + \delta_1^{(2)},\ \delta_2^{(0)} + \delta_2^{(1)}, \ldots, \delta_n^{(0)} + \delta_n^{(1)})$, $u_1^{(2)} = u_n^{(1)} - \delta_1^{(2)} v_1^{(2)}$, $v_2^{(2)} = y_2$, calculate $\Delta_1^{(2)} = \|u_1^{(2)}\|^2$ and determine $\delta_2^{(2)}$ from

$$\delta_2^{(2)} = \frac{(u_1^{(1)}, v_2^{(2)})}{\|v_2^{(2)}\|^2} \quad ,$$

and so on. The calculations are concluded as soon as

$$\Delta_n^{(q)} - \Delta_n^{(q+1)} < \mu \Delta_n^{(0)} \quad , \tag{17.32}$$

where μ is a given constant. The required solution is taken to be the point $P_n^{(q+1)}(\delta_1, \delta_2, \ldots, \delta_n)$ or

$$\phi_j'(E) = \mathrm{Re}[(1-t) \sum_{p=1}^{N} (\delta_{10(2p-2)+j}\, t^{p-1} + \delta_{10(2p-1)+j}\, t^{-p}] \ , \tag{17.33}$$

$$j = 1,2,\ldots,10 \ , \ t = (E-i)/(E+i) \ .$$

From (17.4) the final result is given by the functions

$$\begin{aligned} D_j'(E) &= [F_j(E) + \phi_j'(E)]/E \ , \\ A_j'(E) &= \sum_{s=1}^{10} b_{js}(E) D_s'(E) \ , \ j = 1,2,\ldots,10 \end{aligned} \tag{17.34}$$

which are represented in graphical or tabular form by the values

$$\begin{aligned} D_j'(\epsilon_r) &= [F_j(\epsilon_r) + \phi_j'(\pi r/m)]/\epsilon_r \ , \\ A_j(\epsilon_r) &= \sum_{s=1}^{10} b_{js}(\epsilon_r) D_s'(\epsilon_r) \ , \\ \epsilon_r &= -\cot(\pi r/2m) \ ; \ r = -m, -m+1, \ldots, m-1; \ \ j = 1,2,\ldots,10 \ . \end{aligned} \tag{17.35}$$

The solution of the problem comprises four stages:

1) the calculation of the quantities in (17.3) at the points $\epsilon_r = -\cot(\pi r/2m)$, $r = -1, -2, \ldots, -m+1$;
2) the calculation of the matrices b and c from (17.6) at the points ϵ_r (for matrix inversion, the method of A.P. Ershov [49], was used);

3) the determination of $\delta_k (k = 1,2,\ldots,n)$ by the calculation of the coefficients $a_k^{(j)}$, $b_k^{(js)}$ and $c_k^{(js)}$ from equations (17.27), the calculation of y_k from (17.28) with a recalculation after every n steps, and the application of the coordinate-wise descent algorithm;
4) the calculation of the required values according to equations (17.33) and (17.35).

Let us consider some details of the calculation in the case when $\nu_1 = \nu_2 = 0.5$, $m = 10$ and $N = 20$. We choose a comparatively high degree for the approximating polynomials ($m = 10$) since the functions have first derivatives which are Hölder continuous with index 1/2. Consequently the order of approximation is given only by $m^{-3/2} \ln m$, (see §10.1). The time of one iteration (i.e. from $\Delta_n^{(\ell-1)}$ to $\Delta_n^{(\ell)}$) is about seven minutes. It should be noted that an algebraic system of the form (3.5)(with $\alpha = 0$) of order 800, was solved. After 27 iterations the original residual $\Delta_n^{(0)} = 22.2$ was decreased to $\Delta_n^{(27)} = 2.4$. It was found that $\Delta_n^{(26)} - \Delta_n^{(27)} \approx 0.02$. Additional iterations were considered unnecessary because of the low accuracy of the input data.

The same problem has also been solved by Kim Ze Pkhen [81], using a different method. A comparison of both solutions showed close agreement.

Chapter 4

NUMERICAL SOLUTION OF SINGULAR INTEGRAL EQUATIONS OF CONVOLUTION TYPE†

§18 THE GENERAL FORM OF SINGULAR INTEGRAL EQUATIONS OF CONVOLUTION TYPE

Let us consider the space of m-dimensional vector functions in $L_1(-\infty,\infty)$ and $L_2(-\infty,\infty)$. We shall denote the elements of these spaces by small letters and their corresponding Fourier transforms by capital letters. If f is any element which belongs to L_1 or L_2, then its Fourier transform F is defined by

$$F(\omega) = (2\pi)^{-\frac{1}{2}} \int_{-\infty}^{\infty} f(x)e^{i\omega x}\, dx \quad . \tag{18.1}$$

We shall assume that the space $L_2(-\infty,\infty)$ is a Hilbert space. If $f^{(1)} = (f_1^{(1)},f_2^{(1)},\dots,f_m^{(1)})$ and $f^{(2)} = (f_1^{(2)},f_2^{(2)},\dots,f_m^{(2)})$ are any two elements in $L_2(-\infty,\infty)$, then their scalar product is defined by

$$(f^{(1)}, f^{(2)}) = \frac{1}{m} \sum_{j=1}^{m} \int_{-\infty}^{\infty} f_j^{(1)}(x)\, \overline{f_j^{(2)}(x)}\, dx \ . \tag{18.2}$$

As usual, the norm of f is defined by $\|f\| = (f,f)^{\frac{1}{2}}$. It is well known (see [95]) that if $f^{(1)}, f^{(2)} \in L_2(-\infty,\infty)$ then $F^{(1)}, F^{(2)} \in L_2(-\infty,\infty)$ and also

$$(f^{(1)}, f^{(2)}) = (F^{(1)}, F^{(2)}) \quad . \tag{18.3}$$

If $f \in L_1(-\infty,\infty)$ then $F \in W(-\infty,\infty)$ i.e. F is continuous, is zero at infinity and in addition the functions $F^{\pm}$, where

$$F^{\pm}(\omega) = (2\pi)^{-\frac{1}{2}} \int_{\pm\infty}^{0} f(x)e^{i\omega x}\, dx \quad , \tag{18.4}$$

are analytic in the upper (+) and lower (-) half planes respectively and are continuous on the real axis.

Let us now consider a system of integral equations of convolution type of the general form which was first introduced by Yu. I. Cherskiy [147]:

---

† For a recent summary of the theory for integral equations with convolution kernels see Chapter VIII of P.P. Zabreyko et al., *Integral Equations - a reference text* (a translation from the Russian) Noordhoff International Publishing, Leyden (1975). (Eds.)

$$n\phi \equiv \delta\phi(x) + \eta \operatorname{sgn} x\, \phi(x) + (2\pi)^{-\frac{1}{2}} \int_{-\infty}^{\infty} a_1(x-u)\phi(u)du +$$

(18.5)
$$+ (2\pi)^{-\frac{1}{2}} \int_{-\infty}^{\infty} a_2(x-u)\phi(u) \operatorname{sgn} u\, du + \int_{-\infty}^{\infty} n(u,x)\phi(u)du = f(x) ,$$

$x \in (-\infty,\infty)$. The following assumptions are made concerning the functions occurring in (18.5):

1) the matrices a_1, a_2 are of order m and are in $L_1(-\infty,\infty)$ i.e. each element of these matrices is integrable on the real axis;

2) $f = (f_1, f_2, \ldots, f_m)$ and $f \in L_2(-\infty, \infty)$;

3) $\int_{-\infty}^{\infty} \int_{-\infty}^{\infty} |n_{j\nu}(u,x)|^2 dudx < \infty$ for $j,\nu = 1,2,\ldots,m$,

   where $n(u,x) \equiv (n_{j\nu}(u,x))$.

4) δ and η are constant matrices.

If $\det(\delta \pm \eta) \neq 0$ then equation (18.5) is said to be an equation of the *second kind*. If either $\det(\delta + \eta) = 0$ or $\det(\delta - \eta) = 0$ then (18.5) is said to be an equation of the *first kind*.

We take the Fourier transform of each side of equation (18.5). Recalling that the Fourier transform of the convolution of two functions is equal to the product of their Fourier transforms and that the Fourier transform of ϕ sgn x is given by the singular integral of Φ, we obtain on using (18.3) that

(18.6)
$$N\Phi \equiv [\delta + A_1(\omega)]\Phi(\omega) + \frac{\eta + A_2(\omega)}{\pi i} \int_{-\infty}^{\infty} \frac{\Phi(v)dv}{v - \omega} + \int_{-\infty}^{\infty} N(-v,\omega)\Phi(v)dv = F(\omega) ,$$

where

(18.7)
$$N(-v,\omega) = \frac{1}{2\pi} \int_{-\infty}^{\infty} \int_{-\infty}^{\infty} n(u,x) \exp[i(\omega x - vu)]dudx .$$

If we make the substitutions

(18.8)
$$\omega = i\,\frac{1+t}{1-t} , \quad v = i\,\frac{1+\tau}{1-\tau} ,$$

equation (18.6) becomes

(18.9)
$$K\tilde{\phi} \equiv a(t)\tilde{\phi}(t) + \frac{b(t)}{\pi i} \int_{\gamma} \frac{\tilde{\phi}(\tau)d\tau}{\tau - t} + \int_{\gamma} k(\tau,t)\tilde{\phi}(\tau)d\tau = \tilde{f}(t) ,$$

where

$$a(t) = \delta + A_1\left(i\,\frac{1+t}{1-t}\right) \ , \quad b(t) = \eta + A_2\left(i\,\frac{1+t}{1-t}\right) \ ,$$

$$(18.10) \qquad k(\tau,t) = N\left(-i\,\frac{1+\tau}{1-\tau} \ , \ i\,\frac{1+t}{1-t}\right) \frac{2i}{(1-t)(1-\tau)} \ ,$$

$$\tilde{\phi}(t) = \Phi\left(i\,\frac{1+t}{1-t}\right)\frac{1}{1-t} \ , \quad \tilde{f}(t) = F\left(i\,\frac{1+t}{1-t}\right)\frac{1}{1-t} \ .$$

We shall now examine the properties of a, b, $\tilde{f}$ and k. Since A_1, $A_2 \in W(-\infty,\infty)$, it follows that a and b are continuous. In addition, if $\det(a\pm b) \neq 0$ then from the Wiener-Levy theorem (see, for example, [2]) it follows that $\ln[(a+b)^{-1}(a-b)]$ is continuous. From conditions 2) and 3) on f and n respectively, it follows that $\tilde{f} \in L_2(\gamma)$ and

$$\int_\gamma |k_{j\nu}(\tau,t)|^2 |dt||d\tau| < \infty \ , \quad k(\tau,t) \equiv (k_{j\nu}(\tau,t)) \ .$$

Finally we note that when $t = 1$, $a\pm b = \delta\pm\eta$. Thus in the case when equation (18.5) is of the first kind, we have the exceptional case of equation (18.9). The theory of equations (18.5) and (18.9) is now more complicated (see §8.2, §14.1 and §19 as well as the paper by Gakhov and Smagina [34]) but this poses no problems for the least squares method as described in §3 and §6.2. In general, to solve equation (18.9) we can apply all the methods developed in the previous chapters.(1)

Let us assume that equation (18.9) has been solved and that it is required to find the solution of the original equation (18.5). From equations (18.8) and (18.10) we obtain

$$(18.11) \qquad \Phi(\omega) = \tilde{\phi}\left(\frac{\omega-i}{\omega+i}\right)\frac{2i}{\omega+i} \ .$$

Applying the inverse Fourier transform we obtain

$$(18.12) \qquad \phi(x) = \frac{1}{(2\pi)^{\frac{1}{2}}}\int_{-\infty}^{\infty} \tilde{\phi}\left(\frac{\omega-i}{\omega+i}\right)\frac{2i}{\omega+i}\,e^{-i\omega x}\,d\omega \ .$$

Thus a new computational problem arising in the numerical solution of systems of integral equations of convolution type, is that of the approximate evaluation of Fourier integrals. This problem is examined in §20.1.

In conclusion we note that an alternative theory for equation (18.5) has been developed by Gakhov and Cherskiy [35], under assumptions that are different from those given by conditions 1)-3), above.

---

(1) On the other hand, in a series of applications, S.M. Belonosov [6] has shown that it is advantageous to use the Laplace and Riemann-Mellin transformations to change boundary value problems and singular integral equations into functional equations which are similar to equations of convolution type.

§19 WIENER-HOPF EQUATIONS OF THE SECOND KIND

Equations of the form

$$B\phi \equiv \phi(x) + \int_0^\infty c(x-u)\phi(u)du = f(x) \;, \quad x > 0 \;, \tag{19.1}$$

are usually called *Wiener-Hopf equations.*(2) The theory of these equations has been discussed in detail by M.G. Krein [85]. It is important that equation (19.1) be considered as a specific case of equation (18.5). Now for $x > 0$, the solution of (19.1) will coincide with the solution of

$$\phi(x) + \int_0^\infty c(x-u)\phi(u)du = f(x) \;, \quad -\infty < x < \infty \;, \tag{19.2}$$

where f denotes a continuation of the right hand side of equation (19.1) to the entire real axis. On the other hand, if in equation (18.5) we put

$$a_1(x) = a_2(x) = (\pi/2)^{\frac{1}{2}}\, c(x) \;, \; n(u,x) \equiv 0 \;, \; \delta = I, \; \eta = 0 \;, \tag{19.3}$$

then we obtain equation (19.2).

In the previous section we showed how the solution of equation (18.5) could be reduced to that of the singular integral equation (18.9). In the case given by equations (19.3), this latter equation takes the form

$$K^0\tilde{\phi} \equiv a(t)\tilde{\phi}(t) + \frac{b(t)}{\pi i}\int_\gamma \frac{\tilde{\phi}(\tau)d\tau}{\tau - t} = \tilde{f}(t) \tag{19.4}$$

where

$$a(t) = I + (\pi/2)^{\frac{1}{2}}\, C\left(i\,\frac{1+t}{1-t}\right) \;, \; b(t) = (\pi/2)^{\frac{1}{2}}\, C\left(i\,\frac{1+t}{1-t}\right) \;,$$
$$\tilde{f}(t) = F\left(i\,\frac{1+t}{1-t}\right)\frac{1}{1-t} \;. \tag{19.5}$$

Having solved equation (19.4) in closed form (see §8), we obtain the solution of equation (19.1) as

$$\phi(x) = \frac{1}{(2\pi)^{\frac{1}{2}}}\int_{-\infty}^\infty \tilde{\phi}\left(\frac{\omega-i}{\omega+i}\right)\frac{2i}{\omega+i}\; e^{-i\omega x}\, d\omega \;, \; x > 0 \;. \tag{19.6}$$

Since

$$(f^{(1)}, f^{(2)})_{L_2(-\infty,\infty)} = (F^{(1)}, F^{(2)})_{L_2(-\infty,\infty)} = 4\pi(\tilde{f}^{(1)}, \tilde{f}^{(2)})_{L_2(\gamma)} \;,$$
$$\text{where} \quad \tilde{f}^{(1)}(t) = F^{(1)}\left(i\,\frac{1+t}{1-t}\right)\frac{1}{1-t} \;, \; \tilde{f}^{(2)}(t) = F^{(2)}\left(i\,\frac{1+t}{1-t}\right)\frac{1}{1-t} \;, \tag{19.7}$$

---

(2) N. Wiener and E. Hopf [171] were the first to give solutions in closed form of equation (19.1), in certain cases.

it follows that all results concerning the method of least squares (see §3) and the Ritz-Galerkin method (see [58, 117]) which were obtained for singular integral equations of the form (18.9) and (19.4), will be valid for equations (18.5) and (19.2). In other words, these methods can be applied directly to equations (18.5) and (19.2) without first reducing them to singular integral equations with Cauchy kernel.

Let us illustrate this by applying the Galerkin method to the equation

$$\phi(x) - \frac{\lambda}{\pi}\int_0^\infty K_0(|x-u|)\phi(u)du = e^{-x}, \quad x > 0,$$

(19.8)

$$\lambda \notin (1,\infty), \quad K_0(x) = \int_0^\infty \frac{\cos \omega x}{(1+\omega^2)^{\frac{1}{2}}}\, d\omega, \quad x > 0,$$

which was obtained by Grinberg and Fok [45] in their study of the refraction of plane electromagnetic waves. We introduce the equation

$$G\phi \equiv \phi(x) - \frac{\lambda}{\pi}\int_0^\infty K_0(|x-u|)\phi(u)du = \begin{cases} e^{-x}, & x \geq 0 \\ 0, & x < 0. \end{cases} \tag{19.9}$$

The Fourier transform of $K_0(|x|)$ is given by

$$(2\pi)^{-\frac{1}{2}}\int_{-\infty}^\infty K_0(|x|)e^{i\omega x}\, dx = (\pi/2)^{\frac{1}{2}}(1+\omega^2)^{-\frac{1}{2}}. \tag{19.10}$$

Therefore, in the notation of equation (18.6), corresponding to equation (19.9) we have

$$1 + A_1 + A_2 = 1 - \lambda(1+\omega^2)^{-\frac{1}{2}} \neq 0,$$

(19.11)

$$1 + A_1 - A_2 = 1,$$

withthe index $\varkappa$ of this equation (which is simultaneously the index of equations (19.8) and (19.9)) equal to zero. Accordingly we seek an approximate solution to ϕ in the form

$$\phi_n(x) = \sum_{k=-n}^{n} \alpha_k\psi_k(x), \tag{19.12}$$

where for $k \geq 0$,

$$\psi_k(x) = \begin{cases} e^{-x}P_k(x), & x > 0, \\ 0, & x < 0; \end{cases}$$

(19.13) and for $k < 0$,

$$\psi_k(x) = \begin{cases} 0, & x > 0, \\ e^{x}P_{-k-1}(-x), & x < 0. \end{cases}$$

The polynomials P_k are chosen to satisfy the condition

$$\int_0^\infty e^{-2x} P_k(x) P_j(x)\,dx = \begin{cases} 1, & k = j, \\ 0, & k \neq j. \end{cases} \tag{19.14}$$

The unknowns α_k are then given as solutions of the algebraic system

$$\sum_{k=-n}^{n} \alpha_k (G\psi_k, \psi_j) = \begin{cases} 1/\sqrt{2}, & j = 0, \\ 0, & j = \pm 1, \pm 2, \ldots, \pm n. \end{cases} \tag{19.15}$$

Since we only need α_k for $k \geq 0$, we shall consider the system of equations

$$\alpha_j + \sum_{k=0}^{n} \alpha_k \left(-\frac{\lambda}{\pi} \int_0^\infty K_0(|x-u|)\psi_k(u)\,du \,,\, \psi_j \right) = \begin{cases} 1/\sqrt{2}, & j = 0 \\ 0, & j = 1, 2, \ldots, n. \end{cases} \tag{19.16}$$

Applying equation (18.3) we obtain

$$\begin{aligned} &\left(-\frac{\lambda}{\pi} \int_0^\infty K_0(|x-u|)\psi_k(u)\,du,\ \psi_j \right) \\ &= -\frac{\lambda}{\pi} \left(\left(\frac{\omega-i}{\omega+i}\right)^k \left(\frac{i}{\omega+i}\right) \left(\frac{\pi}{1+\omega^2}\right)^{\frac{1}{2}} ,\ \frac{1}{\pi^{\frac{1}{2}}} \left(\frac{\omega-i}{\omega+i}\right)^j \left(\frac{i}{\omega+i}\right) \right) \\ &= -\frac{\lambda}{\pi} \int_{-\infty}^{\infty} \frac{1}{(1+\omega^2)^{\frac{1}{2}}} \left(\frac{\omega-i}{\omega+i}\right)^{k-j} \frac{d\omega}{1+\omega^2} \\ &= \frac{2\lambda}{\pi} \cdot \frac{1}{4(k-j)^2 - 1} . \end{aligned} \tag{19.17}$$

Thus, for $k = 0, 1, \ldots, n$, the quantities α_k satisfy the equations

$$\alpha_j + \frac{2\lambda}{\pi} \sum_{k=0}^{n} \frac{\alpha_k}{4(k-j)^2 - 1} = \begin{cases} 1/\sqrt{2}, & j = 0, \\ 0, & j = 1, 2, \ldots, n. \end{cases} \tag{19.18}$$

The case of small λ is of considerable practical interest. Since

$$\frac{2|\lambda|}{\pi} \sum_{k=0}^{n} \left| \frac{1}{4(k-j)^2 - 1} \right| < \frac{4|\lambda|}{\pi} , \tag{19.19}$$

iterative methods can be applied when $|\lambda| < \pi/4$. In particular Seidel's method is effective and has been used to obtain an approximate solution of equation (19.1) on the BESM computer (see [64]).

§20 WIENER-HOPF EQUATIONS OF THE FIRST KIND

Equations of the form

$$A\phi \equiv \int_0^\infty c(x-u)\phi(u)du = f(x) \ , \ x > 0 \ , \tag{20.1}$$

are usually called Wiener-Hopf equations of the *first kind*.

As indicated by Krein [85], the theory of these equations is far from complete. However, in a series of important cases (see [48, 61, 77]) equation (20.1) has been thoroughly investigated and its solution obtained in closed form. Unlike the case of equations of the second kind, it is obvious from equation (20.1) that we cannot arbitrarily continue the definition of $f(x)$ for $x \leq 0$. Nevertheless equation (20.1) may be looked upon as a special case of equation (18.5) if we assume that

$$\begin{gathered} a_1(x) = a_2(x) = (\pi/2)^{\frac{1}{2}} c(x) \ , \ n(u,x) = 0 \ , \\ \delta + \eta = 0 \ , \ \ \delta - \eta = I \ . \end{gathered} \tag{20.2}$$

Then equation (18.5) becomes

$$\begin{gathered} \int_0^\infty c(x-u)\phi(u)du = f(x) \ , \ x > 0 \ , \\ \phi(x) + \int_0^\infty c(x-u)\phi(u)du = f(x) \ , \ x \leq 0 \ . \end{gathered} \tag{20.3}$$

If ϕ is the solution of equation (20.3), then obviously for $x \geq 0$ it is also a solution of equation (20.1). Conversely, if the solution of equation (20.1) is determined, then the solution of equation (20.3) is found trivially. The singular integral equation (18.9) will now be written as

$$K^0\tilde{\phi} \equiv a(t)\tilde{\phi}(t) + \frac{b(t)}{\pi i} \int_\gamma \frac{\tilde{\phi}(\tau)d\tau}{\tau - t} = \tilde{f}(t) \ , \tag{20.4}$$

where

$$\begin{gathered} a(t) = \frac{1}{2} I + (\pi/2)^{\frac{1}{2}} C\left(i\ \frac{1+t}{1-t}\right) , \ b(t) = -\frac{1}{2} I + (\pi/2)^{\frac{1}{2}} C\left(i\ \frac{1+t}{1-t}\right) , \\ \tilde{\phi}(t) = \Phi\left(i\ \frac{1+t}{1-t}\right) \frac{1}{1-t} \ , \ \ \tilde{f}(t) = F\left(i\ \frac{1+t}{1-t}\right) \frac{1}{1-t} \ . \end{gathered} \tag{20.5}$$

If $C(\infty) = 0$, then it is obvious that $\det[a(1) + b(1)] = 0$. Thus, the reduction of the Wiener-Hopf equation of the first kind to a singular integral equation with Cauchy kernel gives the exceptional case of a singular integral equation.

In §8.2 an equation of the form (20.4) with $m = 1$, has been solved in closed form under the assumption that

$$G(t) \equiv (a(t) + b(t))^{-1}(a(t) - b(t)) = \frac{\prod_{k=1}^{\mu} (t-\alpha_k)^{m_k}}{\prod_{j=1}^{\nu} (t-\beta_j)^{p_j}} \, G_1(t) \quad , \tag{20.6}$$

where $\alpha_k \neq \beta_j$; $\alpha_k, \beta_j \in \gamma$; m_k, p_j are positive integers; and $G_1(t) \neq 0$ or ∞, for $t \in \gamma$. Keeping in mind the applications to follow, we shall now examine in more detail the particular case when

$$G(t) = (1-t)^{-s} G_1(t) = [(2\pi)^{\frac{1}{2}} \, C(\omega)]^{-1} \quad , \quad t = \frac{\omega - i}{\omega + i} \quad , \quad s > 0 \ . \tag{20.7}$$

As in §8.2[3], we assume that a, b and $\tilde{f}$ are sufficiently smooth. Let $\varkappa$ be the index of G_1 i.e.

$$\varkappa = \frac{1}{2\pi} \, [\arg G_1(t)]_\gamma = \frac{1}{2\pi} \left[\arg \left\{ \left(\frac{2i}{\omega + i} \right)^s \frac{1}{C(\omega)} \right\} \right]_{-\infty}^{\infty} \tag{20.8}$$

Writing G_1 in the form

$$G_1(t) = t^{\varkappa} \, \frac{\exp[\psi_1^+(t)]}{\exp[\psi_1^-(t)]} \quad ,$$
$$\psi_1^{\pm}(t) = \pm \frac{1}{2} \, \ln \left[G_1(t) t^{-\varkappa} \right] + \frac{1}{2\pi i} \int_\gamma \frac{\ln[G_1(\tau)\tau^{-\varkappa}] d\tau}{\tau - t} \quad , \tag{20.9}$$

and substituting (20.7) for G in the boundary condition

$$G^{-1} \tilde{\phi}^+ = \tilde{\phi}^- + \tilde{f} \ , \tag{20.10}$$

we obtain after obvious transformations

$$(1-t)^s \, \frac{\tilde{\phi}^+(t)}{\exp[\psi_1^+(t)]} - g_1^+(t) = t^{\varkappa} \, \frac{\tilde{\phi}^-(t)}{\exp[\psi_1^-(t)]} - g_1^-(t) \ , \tag{20.11}$$

where

$$g_1^{\pm}(t) = \pm \frac{1}{2} \, \frac{t^{\varkappa} \tilde{f}(t)}{\exp[\psi_1^-(t)]} + \frac{1}{2\pi i} \int_\gamma \frac{\tau^{\varkappa} \tilde{f}(\tau) d\tau}{\exp[\psi_1^-(\tau)](\tau - t)} \ . \tag{20.12}$$

From theorems on analytic continuation, the generalized Liouville theorem and equation (20.11), it follows that

---

(3) For simplicity we restrict attention to the case of a single equation (m=1). Recalling the results of §8.4, under the given conditions, the results can be generalized to the case of arbitrary m.

$$\tilde{\phi}^+(t) = \frac{\exp[\psi_1^+(t)]}{(1-t)^s} [g_1^+(t) + P_{\mathcal{H}-1}(t)] ,$$

(20.13)
$$\tilde{\phi}^-(t) = \frac{\exp[\psi_1^-(t)]}{t^{\mathcal{H}}} [g_1^-(t) + P_{\mathcal{H}-1}(t)] ,$$

and
$$\tilde{\phi}(t) = \tilde{\phi}^+(t) - \tilde{\phi}^-(t) ,$$

where $P_{\mathcal{H}-1}$ is a polynomial of degree $\mathcal{H}-1$, and $P_{\mathcal{H}-1} \equiv 0$ for $\mathcal{H}-1 < 0$.

In order that equations (20.13) should yield an integrable solution of equation (20.4) it is necessary and sufficient that the function $g_1^+ + P_{\mathcal{H}-1}$ have a zero of order s at the point $t = 1$, and in addition when $\mathcal{H}-1 < 0$, that g_1^- should have a zero of order greater than $-\mathcal{H}+1$ at infinity.

Let us consider three cases in turn.

1. $\mathcal{H} \geq s$. We construct the Hermite polynomial such that

(20.14)
$$Q_{s-1}^{(\ell)}(1) = - g_1^{+(\ell)}(1) , \quad \ell = 0,1,\dots,s-1 .$$

It is obvious that

(20.15)
$$Q_{s-1}(t) = - g_1^+(1) - g_1^{+\prime}(1)(t-1) - \dots - \frac{g_1^{+(s-1)}(1)}{(s-1)!} (t-1)^{s-1} ,$$

and that an arbitrary polynomial $P_{\mathcal{H}-1}$ which also satisfies conditions (20.14), has the form

(20.16)
$$P_{\mathcal{H}-1}(t) = Q_{s-1}(t) + (1-t)^s P_{\mathcal{H}-s-1}(t) ,$$

where $P_{\mathcal{H}-s-1}$ is an arbitrary polynomial of degree $\mathcal{H}-s-1$, with $P_{\mathcal{H}-s-1} \equiv 0$ when $\mathcal{H} = s$. Substituting the right-hand side of equation (20.16) into (20.13) gives

$$\tilde{\phi}^+(t) = \frac{\exp[\psi_1^+(t)]}{(1-t)^s} [g_1^+(t) + Q_{s-1}(t)] + \exp[\psi_1^+(t)] P_{\mathcal{H}-s-1}(t)$$

(20.17)
$$\tilde{\phi}^-(t) = \frac{\exp[\psi_1^-(t)]}{t^{\mathcal{H}}} [g_1^-(t) + Q_{s-1}(t)] + \frac{\exp[\psi_1^-(t)](1-t)^s}{t^{\mathcal{H}}} P_{\mathcal{H}-s-1}(t),$$

and
$$\tilde{\phi}(t) = \tilde{\phi}^+(t) - \tilde{\phi}^-(t) .$$

Thus equation (20.4) always has $\mathcal{H}-s+1$ linearly independent solutions. Hence the original equation (20.1) has, when $\mathcal{H} \geq s$, $\mathcal{H}-s+1$ linearly independent solutions which can be obtained from

$$\phi(x) = \frac{1}{(2\pi)^{\frac{1}{2}}} \int_{-\infty}^{\infty} \tilde{\phi}\left(\frac{\omega-i}{\omega+i} \right) \frac{2i}{\omega+i} e^{-i\omega x}\, d\omega \ , \quad x > 0 \ , \tag{20.18}$$

where $\tilde{\phi}$ is given by equations (20.17).

2. $0 \leq \mathscr{H} < s$. In this case equations (20.13) will be the solution of equation (20.4) only if

$$P_{\mathscr{H}-1}(t) = - g_1^+(1) - g_1^{+\prime}(1)(t-1) - \ldots - \frac{g_1^{+(\mathscr{H}-1)}(1)}{(\mathscr{H}-1)!} (t-1)^{\mathscr{H}-1} \ , \tag{20.19}$$

and

$$g^{+(\ell)}(1) = 0 \quad \text{for } \ell = \mathscr{H}, \mathscr{H}+1, \ldots, s-1 \ . \tag{20.20}$$

3. $\mathscr{H} < 0$. It is easily seen that in this case, equations (20.13) give the solution of equation (20.4) only if

$$g_1^{+(\ell)}(1) = 0 \quad , \quad \ell = 0,1,\ldots,s-1 \quad , \tag{20.21}$$

and

$$\int_\gamma g_1^-(t) t^k dt = 0 \quad , \ k = 0,1,\ldots,-\mathscr{H}-1 \ . \tag{20.22}$$

It is clear that the requirement that $\tilde{\phi}$ be integrable will impose, in many cases, too rigid a constraint on the differentiability of g_1^+, and the realisation of conditions (20.20) or (20.21). Consequently there arises the question concerning an extension of the class of admissible solutions. An examination of problems from the theory of automatic control (see §21) suggests that a class of admissible solutions for equation (20.1) can be given in terms of generalized functions as

$$\begin{aligned} \phi(x) &= \sum_{j=0}^{p} \sum_{\nu=0}^{r} d_{j\nu}\, \delta^{(\nu)}(x - x_{j\nu}) + \phi_1(x) \\ &\equiv \phi[\phi_1(x);\ d_{00}, d_{10}, \ldots, d_{pr}] \ , \end{aligned} \tag{20.23}$$

where $d_{j\nu}$ are arbitrary constants, $x_{j\nu} \geq 0$, $\phi_1 \in L_2(-\infty,\infty)$. In addition, $\delta^{(\nu)}(x-x_{j\nu})$ denotes the derivative of order ν of the Dirac delta function $\delta(x-x_{j\nu})$ which is defined by

$$\int_a^b f(x)\delta^{(\nu)}(x-x_{j\nu})dx = \lim_{\alpha\to 0} \int_a^b f(x)\delta_x^{(\nu)}(x-x_{j\nu}, \alpha)dx \quad , \tag{20.24}$$

$$\delta(u,\alpha) = \frac{\alpha}{\pi(u^2+\alpha^2)} \quad ,$$

for $a,b \in (-\infty,\infty)$ and f any function defined on $(-\infty,\infty)$. We substitute

(20.23) into (20.1). Assuming that c has sufficiently many continuous derivatives at the points $x_{j\nu}$, we obtain

$$(20.25)\qquad \tilde{A}\phi \equiv \int_0^\infty c(x-u)\phi_1(u)du + \sum_{j=0}^{p}\sum_{\nu=0}^{r} (-1)^\nu d_{j\nu}c^{(\nu)}(x-x_{j\nu}) = f(x) ,$$

for $x > 0$. Let us rewrite this equation as

$$(20.26)\qquad A\phi_1 \equiv \int_0^\infty c(x-u)\phi_1(u)du = f_1(x) = f(x) - \sum_{j=0}^{p}\sum_{\nu=0}^{r} (-1)^\nu d_{j\nu}\, c^{(\nu)}(x-x_{j\nu}).$$

Proceeding as before, we can reduce equation (20.26) to the solution of the equation

$$(20.27)\qquad a(t)\tilde{\phi}_1(t) + \frac{b(t)}{\pi i}\int_\gamma \frac{\tilde{\phi}_1(\tau)d\tau}{\tau - t} = \tilde{f}_1(t) ,$$

where

$$(20.28)\qquad \begin{aligned} \tilde{f}_1(t) &= [(\omega+i)/2i]F_1(\omega) \\ &= [F(\omega) - \sum_{j=0}^{p}\sum_{\nu=0}^{r} d_{j\nu}(i\omega)^\nu \exp(i\omega x_{j\nu})C(\omega)][(\omega+i)/2i] , \\ \omega &= i(1+t)/(1-t). \end{aligned}$$

Since $\tilde{f}_1$ must be integrable on γ, the representation (20.28) is always possible, provided only that $C(\omega)$ has a zero of order greater than r at infinity. In other words, we can choose any value of r in equation (20.23) provided that

$$(20.29)\qquad r \le s ,$$

where s is the order of $C(\omega)$ at infinity.

The constants $d_{j\nu}$ must be determined so that equations (20.20)-(20.22) are satisfied. Thus the quantities $d_{j\nu}$ satisfy an algebraic system of equations, whose solvability is in general very difficult to investigate (see [77]). In some applications, it is possible to avoid this problem (see §21).

A wider class of solutions of equation (20.1) is given by functions of the form

$$(20.30)\qquad \phi(x) = \sum_{\nu=0}^{r} d_\nu x^\nu + \phi_1(x) ,$$

where as before $\phi_1 \in L_2(-\infty,\infty)$.

For the numerical solution of equations (20.1) and (20.25) by the method of least squares, these equations can either be reduced to the corresponding singular integral equations, or considered directly as linear equations in the Hilbert spaces $L_2(0,\infty)$ and $\tilde{L}_2(0,\infty)$ respectively where the inner product is defined by

$$(f^{(1)}, f^{(2)})_{\tilde{L}_2(0,\infty)} = (f_1^{(1)}, f_1^{(2)})_{L_2(0,\infty)} + \sum_{j=0}^{p} \sum_{\nu=0}^{r} d_{j\nu}^{(1)} \bar{d}_{j\nu}^{(2)} ,$$
(20.31)
$$f^{(1)}[f_1^{(1)}; d_{00}^{(1)},\dots,d_{pr}^{(1)}] , f^{(2)}[f_1^{(2)}; d_{00}^{(2)},\dots,d_{pr}^{(2)}] .$$

It is easily seen that, in general, the system of equations (20.1) satisfies

$$\|A\phi\|_{L_2(0,\infty)} \leq \|A\phi\|_{L_2(-\infty,\infty)} = \|C\Phi^+\|_{L_2(-\infty,\infty)}$$
(20.32)
$$\leq \left\{ \max_{\omega} \sum_{j,k=1}^{m} |C_{jk}(\omega)|^2 \right\}^{\frac{1}{2}} \|\phi\|_{L_2(0,\infty)} ,$$

where the matrix $C(\omega) = \{C_{jk}(\omega)\}$ and $C_{jk}(\omega) = (2\pi)^{-\frac{1}{2}} \int_{-\infty}^{\infty} c_{jk}(x) \exp(i\omega x)dx.$

It is readily verified that the conjugate operator is given by

$$A^*\psi = \int_0^{\infty} c^*(u-x)\psi(u)du , \tag{20.33}$$

where c^* denotes the complex conjugate transpose of the matrix c i.e. $c^*(x) = \{\overline{c_{kj}(x)}\}$.

It is also possible to look for an approximate solution of equation (20.1) in the form

$$\phi_n^{(1)}(x) = \sum_{k=0}^{n} \alpha^{(k)}\psi_k(x) , \alpha^{(k)} = (\alpha_1^{(k)},\dots,\alpha_m^{(k)}) , x > 0 , \tag{20.34}$$

where the functions ψ_k are defined by equations (19.13), and to find $\alpha^{(k)}$ such that we minimize $\Delta_n^{(1)}$ defined by

$$\Delta_n^{(1)} = \|A\phi_n^{(1)} - f\|^2_{L_2(0,\infty)} + \alpha\|\phi_n^{(1)}\|^2_{L_2(0,\infty)} , \alpha > 0 . \tag{20.35}$$

In this connection the results of §2.1 and §14.1 should be applied. If equation (20.1) is solvable, then we may also look for an approximate solution in the form

$$\phi_n^{(2)} = \sum_{k=0}^{n} A^*\alpha^{(k)}\psi_k , \tag{20.36}$$

where $\alpha^{(k)}$ are now chosen so as to minimize $\Delta_n^{(2)}$ where

$$\Delta_n^{(2)} = \|\phi - \phi_n^{(2)}\|_{L_2(0,\infty)} , \tag{20.37}$$

ϕ being the exact solution. If c and f are approximated by functions of the form $\sum_{k=-n}^{n} c_k\psi_k$, then it can be shown that for any of the algorithms described in §3 which may be used for the minimization of (20.35) and (20.37), all the scalar products can be determined exactly.

We shall not examine these questions in detail since they represent a repetition of previous analysis. We note only that if it is necessary to seek an approximate solution of equation (20.1) (see (20.25)) in the form

(20.38)
$$\phi_n^{(3)}(x) = \sum_{j=0}^{p} \sum_{\nu=0}^{r} d^{(j\nu)}\, \delta^{(\nu)}(x-x_{j\nu}) + \sum_{k=0}^{n} \alpha^{(k)}\psi_k(x) ,$$
$$d^{(j\nu)} = (d_1^{(j\nu)}, d_2^{(j\nu)}, \ldots, d_m^{(j\nu)}) ,$$

then we can choose $d^{(j\nu)}$ and $\alpha^{(k)}$ so as to minimize $\Delta_n^{(3)}$, where[4]

(20.39)
$$\Delta_n^{(3)} = \|\tilde{A}\phi_n^{(3)} - f\|^2_{L_2(0,\infty)} + \alpha\|\phi_n^{(3)}\|^2_{\tilde{L}_2(0,\infty)} .$$

The choice of the form (20.38) arises from the expected character of the solution. As we shall explain in the next section, in certain cases we know that the components of the solution involve $\delta(x-x_{j\nu})$ and its derivatives. The determination of an approximate solution in the form given by equation (20.38), then becomes admissable.

In conclusion, we note that for sufficiently large m, effective algorithms are given by the iterative methods described in §5 and §15.3.

---

(4) Obviously we must substitute $d^{(j\nu)}$ for $d_{j\nu}$ in equation (20.25).

§21 APPLICATIONS IN THE THEORY OF AUTOMATIC CONTROL(5)

§21.1 The Approximate Calculation of Fourier Transforms

We examine two methods:

1. Let f be a function which is approximated by zero outside an interval $[T_1, T_2]$ and is defined by a table of values at the points

$$x_1 = T_1, \quad x_2 = T_1 + \Delta x, \quad x_3 = T_1 + 2\Delta x, \ldots, x_n = T_1 + (n-1)\Delta x = T_2 . \tag{21.1}$$

Then the Fourier transform of this function is given by

$$\begin{aligned} F(\omega) &= (2\pi)^{-\frac{1}{2}} \int_{-\infty}^{\infty} f(x) e^{i\omega x} dx \\ &\approx (2\pi)^{-\frac{1}{2}} \int_{T_1}^{T_2} f(x) e^{i\omega x} dx \\ &= (2\pi)^{-\frac{1}{2}} \sum_{j=1}^{n-1} \int_{x_j}^{x_{j+1}} f(x) e^{i\omega x} dx \\ &\approx (2\pi)^{-\frac{1}{2}} \sum_{j=1}^{n-1} f(x_j) \int_{x_j}^{x_{j+1}} e^{i\omega x} dx \\ &= (2\pi)^{-\frac{1}{2}} \sum_{j=1}^{n-1} f(x_j) e^{i\omega x_j} [(e^{i\omega\Delta x} - 1)/i\omega] . \text{ (6)} \end{aligned} \tag{21.2}$$

It follows that

$$\begin{aligned} \Delta F &= \left| F(\omega) - (2\pi)^{-\frac{1}{2}} \sum_{j=1}^{n-1} f(x_j) e^{i\omega x_j} [(e^{i\omega\Delta x} - 1)/i\omega] \right| \\ &\leq (2\pi)^{-\frac{1}{2}} \left[\int_{T_2}^{\infty} |f(x)| dx + \int_{-\infty}^{T_1} |f(x)| dx + \omega_f(\Delta x)(T_2 - T_1) \right] , \end{aligned} \tag{21.3}$$

where $\omega_f(\delta) = \max_x |f(x+\delta) - f(x)|$, is the modulus of continuity of f. If we know that

(5) Many of the algorithms described below have been published (see [62]) as standard subroutines.

(6) For $\omega = 0$, we write

$$F(0) = \lim_{\omega \to 0} F(\omega) \approx (2\pi)^{-\frac{1}{2}} \Delta x \sum_{j=1}^{n-1} f(x_j) .$$

(21.4) $$|f(x)| < C_1|x|^{-r} \ , \ r > 1 \ , \ x \notin [T_1, T_2] \ ,$$

then it follows that

(21.5) $$\int_{T_2}^{\infty} |f(x)|dx + \int_{-\infty}^{T_1} |f(x)|dx < C_1(|T_2|^{-r+1} + |T_1|^{-r+1})/(r-1) \ .$$

This method is not new and, for example, is similar to methods suggested by Solodovnikov [125]. If all that we know about f is that it is continuous, then the above algorithm is the most appropriate. We shall consider an alternative method when f is sufficiently smooth. An increase in the degree of smoothness of f does not give an increase in the accuracy of (21.2), and in such cases the following method may be more appropriate.

2. Let $F(\omega)$ be a sufficiently smooth function which is defined on the real axis and becomes zero at infinity. We make the substitution

$$\omega = i\,\frac{1+\tau}{1-\tau} \ , \ \tau = e^{i\theta} \ , \ -\pi \leq \theta \leq \pi \ ,$$

and put

(21.6) $$\tilde{f}(\tau) = F(i\,\frac{1+\tau}{1-\tau})\,\frac{1}{1-\tau} \ , \ F(\omega) = \tilde{f}(\frac{\omega-i}{\omega+i})\,\frac{2i}{\omega+i} \ .$$

We approximate to $\tilde{f}(\tau)$ by the interpolation polynomial $U_n(\tilde{f},\tau)$ such that

(21.7) $$\tilde{f}(\tau) \approx U_n(\tilde{f},\tau) = \sum_{k=-n}^{n} \tilde{a}_k \tau^k,$$

where

(21.8)
$$\tilde{a}_k = \frac{1}{(2n+1)} \sum_{j=-n}^{n} \tilde{f}(\tau_j)\tau_j^{-k} = \frac{1}{(2n+1)} \sum_{j=-n}^{n} F(\omega_j)\,\frac{\tau_j^{-k}}{1-\tau_j} \ ,$$
$$\tau_j = \exp[2\pi ij/(2n+1)], \ \omega_j = -\cot[\pi j/(2n+1)] \ .$$

Then we have

(21.9) $$F(\omega) \approx \sum_{k=-n}^{n} \tilde{a}_k \left(\frac{\omega-i}{\omega+i}\right)^k \frac{2i}{\omega+i} \ .$$

<u>Lemma 1</u>. *If* $F(\omega) = (2\pi)^{-\frac{1}{2}} \int_{-\infty}^{\infty} f(x)\exp(i\omega x)dx$, *and f is real, then the coefficients $\tilde{a}_k$ are real.*

In order to prove this we note that $\tilde{f}(\tau) = F(\omega)/(1-\tau)$ where $\omega = i(1+\tau)/(1-\tau)$, is real when τ is real and $-1 \leq \tau \leq 1$. Hence the Laurent series expansion for $\tilde{f}(\tau)$, viz. $\tilde{f}(\tau) = \sum_{s=-\infty}^{\infty} \tilde{\alpha}_s \tau^s$, will have real coefficients $\tilde{\alpha}_s$. It remains to note that

$$\tilde{a}_k = \frac{1}{(2n+1)} \sum_{j=-n}^{n} \sum_{s=-\infty}^{\infty} \tilde{\alpha}_s \tau_j^s \tau_j^{-k} = \sum_{\ell=-\infty}^{\infty} \tilde{\alpha}_{(2n+1)\ell+k} .$$

We now assume that

$$f(x) = (2\pi)^{-\frac{1}{2}} \int_{-\infty}^{\infty} F(\omega) e^{-i\omega x} d\omega \approx (2\pi)^{-\frac{1}{2}} \sum_{k=-n}^{n} \tilde{a}_k \int_{-\infty}^{\infty} \left(\frac{\omega - i}{\omega + i} \right)^k \frac{2i}{\omega + i} e^{-i\omega x} d\omega . \tag{21.10}$$

The calculation of these integrals poses no difficulty since (see [133])

$$\begin{aligned} &\int_{-\infty}^{\infty} \left(\frac{\omega-i}{\omega+i} \right)^k \frac{2i}{\omega+i} e^{-i\omega x} d\omega = \begin{cases} e^{-x} P_k(x) , & x > 0 , \\ 0 , & x < 0 , \end{cases} \quad k \geq 0 , \\ &\int_{-\infty}^{\infty} \left(\frac{\omega-i}{\omega+i} \right)^k \frac{2i}{\omega+i} e^{-i\omega x} d\omega = \begin{cases} 0 , & x > 0 , \\ -e^{x} P_{-k-1}(-x) , & x < 0 , \end{cases} \quad k < 0 , \end{aligned} \tag{21.11}$$

where P_k denote the Laguerre polynomials which are orthogonal on $(0,\infty)$ with respect to the weight function $\exp(-2x)$. For $k = 0,1$ we find

$$\begin{aligned} &\int_{-\infty}^{\infty} \frac{2i}{\omega+i} e^{-i\omega x} d\omega = \begin{cases} 4\pi e^{-x} , & x > 0 , \\ 0 , & x < 0 , \end{cases} \\ &\int_{-\infty}^{\infty} \frac{\omega-i}{(\omega+i)^2} 2i e^{-i\omega x} d\omega = \begin{cases} 4\pi e^{-x}(1-2x) , & x > 0 , \\ 0 , & x < 0 . \end{cases} \end{aligned} \tag{21.12}$$

Consequently $P_0(x) = 4\pi$, $P_1(x) = 4\pi(1-2x)$. Polynomials of higher degree may be obtained from the recurrence equation (see [133]),

$$P_{k+1}(x) = \frac{2k+1-2x}{k+1} P_k(x) - \frac{k}{k+1} P_{k-1}(x) . \tag{21.13}$$

Thus we have

$$\begin{aligned} f(x) &= (2\pi)^{-\frac{1}{2}} \int_{-\infty}^{\infty} F(\omega) e^{-i\omega x} d\omega \\ &\approx \begin{cases} (2\pi)^{-\frac{1}{2}} \sum_{k=0}^{n} \tilde{a}_k e^{-x} P_k(x) , & x > 0 , \\ -(2\pi)^{-\frac{1}{2}} \sum_{k=-n}^{-1} \tilde{a}_k e^{x} P_{-k-1}(-x) , & x < 0 . \end{cases} \end{aligned} \tag{21.14}$$

Let us denote by Δ_2 the absolute error in the approximations (21.14) with respect to the L_2 norm.

Theorem 21.1 *The estimate*

$$\Delta_2 \le 8\pi E_n(\tilde{f}, \gamma) \tag{21.15}$$

is valid.

The proof readily follows from the equations $\|f\|_{L_2(-\infty,\infty)} = \|F\|_{L_2(-\infty,\infty)} = 4\pi\|\tilde{f}\|_{L_2(\gamma)}$, and inequality (10.26).

In order to obtain a similar estimate in the space $C(-\infty,\infty)$, we introduce the function f_1, where $f_1(\tau) = F(i\,\frac{1+\tau}{1-\tau})/(1-\tau)^2$, and construct the interpolation polynomial $U_n(f_1,\tau)$ given by

$$U_n(f_1,\tau) = \sum_{k=-n}^{n} \gamma_k\tau^k \;, \quad \gamma_k = \frac{1}{2n+1} \sum_{j=-n}^{n} f_1(\tau_j)\tau_j^{-k} \;.$$

Since $\tilde{f}(\tau) = (1-\tau)f_1(\tau)$ then we have $\tilde{f}(\tau) \approx \sum_{k=-n}^{n+1} \tilde{b}_k\tau^k$ where $\tilde{b}_k = \gamma_k - \gamma_{k-1}$ for $k = -n+1,-n+2,\ldots,n$ and $\tilde{b}_{-n} = \gamma_{-n}$, $\tilde{b}_n = \gamma_n$. As a result we have

$$f(x) = (2\pi)^{-\frac{1}{2}} \int_{-\infty}^{\infty} F(\omega)e^{-i\omega x}d\omega$$

$$\approx \begin{cases} (2\pi)^{-\frac{1}{2}} \sum\limits_{k=0}^{n+1} \tilde{b}_k\, e^{-x}P_k(x) \;, & x > 0 \;, \\ -(2\pi)^{-\frac{1}{2}} \sum\limits_{k=-n}^{-1} \tilde{b}_k\, e^{x}P_{-k-1}(-x) \;, & x < 0 \;. \end{cases} \tag{21.16}$$

Let us denote by Δ_C the absolute error in the approximations (21.16) with respect to the uniform norm.

Theorem 21.2. *The estimate*

$$\Delta_C \le 4(2\pi)^{\frac{1}{2}}\, E_n(f_1,\gamma) \tag{21.17}$$

is valid.

The proof follows from the relation

$$\Delta_C \le (2\pi)^{-\frac{1}{2}} \int_{-\infty}^{\infty} \left| F(\omega) - \sum_{k=-n}^{n+1} \tilde{b}_k \tau^k (1-\tau) \right| d\omega$$

$$= \frac{2}{(2\pi)^{\frac{1}{2}}} \int_{\gamma} \left| \frac{F(\omega)}{(1-\tau)^2} - \sum_{k=-n}^{n} \gamma_k \tau^k \right| |d\tau|$$

$$\le 2(2\pi)^{\frac{1}{2}} \|f_1(\tau) - U_n(f_1,\tau)\|_{L_2(\gamma)} ,$$

and inequality (10.26).

We note that in the more general case when

$$\text{(21.18)} \qquad \begin{aligned} F(\omega) &= b_{-1}\omega^{-1} + b_0 + b_1\omega + \ldots + b_p\omega^p + F_1(\omega) , \\ F_1(\omega) &= O(1/\omega^2) , \end{aligned}$$

for large $|\omega|$, it is necessary to apply the approximation (21.16) to the function F_1. As a result we obtain

$$\text{(21.19)} \qquad \begin{aligned} f(x) &= (2\pi)^{-\frac{1}{2}} \int_{-\infty}^{\infty} F(\omega)\, e^{-i\omega x} d\omega \\ &\approx -b_{-1} i(\pi/2)^{\frac{1}{2}} \operatorname{sgn} x + (2\pi)^{\frac{1}{2}} [b_0\delta(x) + ib_1\delta'(x) + \ldots + i^p b_p \delta^{(p)}(x)] \\ &+ \begin{cases} (2\pi)^{-\frac{1}{2}} \sum_{k=0}^{n+1} \tilde{b}_k^{(1)} e^{-x} P_k(x) , & x > 0 , \\ -(2\pi)^{-\frac{1}{2}} \sum_{k=-n}^{-1} \tilde{b}_k^{(1)} e^{x} P_{-k-1}(-x) , & x < 0 , \end{cases} \end{aligned}$$

where $\delta(x)$ denotes the Dirac delta function.

§21.2 Approximation of the Transfer Function

It is known (see [125]) that the *transfer function*(7) $\Phi(i\omega)$ is defined by the equation

$$\text{(21.20)} \qquad \Phi(i\omega) = \int_0^{\infty} k(t) e^{i\omega t} dt ,$$

in which k is the so-called *impulse transfer function* of a linear filter with constant parameters.(8) If $k \in L_1(0,\infty)$ then (see [133]) $\Phi(i\omega)$ is analytic in

(7) Often $\Phi(-i\omega)$ is taken as the transfer function rather than $\Phi(i\omega)$.

(8) The function k describes the reaction of the filter to an initial perturbation in the form of a Dirac delta function (with zero initial conditions).

the upper half plane and continuous on the real axis. In addition, it is obvious that Φ is real on the upper imaginary axis. If we make the substitution $\omega = i(1+\tau)/(1-\tau)$, then the function

(21.21) $$\phi(\tau) = \Phi\left(\frac{\tau+1}{\tau-1} \right)$$

is analytic inside the unit circle γ, is continuous on γ, and is real on the interval [-1,1]. It then follows that the Laurent expansion

(21.22) $$\phi(\tau) = \sum_{k=0}^{\infty} \alpha_k \tau^k$$

coincides with the power series expansion of ϕ (and with the Fourier series for $\phi(e^{i\theta})$), and has real coefficients α_k. If $\sum_{k=0}^{n} \alpha_k^* \tau^k$ denotes the polynomial of best approximation to ϕ, then we may assume that the coefficients α_k^* are real also. Thus we have, for example, the possibility of obtaining estimates of the form

(21.23) $$\left| \phi(\tau) - \sum_{k=0}^{n} \alpha_k^* \tau^k \right| \leq \frac{\pi}{2} \, \frac{\max_\theta |\phi_\theta^{(r)}(e^{i\theta})|}{(n+1)^r} ,$$

with real α_k^*. With respect to the transfer function, this estimate can be written as

(21.24) $$\left| \Phi(i\omega) - \sum_{k=0}^{n} \alpha_k^* \left(\frac{i\omega+1}{i\omega-1}\right)^k \right| \leq \frac{\pi}{2} \, \frac{\max_\theta |\phi_\theta^{(r)}(e^{i\theta})|}{(n+1)^r} ,$$

where ϕ is defined by equation (21.21). The evaluation of α_k^* is computationally difficult. Let us introduce the interpolation polynomial $\tilde{U}_n$ where

$$\tilde{U}_n(\phi,\tau) = \sum_{k=0}^{n} b_k \tau^k ,$$

(21.25) $$b_k = \frac{1}{n+1} \sum_{j=0}^{n} \phi(\tau_j)\tau_j^{-k} = \frac{1}{n+1} \sum_{j=0}^{n} \Phi(i\omega_j)\tau_j^{-k} ,$$

$$\tau_j = \exp(2\pi ij/(n+1)), \; \omega_j = -\cot(\pi j/(n+1)) .$$

It is easily verified that $\tilde{U}_n(\phi,\tau_j) = \phi(\tau_j)$. By analogy with the analysis for the polynomials $U_n(\phi,\tau)$ (see [63]), it is possible to prove the following theorem.

Theorem 21.3. *The inequality*

$$|\phi(\tau) - \tilde{U}_n(\phi,\tau)| \leq \{2 + \frac{2}{\pi} + \frac{2}{\pi} \ln [\frac{2}{\pi}(n+1)]\} E_n(\phi,\gamma) \tag{21.26}$$

is valid, where $E_n(\phi,\gamma)$ *denotes the error in the best approximation to* ϕ *on* γ *by polynomials of the form* $\sum_{k=0}^{n} \alpha_k \tau^k$.

If ϕ has a derivative of order r on γ, it follows from (21.26) that

$$|\phi(\tau) - \tilde{U}_n(\phi,\tau)| \leq \{\pi + 1 + \ln[2(n+1)/\pi]\} \frac{\max_\tau |\phi_\theta^{(r)}(\tau)|}{(n+1)^r} \tag{21.27}$$

Recalling equation (21.21) it is easy to obtain corresponding results for the approximation to $\Phi(i\omega)$. We have

$$\Phi(i\omega) \approx \sum_{k=0}^{n} b_k \left(\frac{i\omega+1}{i\omega-1}\right)^k = \Phi_n(i\omega) \quad , \tag{21.28}$$

where the real coefficients b_k (see Lemma 1) are calculated using equations (21.25).

Let us examine the problem of the practical realization of a linear filter with transfer function $\Phi_n(i\omega)$. To do this (see [125]) we write $\Phi_n(i\omega)$ in the form

$$\begin{aligned}\Phi_n(i\omega) &= \frac{b_0(i\omega-1)^n + b_1(i\omega-1)^{n-1}(i\omega+1) + \ldots + b_n(i\omega+1)^n}{(i\omega-1)^n} \\ &= \frac{\tilde{b}_0 + \tilde{b}_1(i\omega) + \ldots + \tilde{b}_n(i\omega)^n}{(i\omega-1)^n} .\end{aligned} \tag{21.29}$$

If we put

$$(i\omega-1)^n Y_n(i\omega) = [\tilde{b}_0 + \tilde{b}_1(i\omega) + \ldots + \tilde{b}_n(i\omega)^n]X_n(i\omega) \; , \tag{21.30}$$

where we assume

$$X_n(i\omega) = \int_{-\infty}^{\infty} x_n(t)e^{i\omega t}dt \; , \quad Y_n(i\omega) = \int_{-\infty}^{\infty} y_n(t)e^{i\omega t}dt \quad , \tag{21.31}$$

then x_n and y_n will be the *input* and *output* respectively of the filter. They are related by the differential equation

$$\begin{aligned}(-1)^n(\frac{d}{dt} + 1)^n y_n(t) &= [b_0(-\frac{d}{dt} - 1)^n + b_1(-\frac{d}{dt} - 1)^{n-1}(-\frac{d}{dt} + 1) + \ldots \\ &\quad \ldots + b_n(-\frac{d}{dt} + 1)^n] x_n(t) \\ &= [\tilde{b}_0 - \tilde{b}_1 \frac{d}{dt} + \ldots + (-1)^n \tilde{b}_n \frac{d^n}{dt^n}] x_n(t) .\end{aligned} \tag{21.32}$$

From equation (21.32), it follows that a filter with transfer function $\Phi_n(i\omega)$ may be constructed from differentiation, amplifier and adder networks in order to give the right hand side of (21.32), and also from n passive networks in order to give $y_n(t)$. In many control systems, differentiation of the input signal x_n is undesirable because in general this signal is not sufficiently smooth and differentiation can give rise to large errors. Therefore it is more appropriate to rewrite equations (21.29) and (21.30) as follows:

$$\Phi_n(i\omega) = B_0 + \frac{B_1}{(i\omega-1)} + \frac{B_2}{(i\omega-1)^2} + \ldots + \frac{B_n}{(i\omega-1)^n} \tag{21.33}$$

and

$$Y_n(i\omega) = \Phi_n(i\omega)\, X_n(i\omega) = \tilde{Y}_0 + \tilde{Y}_1 + \ldots + \tilde{Y}_n \ , \tag{21.34}$$

where

$$\tilde{Y}_j = \frac{B_j}{(i\omega-1)^j}\, X_n \ , \ j = 0,1,\ldots,n \ . \tag{21.35}$$

From (21.34) and (21.35) it follows that the filter with transfer function $\Phi_n(i\omega)$ can be constructed from passive networks (which correspond to (21.35) with $j \neq 0$), an amplifier $\tilde{Y}_0 = B_0X_n$, and an adder $\tilde{Y}_n = \tilde{Y}_0 + \tilde{Y}_1 + \ldots + \tilde{Y}_n$. We note that

$$B_0 = \tilde{b}_n = b_0 + b_1 + \ldots + b_n = \phi(1) = \Phi(\infty) \ . \tag{21.36}$$

If $\Phi(\infty) = 0$, which occurs for example when $k \in L_2(0,\infty)$, then $B_0 = \tilde{b}_n = 0$.

In the case when $\tilde{\phi}(\tau) = \Phi\left(\frac{\tau+1}{\tau-1}\right)\left(\frac{1}{1-\tau}\right)$ is differentiable with respect to τ and $\tilde{\phi}(1) = 0$, it is better to approximate $\Phi(i\omega)$ by

$$\Phi(i\omega) \approx \sum_{k=0}^{n} \tilde{b}_k \left(\frac{i\omega+1}{i\omega-1}\right)^k \left(\frac{-2}{i\omega-1}\right) \ , \tag{21.37}$$

where

$$\begin{gathered} \tilde{b}_k = \frac{1}{(n+1)} \sum_{j=0}^{n} \tilde{\phi}(\tau_j)\tau_j^{-k} = \frac{1}{(n+1)} \sum_{j=0}^{n} \Phi(i\omega_j)\, \frac{\tau_j^{-k}}{1-\tau_j} \ , \\ \tau_j = \exp\,[2\pi ij/(n+1)],\ \omega_j = -\cot(\pi j/(n+1)) \ . \end{gathered} \tag{21.38}$$

We now have

$$(21.39)\quad \begin{aligned} \Big| \Phi(i\omega) - \sum_{k=0}^{n} \tilde{b}_k \left(\frac{i\omega+1}{i\omega-1}\right)^k \left(\frac{-2}{i\omega-1}\right) \Big| &\le \frac{2}{(1+\omega^2)^{\frac12}} |\tilde{\phi}(\tau) - \tilde{U}_n(\tilde{\phi},\tau)| \\ &\le \frac{2}{(1+\omega^2)^{\frac12}} \{ 2 + \frac{2}{\pi} + \frac{2}{\pi} \ln [\frac{2}{\pi} (n+1)] \} \rho_n(\tilde{\phi},\gamma) . \end{aligned}$$

The advantages of this latter method of approximation become obvious if it is necessary to calculate $k(t)$ when $\Phi(i\omega)$ is known. We have (recall method 2),

$$(21.40)\qquad k(t) = \frac{1}{2\pi} \int_{-\infty}^{\infty} \Phi(i\omega) e^{-i\omega t} d\omega \approx \begin{cases} \frac{1}{2\pi} \sum_{k=0}^{n} b_k e^{-t} P_k(t), & t > 0 , \\ 0 , & t < 0 . \end{cases}$$

It is not difficult to obtain good estimates of the absolute errors in this approximation with respect to both the uniform norm and the $L_2(0,\infty)$ norm.

§21.3 The Determination of the Dynamic Characteristics of Linear Filters with Constant Parameters

We shall examine the approximate calculation of the impulse transfer function k of an isolated stable linear system with constant parameters, assuming that the input x and output y of the system are stationary random ergodic functions of time (see [120]) which are known.[9] The input and output functions are related by

$$(21.41)\qquad y(t) = \int_0^{\infty} x(t-\theta)k(\theta)d\theta + n(t) , \quad t \in (-\infty,\infty) ,$$

where the unknown noise function n is independent of x. Multiplying each side of equation (21.41) by $x(t-u)$ and averaging the results with respect to t, we obtain the equation

$$(21.42)\qquad R_{yx}(u) = \int_0^{\infty} R_{xx}(u-\theta)k(\theta)d\theta , \quad u \in (-\infty,\infty) ,$$

in which R_{xx} and R_{yx} are autocorrelation and cross-correlation functions respectively, defined by

$$(21.43)\qquad \begin{aligned} R_{xx}(u) &= \lim_{T\to\infty} \frac{1}{2T} \int_{-T}^{T} x(t)x(t-u)dt , \\ R_{yx}(u) &= \lim_{T\to\infty} \frac{1}{2T} \int_{-T}^{T} y(t)x(t-u)dt . \end{aligned}$$

(9) Without loss of generality, we can assume that the means of x and y are zero.

The term $R_{nx}(u)$ vanishes since we assume that n and x are not correlated. From the properties of the Fourier transform, we define

(21.44) $$\Phi(i\omega) = S_{yx}(i\omega)/S_{xx}(i\omega)$$

and

(21.45) $$k(t) = \frac{1}{2\pi} \int_{-\infty}^{\infty} \Phi(i\omega) e^{-i\omega t} d\omega \quad ,$$

where

(21.46) $$\begin{aligned} S_{yx}(i\omega) &= \int_{-\infty}^{\infty} R_{yx}(u) e^{i\omega u} du \quad , \\ S_{xx}(i\omega) &= \int_{-\infty}^{\infty} R_{xx}(u) e^{i\omega u} du \quad . \end{aligned}$$

Equations (21.41)-(21.46) are derived in more detail in the book by Solodovnikov [125]. From the results of the two preceding sections and of §2, we shall examine an algorithm for the approximate evaluation of the dynamic characteristics, using the above equations.

Questions concerning the calculation of correlation functions are examined in detail by Solodovnikov [125]; we shall not consider them here. We note only (see [120]) that the smoothness of the correlation function is about the same as that of the corresponding random processes. For the smoothness of the latter, it is generally impossible to assume anything more than continuity. Therefore to calculate S_{yx} and S_{xx} (see equation (21.46)) it is appropriate to apply the first method for the approximate calculation of Fourier transforms. Another important property of correlation functions is that they rapidly tend to zero. For Fourier transforms of correlation functions this implies (see [120]) that they are sufficiently smooth functions everywhere except perhaps at infinity. In connection with this, it is often appropriate to approximate Φ in equation (21.44) by the form given in equation (21.37), and to calculate the impulse transfer function k by means of (21.40).

We consider now a series of cases where it is necessary to approximate Φ in a rather different way.

1. Suppose that the filter has an observable delay and an estimate T_1 (>0) of the time of this delay is known. We assume

(21.47) $$\Phi(i\omega) \approx e^{-i\omega T_1} \sum_{k=0}^{n} a_k \left(\frac{i\omega+1}{i\omega-1}\right)^k \left(\frac{-2}{i\omega-1}\right) \quad ,$$

and we calculate a_k from

$$a_k = \frac{1}{(n+1)} \sum_{j=0}^{n} \Phi(i\omega_j) e^{i\omega_j T_1} \frac{\tau_j^{-k}}{1-\tau_j} ,$$

(21.48)

$$\tau_j = \exp [2\pi ij/(n+1)],\ \omega_j = -\cot[\pi j/(n+1)] .$$

2. For the calculation of $\Phi(i\omega)$ using (21.44) when S_{yx} and S_{xx} are calculated using (21.2), it can happen that $|\Phi(i\omega)|$ increases indefinitely or approaches a non-zero value as $\omega \to \pm\infty$. In this case, it is necessary to select beforehand a polynomial $\alpha_0+\alpha_1(i\omega)+ \ldots + \alpha_m(i\omega)^m$ (usually of not too high a degree), which approximates $\Phi(i\omega)$ on the real axis outside some interval $[-N,N]$, where N is large. We may, for example, use the method of least squares by calculating $\Phi(i\omega)$ at a set of points $\omega_s \notin [-N,N]$ and determining coefficients $\mathring{\alpha}_k$ by minimizing

$$\Delta(\alpha_0,\alpha_1,\ldots,\alpha_m) = \sum_s \left[\Phi(i\omega_s) - \sum_{k=0}^{m} \alpha_k (i\omega_s)^k \right]^2 . \tag{21.49}$$

Then it is possible to write $\Phi(i\omega)$ in the form

$$\Phi(i\omega) \approx \sum_{k=0}^{m} \mathring{\alpha}_k (i\omega)^k + \sum_{k=0}^{n} \beta_k \left(\frac{i\omega+1}{i\omega-1}\right)^k \left(\frac{-2}{i\omega-1}\right) , \tag{21.50}$$

and to define β_k by

$$\beta_k = \frac{1}{(n+1)} \sum_{j=0}^{n} \left[\Phi(i\omega_j) - \sum_{k=0}^{m} \mathring{\alpha}_k (i\omega_j)^k\right] \frac{\tau_j^{-k}}{1-\tau_j} , \tag{21.51}$$

where τ_j and ω_j are as defined in (21.48). As a result we obtain

$$k(t) = \frac{1}{2\pi} \int_{-\infty}^{\infty} \Phi(i\omega) e^{-i\omega t} d\omega$$

(21.52)

$$\approx 2 \sum_{k=0}^{m} (-1)^k \mathring{\alpha}_k \delta^{(k)}(t) + \frac{1}{2\pi} \sum_{k=0}^{n} \beta_k e^{-t} P_k(t), \quad t > 0 ,$$

where δ denotes the Dirac delta function. It is also possible to determine α_k and β_k simultaneously using the method of least squares.

3. Suppose on the upper imaginary axis, $\Phi(i\omega)$ possesses a Dirac delta function near $\omega = 0$. Then

$$\Phi(i\omega) \approx A\delta(i\omega) + \sum_{k=0}^{n} \gamma_k \left(\frac{i\omega+1}{i\omega-1}\right)^k \left(\frac{-2}{i\omega-1}\right) . \tag{21.53}$$

In practice, in place of $\delta(i\omega)$ we may take, for example, the function

(21.54) $$\delta(i\omega,\alpha) = \frac{\alpha}{\pi[(i\omega)^2 + \alpha^2]} ,$$

and select A, α so that $|\Phi(i\omega) - A\delta(i\omega,\alpha)|$ is close to a minimum in a neighbourhood of $\omega = 0$ on the upper imaginary axis. Then we define

(21.55) $$\gamma_k = \frac{1}{(n+1)} \sum_{j=0}^{n} \left[\Phi(i\omega_j) - A\delta(i\omega_j,\alpha)\right] \frac{\tau_j^{-k}}{1-\tau_j} ,$$

where τ_j and ω_j are defined in (21.48). From (21.53) it follows that

(21.56) $$k(t) = \frac{A}{2\pi} + \frac{1}{2\pi} \sum_{k=0}^{n} \gamma_k e^{-t} P_k(t) , \quad t > 0 .$$

4. $\Phi(\pm i\infty) = \alpha_0$, either $\alpha_0 \neq 0$ or $\alpha_0 = 0$, but $\tilde{\phi}(\tau) = \Phi(\frac{\tau+1}{\tau-1})(\frac{1}{1-\tau})$ does not vanish at $\tau = 1$. Then it is better to approximate $\Phi(i\omega)$ as

(21.57) $$\Phi(i\omega) \approx \sum_{k=0}^{n} b_k \left(\frac{i\omega+1}{i\omega-1}\right)^k = \alpha_0 + \sum_{k=0}^{n-1} c_k \left(\frac{i\omega+1}{i\omega-1}\right)^k \left(\frac{-2}{i\omega-1}\right) ,$$

where the b_k are given by equations (21.25). It is easy to show that

(21.58) $$\alpha_0 = b_0 + b_1 + \dots + b_n ,$$
$$c_k = b_0 + b_1 + \dots + b_k .$$

As an example, let us consider equation (21.42) in which

(21.59)
$$R_{xx}(u) = \begin{cases} \exp(-c^2 u)\cos(bu) , & u > 0 , \\ \exp(c^2 u)\cos(bu) , & u < 0 ; \end{cases}$$
$$R_{yx}(u) = \frac{2cc_1}{c^2+c_1^2} \begin{cases} \exp(-c_1^2 u)\cos(b_1 u), & u > 0 , \\ \exp(c^2 u)\cos(bu) , & u < 0 ; \end{cases}$$

where c, b, c_1 and b_1 are real parameters satisfying the equation $c_1^4+b_1^2 = c^4+b^2$. It is not difficult to show that

(21.60)
$$S_{xx}(i\omega) = \frac{2c^2(c^4+b^2+\omega^2)}{(c^2+ib-i\omega)(c^2-ib-i\omega)(c^2-ib+i\omega)(c^2+ib+i\omega)} ,$$
$$S_{yx}(i\omega) = \frac{2cc_1(c^4+b^2+\omega^2)}{(c^2+ib+i\omega)(c^2-ib+i\omega)(c_1^2-ib_1-i\omega)(c_1^2+ib_1-i\omega)} ,$$

and consequently

$$\Phi(i\omega) = \frac{c_1}{c} \cdot \frac{(c^2-ib-i\omega)(c^2+ib-i\omega)}{(c_1^2-ib_1-i\omega)(c_1^2+ib_1-i\omega)} . \tag{21.61}$$

Selecting from this rational function the polynomial part, expanding the remainder as a sum of partial fractions, and using the result

$$\frac{1}{2\pi i} \int_{-\infty}^{\infty} \frac{e^{-i\omega t}}{\omega - \alpha} d\omega = \begin{cases} -e^{-i\alpha t} , & t > 0 , \\ 0 , & t < 0 , \end{cases} \qquad \operatorname{Im} \alpha < 0 ,$$

we obtain the exact solution

$$k(t) = \frac{c_1}{c} \delta(t) + 2(c_1^2-c^2)\left[\frac{c_1^3}{cb_1} \sin(b_1 t) - \frac{c_1}{c} \cos(b_1 t)\right] \exp(-c_1^2 t). \tag{21.62}$$

This example has been solved numerically using an approximation of the type (21.57) for $\Phi(i\omega)$.(10) The results of the calculation have been given in [65]. The agreement between the calculated and exact values of c_1/c and k (in the cases when $c_1 = 2$, $c = 1$, $n = 20$, $b_1 = 1/3, 1, 2$ and $t = 1/4, 1/2, 1$) was better than two significant digits.

The problem of the determination of the dynamic characteristics of isolated stable linear systems in the case of m inputs and n outputs can be reduced to an examination of an analogous system with m cross-correlated inputs and one output (see [125]). To determine the m impulse transfer functions $k_1, k_2, \dots, k_m$ corresponding to the stationary ergodic inputs $x_1, x_2, \dots, x_m$ and a stationary ergodic output y, we set up the system of m linear integral equations given by

$$R_{yx_s}(u) = \sum_{\ell=1}^{m} \int_0^{\infty} R_{x_\ell x_s}(u-\theta) k_\ell(\theta) d\theta , \tag{21.63}$$

$u \in (-\infty,\infty)$, $s = 1,2,\dots,m$ in which

$$R_{x_\ell x_s}(u) = \lim_{T\to\infty} \frac{1}{2T} \int_{-T}^{T} x_\ell(t) x_s(t-u) dt ,$$

(21.64) and

$$R_{yx_s}(u) = \lim_{T\to\infty} \frac{1}{2T} \int_{-T}^{T} y(t) x_s(t-u) dt .$$

Applying the Fourier transform to each equation of this system, we obtain a linear algebraic system of equations for $\Phi_\ell(i\omega)$ given by

(10) By E.K. Yadrenko.

(21.65) $$S_{yx_s}(i\omega) = \sum_{\ell=1}^{m} S_{x_\ell x_s}(i\omega)\Phi_\ell(i\omega) \ , \ s = 1,2,\ldots,m$$

where

(21.66) $$S_{x_\ell x_s}(i\omega) = \int_{-\infty}^{\infty} R_{x_\ell x_s}(u)e^{i\omega u}du \ ,$$
$$S_{yx_s}(i\omega) = \int_{-\infty}^{\infty} R_{yx_s}(u)e^{i\omega u}du \ .$$

Having found $\Phi_\ell(i\omega)$, we take their inverse Fourier transforms to obtain exact expressions for the impulse transfer functions, k_ℓ.

In order to compute k_ℓ, it is necessary to develop a program for the solution of the system of equations (21.65) at the points $\omega_j = -\cot(\pi j/(n+1))$, $j = 0,1,\ldots,n$. To this end we may use the method of Gaussian elimination with pivoting as described in §3.3. It is not difficult to show that the total computing time is approximately m^2 times that for the problem of one input and output, and that approximately m times the storage is required (see [65]).

It should be noted that the problem of the solution of equation (21.42) (and the system of equations (21.63)) is not properly posed. Arbitrarily large variations in Φ with respect to (21.44) can correspond to arbitrarily small variations in R_{yx} and R_{xx}. In fact if S_{xx} and S_{yx} become zero at some finite values of ω, then their corresponding approximations $\tilde{S}_{xx}$ and $\tilde{S}_{yx}$ respectively, will not necessarily satisfy this condition. This can be the cause of an arbitrarily large error in the solution arising from an arbitrarily small error in the input data. Even in the case when the approximations $\tilde{S}_{xx}$ and $\tilde{S}_{yx}$ do have zeros at the same points as the exact functions, and of the same multiplicity, the errors in the solution may still be significant since the closeness of function values does not imply closeness of their derivatives. It is not difficult to see that for arbitrarily small variations in S_{xx} and S_{yx}, the function Φ can change its order at infinity and may become non-analytic in the upper half plane.

These difficulties may be overcome by using either regularization or the method of quasi-solutions (see §2). However in the case of the simplest integral equations of the first kind, it is not difficult to transform the method described above so that it will yield the required results.

Theorem 21.4. *Suppose*

$$(21.67)\qquad \begin{aligned} |\tilde{S}_{xx} - S_{xx}| &\le \int_{-\infty}^{\infty} |\tilde{R}_{xx}(u) - R_{xx}(u)|du \le \epsilon \,, \\ |\tilde{S}_{yx} - S_{yx}| &\le \int_{-\infty}^{\infty} |\tilde{R}_{yx}(u) - R_{yx}(u)|du \le \epsilon \,, \end{aligned}$$

and let

$$(21.68)\qquad \tilde{\Phi}(i\omega) = \tilde{S}_{yx}(i\omega)/\tilde{S}_{xx}(i\omega) \quad .$$

Then at those values of ω *for which* $|\tilde{S}_{xx}(i\omega)| \ge \epsilon/\delta$, $\delta < 1$, *we have*

$$(21.69)\qquad |\tilde{\Phi}(i\omega) - \Phi(i\omega)| \le \frac{1 + |\tilde{\Phi}(i\omega)|}{|\tilde{S}_{xx}(i\omega)|-\epsilon}\,\epsilon \le \frac{1 + |\tilde{\Phi}(i\omega)|}{1 - \delta}\,\delta \quad .$$

For the proof we note that

$$\begin{aligned} |\tilde{\Phi} - \Phi| &= \left| \frac{\tilde{S}_{yx}}{\tilde{S}_{xx}} - \frac{S_{yx}}{\tilde{S}_{xx}} + \frac{S_{yx}}{\tilde{S}_{xx}} - \frac{S_{yx}}{S_{xx}} \right| \\ &\le \frac{\epsilon}{|\tilde{S}_{xx}|} + \frac{\epsilon|S_{yx}|}{|\tilde{S}_{xx}|\;|S_{xx}|} \\ &\le \frac{1}{|\tilde{S}_{xx}|}[\epsilon + |\tilde{\Phi} - \Phi|\epsilon + |\tilde{\Phi}|\epsilon] \,, \end{aligned}$$

from which the result follows.

This theorem implies that in order to avoid the complications mentioned above, it is necessary to calculate Φ (using (21.68)) only at those points for which $|S_{xx}(i\omega)| \ge \epsilon/\delta$. At the remaining points we can evaluate Φ by means of an interpolation polynomial using the values $\tilde{\Phi}$ at the points where equation (21.68) is used.

In conclusion, we note that in practice $\tilde{R}_{xx}$ and $\tilde{R}_{yx}$ are random estimates of R_{xx} and R_{yx}, respectively. Consequently inequalities (21.67) and (21.69) are only valid with a probability p.

§21.4 The Numerical Solution of Statistical Linearization and the Synthesis of Optimum Linear Systems

Let the functions x and y denote representations of stationary ergodic random processes for the input and output respectively of a filter. The problem of statistical linearization (or linear synthesis) of the filter consists of finding the so-called optimal impulse transfer function k^* (where $k^*(t) = 0$ for $t < 0$) which minimises I, the mathematical expectation of the square of the deviation, defined by

$$
(21.70) \qquad \begin{aligned} I(k) &= E\{y(t) - \int_0^\infty x(t-\theta)k(\theta)d\theta\}^2 \\ &= \lim_{T\to\infty} \frac{1}{2T} \int_{-T}^{T} \left(y(t) - \int_0^\infty x(t-\theta)k(\theta)d\theta\right)^2 dt \,. \end{aligned}
$$

Thus the formal difference between the problem of linearization (analysis) of a given filter and the problem of synthesis of the filter consists only in the way in which the output y is represented.

Important special cases of the problem of synthesis are given by: (i) the problem of optimum extrapolation of a signal, where $y(t) = x(t+\tau)$, $\tau > 0$; (ii) the problem of the optimum filtering of a signal where $x(t) = z(t)+n(t)$, $y(t) = z(t)$, where z is the required signal and n is the noise; (iii) the problem of simultaneous filtering and extrapolation, where $x(t) = z(t)+n(t)$, $y(t) = z(t+\tau)$.

Solodovnikov [125] has shown that the required optimum impulse transfer function k^* is given as the solution of a Wiener-Hopf equation of the first kind:

$$
(21.71) \qquad Ak \equiv \int_0^\infty R_{xx}(u-\theta)k(\theta)d\theta = R_{yx}(u) \ , \quad u > 0 \ ,
$$

where R_{xx} and R_{yx} are correlation functions of the signals x and y (see (21.43)). If $R_{xx} \in L_1(-\infty,\infty)$, then A is a bounded, positive definite operator in $L_2(0,\infty)$ ($\|A\| \leq \int_{-\infty}^{\infty} |R_{xx}(\theta)|d\theta$), and this implies that (21.71) has a unique solution in $L_2(0,\infty)$. However, (21.71) is not always solvable in $L_2(0,\infty)$.

For the analytic solution of equation (21.71), let us rewrite it in the form

$$
(21.72) \qquad R_{yx}(u) - \int_0^\infty R_{xx}(u-\theta)k_+(\theta)d\theta = \begin{cases} 0 \ , & u > 0 \ , \\ k_-(u) \ , & u \leq 0 \ . \end{cases}
$$

Taking the Fourier transform(11) of each side of (21.72) we obtain

$$
(21.73) \qquad S_{yx}(i\omega) - S_{xx}(i\omega)\Phi^+(i\omega) = \Phi^-(i\omega) \ ,
$$

---

(11) The factor $(2\pi)^{-1/2}$ is omitted.

where

$$\Phi^+(i\omega) = \int_0^\infty k_+(\theta)e^{i\omega\theta}d\theta \quad \text{and} \quad \Phi^-(i\omega) = \int_{-\infty}^0 k_-(\theta)e^{i\omega\theta}d\theta \ .$$

The quantities $\Phi^+(i\omega)$, $\Phi^-(i\omega)$ are the boundary values of functions analytic in the upper and lower half-planes respectively (see [133]), and S_{yx}, S_{xx} are the Fourier transforms of R_{yx}, R_{xx} respectively. Making the transformation $\omega = i(1+\tau)/(1-\tau)$, we obtain the Riemann boundary value problem on the unit circle γ given by

$$\phi^-(\tau) = -S_{xx}(\tau)\phi^+(\tau) + S_{yx}(\tau) \ , \ \tau \in \gamma \ . \tag{21.74}$$

We shall assume that $R_{xx} \in L_1(-\infty,\infty)$. Since, in addition, R_{xx} is an even function, S_{xx} is the spectral density of the signal x(x is a real, continuous and even function which is zero at infinity). Hence $S_{xx}(\tau) \equiv S_{xx}(i\omega)$, $\omega = i(1+\tau)/(1-\tau)$, will be zero at $\tau = 1$. Thus we are dealing with the exceptional case of a Riemann boundary value problem (see §8.2, §14.1).

Assuming always that $S_{yx} \in L_2(\gamma)$, we shall examine this problem under particular assumptions concerning S_{xx} (cf. §19).

A. Suppose

$$-S_{xx}(\tau) = \left[\frac{(\tau-1)^2}{\tau}\right]^s G(\tau) \ , \tag{21.75}$$

where s is a non-negative integer and $G(\tau) \neq 0$ for $\tau \in \gamma$. It follows since G is real that its index is zero, so using the Sokhotski-Plemelj formulae for $\ln G$ we find (see equation (8.11))

$$G(\tau) = \frac{\exp[\psi^+(\tau)]}{\exp[\psi^-(\tau)]} \ , \quad \psi^\pm(\tau) = \pm\frac{1}{2}\ln G(\tau) + \frac{1}{2\pi i}\int_\gamma \frac{\ln G(u)du}{u-\tau} \ . \tag{21.76}$$

Substituting this representation for G into equation (21.74) we obtain

$$\phi^-(\tau)\exp[\psi^-(\tau)] + (S_{yx}\exp\psi^-)^-(\tau) = [\tau^{-1}(\tau-1)^2]^s\,\phi^+(\tau)\exp[\psi^+(\tau)] + (S_{yx}\exp\psi^-)^+(\tau) \ , \tag{21.77}$$

where

$$(S_{yx}\exp\psi^-)^\pm(\tau) = \pm\frac{1}{2}S_{yx}(\tau)\exp[\psi^-(\tau)] + \frac{1}{2\pi i}\int_\gamma \frac{S_{yx}(u)\exp[\psi^-(u)]du}{u-\tau} \ . \tag{21.78}$$

Recalling that $\phi^-(\infty) = \Phi^-(1) = \int_{-\infty}^{0} k_-(\theta)e^{\theta}d\theta < \infty$, we look for a solution of the Riemann boundary value problem (21.74) which is bounded at infinity. Using Liouville's theorem, it follows from equation (21.77) that

$$\left[\frac{(\tau-1)^2}{\tau}\right]^s \phi^+(\tau)\exp[\psi^+(\tau)] + (S_{yx}\exp\psi^-)^+(\tau) \equiv P_s\left(\frac{1}{\tau}\right) \quad, \tag{21.79}$$

where P_s is an arbitrary real polynomial of degree s. We assume P_s is real since for real τ the left-hand side of (21.79) is real. Thus

$$\phi^+(\tau) = \frac{\tau^s}{(\tau-1)^{2s}\exp[\psi^+(\tau)]}\{P_s\left(\frac{1}{\tau}\right) - (S_{yx}\exp\psi^-)^+(\tau)\} \quad. \tag{21.80}$$

From (21.80) we conclude that in general a solution of the boundary value problem (21.74) which is integrable on γ does not exist. This implies that the minimum of I is not attained from among those functions k which are either integrable or square integrable on $(0,\infty)$. On the other hand if we assume that Φ^+ behaves at infinity like $\omega^{2s+\alpha}$, $0 \leq \alpha < 1$, then (21.73) will always be solvable and its solution will be given by (21.80) with $\tau = (\omega-i)/(\omega+i)$.

This last condition implies that the least value of I is attained on the class of generalized functions (see [36]) whose Fourier transforms are of the form

$$\int_0^{\infty} k(\theta)e^{i\omega\theta}d\theta = \omega^{2s}\chi(\omega) \quad, \tag{21.81}$$

where χ satisfies (because of the restrictions on S_{yx}) the condition

$$\int_{-\infty}^{\infty} \frac{|\chi(\omega)|^2}{1+\omega^2}d\omega < \infty \quad. \tag{21.82}$$

We note that generalized functions of the form given in equation (20.23) are obviously a special case of functions satisfying (21.81) and (21.82).

B. Suppose now that

$$-S_{xx}(\tau) = \exp\left\{-\left(i\frac{1+\tau}{1-\tau}\right)^2\right\}G(\tau),\ G(\tau) \neq 0,\ \tau \in \gamma. \tag{21.83}$$

Using the same transformation as in A, we obtain

$$\begin{aligned}&\phi^-(\tau)\exp[\psi^-(\tau)] + (S_{yx}\exp\psi^-)^-(\tau)\\ &\quad = \exp\left[-\left(i\frac{1+\tau}{1-\tau}\right)^2\right]\phi^+(\tau)\exp[\psi^+(\tau)] + (S_{yx}\exp\psi^-)^+(\tau) \quad,\end{aligned} \tag{21.84}$$

and application of Liouville's theorem gives

$$\phi^+(\tau) = \exp\left[\left(i\,\frac{1+\tau}{1-\tau}\right)^2 - \psi^+(\tau)\right]\left\{C - (S_{yx}\exp\psi^-)^+(\tau)\right\} . \qquad (21.85)$$

It appears that in this case it is necessary to allow Φ^+ to behave like $\exp(\omega^2)$. If we assume that Φ^+ can be written as

$$\Phi^+(i\omega) = \sum_{j=0}^{\infty} a_j(-i\omega)^j + \Phi_1^+(i\omega) \;,\; \Phi_1^+ \in L_2(-\infty,\infty) \;, \qquad (21.86)$$

then

$$k(t) \equiv k_+(t) = \sum_{j=0}^{\infty} a_j\delta^{(j)}(t) + k_1(t) \;,\; k_1 \in L_2(0,\infty) \;, \qquad (21.87)$$

and from (21.41) this implies that

$$y(t) = \sum_{j=0}^{\infty} a_j x^{(j)}(t) + \int_0^{\infty} x(t-\theta)k_1(\theta)d\theta + n(t) . \qquad (21.88)$$

This expression for y is only of theoretical interest since in practice input signals do not possess a large number of derivatives.

C. Suppose

$$\begin{aligned} -S_{xx}(\tau) &= p(\tau)G(\tau) \;,\; G(\tau) \neq 0 \;,\; \tau \in \gamma, \text{ where} \\ p(\tau) &= (\omega^2-\omega_1^2)^{m_1}\,[\tau^{-1}(\tau-1)^2]^{m_2} \;,\; m_1 < m_2. \end{aligned} \qquad (21.89)$$

Proceeding as before, we find

$$\phi^+(\tau) = \frac{1}{p(\tau)\exp[\psi^+(\tau)]}\left[P_{m_2}\left(\frac{1}{\tau}\right) - (S_{yx}\exp\psi^-)^+(\tau)\right] . \qquad (21.90)$$

It follows from (21.90) that the minimum value of I is attained on a class of generalized functions for which

$$\int_0^{\infty} k(\theta)e^{i\omega\theta}d\theta = \frac{\omega^{2m_2}}{(\omega^2-\omega_1^2)^{m_1}}\,\chi(\omega) \;,\; \int_{-\infty}^{\infty}\frac{|\chi(\omega)|^2}{1+\omega^2}d\omega < \infty . \qquad (21.91)$$

We note that the generalized function $(\omega^2-\omega_1^2)^{-m_1}$ in the frequency (ω) domain corresponds in the θ domain to the generalized function

$$k_{m_1}(\theta) = \begin{cases} P^{(1)}_{m_1-1}(\theta)\cos(\omega_1\theta) + P^{(2)}_{m_1-1}(\theta)\sin(\omega_1\theta) \;,\; \theta > 0 \;, \\ 0 \;,\; \theta < 0 \;, \end{cases}$$

where $P^{(1)}_{m_1-1}$ and $P^{(2)}_{m_1-1}$ denote polynomials of degree m_1-1 (see [36]).

It is not difficult to generalize the above results to the case when any finite number of factors becomes zero on the real axis. This implies, in particular, the validity of the following theorem.

Theorem 21.5. *If S_{xx} is a rational function (which is most often encountered in applications) then the minimum value of I will always be attained on a set of generalized functions k such that*

(21.92)
$$k(\theta) = O(\theta^r) \ , \quad r \geq 0 \ ,$$
$$\Phi(i\omega) = \int_0^\infty k(\theta) e^{i\omega\theta} d\theta = O(\omega^s) \ , \quad s \geq 0 \ ,$$

for all sufficiently large r and s.[12]

Before considering numerical methods for the minimization of the functional (21.70) and the solution of equation (21.71), we note that from (21.43), (21.70) and (21.71) we have

(21.93)
$$I(k) = R_{yy}(0) - 2(R_{yx}, k)_{L_2(0,\infty)} + (Ak, k)_{L_2(0,\infty)} \ ,$$
$$I(k^*) = R_{yy}(0) - (R_{yx}, k^*)_{L_2(0,\infty)} \ .$$

Furthermore

(21.94)
$$\begin{aligned} I(k) - I(k^*) &= (Ak_+, k_+)_{L_2(-\infty,\infty)} - 2(R_{yx}, k_+)_{L_2(-\infty,\infty)} + \\ &\qquad + (Ak^*, k^*)_{L_2(-\infty,\infty)} \\ &= (2\pi)^{-1} \left[(S_{xx}\Phi^+, \Phi^+)_{L_2(-\infty,\infty)} - 2(S_{xx}\Phi^*, \Phi^+)_{L_2(-\infty,\infty)} + (S_{xx}\Phi^*, \Phi^*)_{L_2(-\infty,\infty)} \right] \\ &= (2\pi)^{-1} \int_{-\infty}^\infty S_{xx}(i\omega) \ |\Phi^*(i\omega) - \Phi^+(i\omega)|^2 d\omega \ , \end{aligned}$$

where

$$\Phi^+(i\omega) = \int_0^\infty k(\theta) e^{i\omega\theta} d\theta = \int_{-\infty}^\infty k_+(\theta) e^{i\omega\theta} d\theta \ ,$$
$$\Phi^*(i\omega) = \int_0^\infty k^*(\theta) e^{i\omega\theta} d\theta \ , \quad \text{and}$$

k^* is the function (generalized) at which the minimum value of I is attained. We also note that

---

(12) As usual, $g = O(f)$ means that $|g/f| \leq$ constant.

$$\|Ak_+ - Ak^*\|^2_{L_2(-\infty,\infty)} = (2\pi)^{-1}\int_{-\infty}^{\infty} S^2_{xx}(i\omega)\,|\Phi^*(i\omega) - \Phi^+(i\omega)|^2 d\omega \tag{21.95}$$

and

$$\|k - k^*\|^2_{L_2(0,\infty)} = (2\pi)^{-1}\int_{-\infty}^{\infty} |\Phi^*(i\omega) - \Phi^+(i\omega)|^2 d\omega \quad . \tag{21.96}$$

From equations (21.94)-(21.96) the following important result is obtained.

Theorem 21.6. *The inequalities*

$$\|Ak - Ak^*\|^2_{L_2(0,\infty)} \le \|Ak_+ - Ak^*\|^2_{L_2(-\infty,\infty)}$$

$$\le \{\max_{\omega} |S^2_{xx}(i\omega)|\}\ \|k - k^*\|^2_{L_2(0,\infty)} , \tag{21.97}$$

$$I(k) - I(k^*) \le \{\max_{\omega}|S_{xx}(i\omega)|\} \cdot \|k - k^*\|^2_{L_2(0,\infty)} ,$$

are valid.

This implies that from the convergence of k to k^* in the $L_2(0,\infty)$ norm there follows both $I(k) - I(k^*) \to 0$ and $\|Ak - Ak^*\|_{L_2(0,\infty)} \to 0$. The converse however is in general not true since S_{xx} can become zero for real ω.

Let $k_+ = k_n$ be an approximate solution of (21.71) such that

$$\|Ak_n - Ak^*\|_{L_2(-\infty,\infty)} \to 0 \text{ as } n \to \infty . \tag{21.98}$$

Thus we shall also have

$$\|Ak_n - R_{yx}\|_{L_2(0,\infty)} \to 0 \text{ as } n \to \infty . \tag{21.99}$$

However it does not follow that $I(k_n)$ tends to the minimum value of I as $n \to \infty$. This occurs, for example, when $\Phi_n^+ = \int_0^{\infty} k_n(\theta)e^{i\omega\theta}d\theta$ approximates arbitrarily closely to Φ^* in the metric C on any finite part of the real axis, but has at infinity a pole of order p_n such that

$$r - 1 \le 2p_n \le 2r - 2 , \tag{21.100}$$

where r is the order of the zero of S_{xx} at infinity. From (21.94) it follows that

$$I(k_n) = \infty . \tag{21.101}$$

In order to overcome this difficulty, it is necessary to obtain an approximate solution of equation (21.71) in $L_2(0,\infty)$ by applying either Tikhonov's regularization method or Ivanov's method of quasi-solutions (see §2, §3, §5, §6, §20). It is also necessary to introduce an extended space of generalized functions $\tilde{L}_2(0,\infty)$ instead of the space $L_2(0,\infty)$ (see §20). We shall not however discuss these problems any further here.

For a numerical solution of the problems of linearization and synthesis, it is necessary to adapt algorithms which minimize directly the functional I (see [65]). Now the observation of x and y can obviously only be done for a finite length of time. Therefore in practice we can minimize only

$$I_T(k) = \frac{1}{T} \int_{T_2}^{T+T_2} \left| y(t) - \int_{T_1}^{T_2} x(t-\theta)k(\theta)d\theta \right|^2 dt \quad , \tag{21.102}$$

where T_1, T_2 are known values of the delay time and the end of the transfer process, respectively. (For $\theta \leq T_1$ and $\theta \geq T_2$, the impulse transfer function k is assumed to be zero). It should be noted that I_T is a random variable, estimates of which can be obtained using methods described by Pugachev [120].

It is possible to pose the problem of the minimization of I_T on any set of functions for which I_T is defined. Obviously the extension of the admissible set of functions can lead only to a decrease in the minimum of I_T i.e. to a more exact linearization or to a better synthesis, and so it is appropriate to consider the largest possible extension of the admissible set. In practice, however, the extension will always be bounded by computer limitations. In addition, for problems of synthesis (and bearing in mind the necessity for the physical realization of the impulse transfer function k), it can be shown that restricting attention to a very narrow set of functions for the minimization of I_T is both practical and economically justifiable.

Let M be a set on which I_T is minimized. We shall give a series of algorithms for the construction of the minimizing sequence $k_1, k_2, \ldots, k_s, \ldots \in M$ such that

$$\begin{gathered} I_T(k_1) \geq I_T(k_2) \geq \ldots \geq I_T(k_s) \geq \ldots , \\ \lim_{s\to\infty} I_T(k_s) = \inf_{k\in M} I_T(k) . \end{gathered} \tag{21.103}$$

1. Suppose M is a finite-dimensional space of functions of the form

$$k_n(\theta) = \sum_{j=1}^{n} a_j \ell_j(\theta) \quad , \tag{21.104}$$

where ℓ_j are certain given functions. It is obvious that $I_T(k_n)$ will be a quadratic function of the a_j, so that the minimizing sequence can be constructed, for example, by the coordinate-wise descent algorithm of §3.4.

The success of the method will depend upon the choice of the coordinate functions ℓ_j. As discussed above, the analytic structure of S_{xx} defines, to a considerable extent, the structure of the solution of equation (21.71) and in turn suggests the choice of the functions ℓ_j. It is appropriate to choose, in addition to the usual complete system in $L_2(0,\infty)$ (or in $L_2(T_1,T_2)$) (§10.3), the generalized functions $\delta(\theta-\theta_j)$, where δ is the Dirac delta function. These latter functions are particularly useful for the solution of synthesis problems, since amplifiers with ideal delay correspond to them (see [125]).

2. Suppose M is a set of arbitrary integrable functions on the interval $[T_1,T_2]$. The problem of the minimization of I_T on M is reduced to that of the minimization of the quadratic function $\tilde{I}_T$, where

$$\tilde{I}_T(k) = \frac{1}{T}\int_{T_2}^{T+T_2} \left| y(t) - \sum_{\nu=N_1}^{N_2-1} x(t-\nu\Delta t) \int_{\nu\Delta t}^{(\nu+1)\Delta t} k(\theta)d\theta \right|^2 dt , \tag{21.105}$$

$$N_1\Delta t = T_1 , \quad N_2\Delta t = T_2 ,$$

with respect to the variables $\int_{\nu\Delta t}^{(\nu+1)\Delta t} k(\tau)d\tau$. The success of this method improves as the number of oscillations of x on the interval of length (T_2-T_1) decreases.

3. Suppose M is the Hilbert space $\tilde{L}_2(T_1,T_2)$ with elements of the form

$$\tilde{k}(\theta) = k(\theta) + \sum_{j=1}^{m} d_j\, \delta(\theta-\theta_j) , \quad k \in L_2(T_1,T_2) , \tag{21.106}$$

and scalar product defined by

$$(\tilde{k}^{(1)}, \tilde{k}^{(2)})_{\tilde{L}_2} = (k^{(1)}, k^{(2)})_{L_2} + \sum_{j=1}^{m} d_j^{(1)} d_j^{(2)} . \tag{21.107}$$

Algorithms for the minimization of I_T reduce to iterative algorithms of steepest descent type (see §5.2).

4. Suppose M is a weakly compact set of functions from $\tilde{L}_2(T_1,T_2)$, defined by the condition that $\|\tilde{k}\|_{\tilde{L}_2} \le C_1$, where C_1 does not depend on $\tilde{k}$. Let $\tilde{k}_n^\circ$ be a function of the form (21.104) which minimizes I_T and satisfies the condition $\|\tilde{k}\| \le C_1$ (see §6.2). Then the sequence $\{\tilde{k}_n^\circ\}$, $n = 1,2,\ldots$ will be

minimizing on M (see §6.2).

Let us now examine the case of an arbitrary filter with n inputs $X = (x_1, x_2, \ldots, x_n)$ and m outputs $Y = (y_1, y_2, \ldots, y_m)$. The problem of linear analysis (or synthesis) consists of determining a matrix $K = (k_{sj})$ of optimum impulse transfer functions, which minimizes the expectations

$$(21.108) \qquad I_s(k) = E\left[y_s(t) - \sum_{j=1}^{m} \int_0^\infty x_j(t-\theta)k_{sj}(\theta)d\theta \right]^2 ,$$

where x_j and y_s denote the inputs and outputs respectively of the filter. The required impulse transfer functions k_{sj} are given by the solution of a system of Wiener-Hopf equations of the first kind (see [125]),

$$(21.109) \qquad R_{y_s x_r}(u) = \sum_{j=1}^{m} \int_0^\infty R_{x_j x_r}(u-\theta)k_{sj}(\theta)d\theta \ , \ u > 0 \ ,$$

$$r = 1,2,\ldots,m \ .$$

As in the case of a single equation, this system of equations is easily reduced to a Riemann boundary value problem for a system of functions. Although the latter problem has received almost no analytic investigation, this does not prevent the application of general methods and algorithms for its solution (see §2, §3, §5, §6, §19, §20 and §15.2, §15.3, §21.1-21.3). In practice we minimize

$$(21.110) \qquad I_{sT}(k) = \frac{1}{T} \int_{T_2}^{T+T_2} \left| y_s(t) - \sum_{j=1}^{m} \int_{T_1^{(j,s)}}^{T_2^{(j,s)}} x_j(t-\theta)k_{sj}(\theta)d\theta \right|^2 dt \ ,$$

where $T_1^{(j,s)}$, $T_2^{(j,s)}$ are estimates of the start and finish respectively of the transfer process along the j,s channel; T_1+T_2 is the total time of observation of x_j and y_s; and $T_2 = \max_{j,s} T_2^{(j,s)}$. The algorithms examined above for the minimization of I_T may be generalised (see §15.2, §15.3, §19.20) to the problem of the minimization of I_{sT} (see [62]).

In conclusion, we note that the minimization of I_s on a compact set ensures that the general problem is well posed. This is essential for problems of analysis. For synthesis problems, there is no need to ensure that they are well posed in the classical sense. In this case any functional matrix K which provides, on the given set M, a value of I_s close to the minimum value will be satisfactory. Furthermore, an improperly posed problem (in the classical sense) can have favourable properties since large changes in the structure of K will be associated with only a small increase of I_s. Accordingly the most economic structure of the system (i.e. the matrix K) can be chosen approximately in conjunction with this type of measure of the quality of the system ($I_s(k)$). Moreover, in practice, the synthesis problem is to calculate K subject to a set of constraints. Once again we obtain the general problem of minimization with constraints (see §6.3).

Chapter 5

NUMERICAL SOLUTION OF OTHER FORMS OF SINGULAR EQUATIONS

§22 INTEGRO-DIFFERENTIAL EQUATIONS

Many important problems from applied mathematics can be reduced to a singular integro-differential equation of the form

$$R\phi \equiv \sum_{r=0}^{m} \left[a_r(t)\phi^{(r)}(t) + \frac{1}{\pi i} \int_\Gamma \frac{k_r(t,\tau)\phi^{(r)}(\tau)d\tau}{\tau - t}\right] = f(t) , \tag{22.1}$$

$t \in \Gamma$, where a_r, k_r, f are given functions and $\phi^{(r)}$ denotes the rth derivative of ϕ, with $\phi^{(0)} \equiv \phi$.

A particular case of this equation is the integro-differential equation of Prandtl (the aerofoil equation)

$$R\Gamma \equiv \frac{\Gamma(x)}{B(x)} - \frac{1}{2\pi} \int_{-1}^{1} \frac{\Gamma'(u)du}{u - x} = f(x) , \ x \in (-1,1) . \tag{22.2}$$

This equation has been discussed in many papers.

The equation as given in (22.1) was first examined by Magnaradze [97]. Assuming that a_r, k_r and f are sufficiently differentiable and that Γ is a simple, closed, sufficiently smooth contour, Magnaradze reduced equation (22.1) to an equivalent singular or regular integral equation.

Following Vekua [22], we shall give one method for the reduction of (22.1) to a singular integral equation.

Let $\Gamma \equiv \widetilde{ab}$ be an open smooth curve. Writing

$$\phi^{(m)}(t) = \mu(t) , \tag{22.3}$$

we obtain

$$\phi^{(k)}(t) = \int_\Gamma \omega_{m-k-1}(t,t_1)\mu(t_1)dt_1 + \sum_{j=0}^{m-k-1} c_{m-k-j}t^j/j\,! \tag{22.4}$$

for $k = 0,1,\ldots,m-1$, where

$$\begin{aligned} \omega_0(t,t_1) &= 1 , \quad \text{if } t_1 \in \widetilde{at} , \\ \omega_0(t,t_1) &= 0 , \quad \text{if } t_1 \notin \widetilde{at} , \\ \omega_{k-1}(t,t_1) &= \int_\Gamma \omega_0(t,t_2)\omega_{k-2}(t_2,t_1)dt_2 , \ k = 2,3,\ldots,m , \end{aligned} \tag{22.5}$$

and $C_1, C_2, \ldots, C_m$ are arbitrary constants. Substituting into (22.1) we obtain a singular integral equation for μ of the form

$$K\mu \equiv a_m(t)\mu(t) + \frac{1}{\pi i}\int_\Gamma \frac{k_m(t,\tau)\mu(\tau)dt}{\tau - t} + \int_\Gamma k(t,\tau)\mu(\tau)d\tau \tag{22.6}$$
$$= f(t) - \sum_{k=1}^{m} C_k\, \chi_k(t) \;,$$

where

$$k(t,\tau) = \sum_{r=0}^{m-1}\left[a_r(t)\omega_{m-r-1}(t,\tau) + \right.$$
$$\left. + \frac{1}{\pi i}\int_\Gamma \frac{k_r(t,u)\omega_{m-r-1}(u,\tau)}{u - t}\, du \right] , \tag{22.7}$$

and

$$\chi_k(t) = \sum_{r=0}^{m-k}\left[a_r t^{m-r-k} + \frac{1}{\pi i}\int_\Gamma \frac{k_r(t,\tau)\tau^{m-r-k}d\tau}{\tau - t}\right] \Big/ (m-r-k)!$$

It follows that if for any values of the constants $C_1, C_2, \ldots, C_m$, the function μ is the solution of equation (22.6), then the function ϕ as given by equation (22.4) will be a solution of the original equation (22.1). Conversely, it is obvious that if ϕ is the solution of equation (22.1) then $\phi^{(m)}(t) = \mu(t)$ gives the solution of equation (22.6) for the specific values of $C_1, C_2, \ldots, C_m$.

It is also possible to apply this method to the case when Γ is a closed smooth curve, but it is now necessary to consider carefully the functions $\phi^{(k)}$, $k = 0,1,\ldots,m-1$ as defined by (22.4), since in general they will not be unique. Therefore, with the help of the solution of equation (22.6) it is possible to obtain solutions of equation (22.1) having discontinuities of the first kind at certain fixed points of Γ. Moreover, with the help of the solution of equation (22.6), it is always possible to obtain single valued continuous solutions of equation (22.1), provided that such solutions exist.

It can be seen that for the Cauchy problem, when the values of $\phi^{(k)}(a)$, $k = 0,1,\ldots,m-1$, are given, we have

$$C_k = \sum_{j=0}^{k-1} (-1)^j\, \phi^{(m-j)}(a)/j! \quad \text{for } k = 1,2,\ldots,m \; . \tag{22.8}$$

By the same token, equation (22.6) becomes completely defined and all the methods and algorithms of the previous sections may be applied for its numerical solution.

In the case of boundary value or mixed problems for the integro-differential equation (22.1), the constants C_k cannot be so simply determined. For the numerical solution of these problems, it is appropriate to apply the methods and algorithms which were described in §2.1 and §6.2 (the case of equations $Ax = y$ with an unbounded linear operator A) directly to equation (22.1). It is convenient to take as the domain of the operator R a subspace of those functions from $L_2(\Gamma)$ which satisfy the given boundary conditions.

§23 MULTI-DIMENSIONAL EQUATIONS†

A detailed theory of multi-dimensional equations can be found in the book by Mikhlin [104]. We shall state here some of the simpler properties.

1. Let f be a function defined almost everywhere on a manifold Γ in which a metric is defined, and let x_0 be a point on Γ. We delete from the manifold Γ those points which lie within a sphere of radius ϵ and centre at x_0. We assume that on the remaining part Γ_ϵ of the manifold Γ, the function f is integrable for all $\epsilon > 0$. If the limit

$$(23.1) \qquad \lim_{\epsilon\to 0} \int_{\Gamma_\epsilon} f(x)dx$$

exists, then it is called the *singular integral* of the function f on the manifold Γ. In certain cases Γ may coincide with the Euclidean space E_m, of dimension m, or with some of its subregions.

2. Consider the singular integral of the form

$$(23.2) \qquad V(x) = \int_{E_m} \frac{f(x,\Theta)}{r^m}\, u(y)dy ,$$

where x and y are points of E_m, $r = \|y-x\|$, and $\Theta = (y-x)/2$. The point x is called the *pole*, the function f the *characteristic*, and the function u the *density* of the singular integral (23.2). Let us assume that the density and characteristic satisfy the following conditions: 1) in any bounded part of the space E_m the density u is Hölder continuous; 2) at infinity $u(x) = O(\|x\|^{-k})$, $k > 0$; 3) the characteristic is bounded and, for a fixed x, is continuous with respect to Θ.

Theorem 23.1 (Tricomi [168]). *If conditions* 1)-3) *are satisfied, then for the existence of the singular integral* (23.2) *it is necessary and sufficient that*

$$(23.3) \qquad \int_S f(x,\Theta)dS = 0$$

where S is the unit sphere over which Θ is taken.

3. Theorem 23.2. (Calderon and Zygmund [154]).

If

$$(23.4) \qquad \int_S |f(x,\Theta)|^{p'}\, dS \le \text{const.} \quad \textit{and} \quad \frac{1}{p} + \frac{1}{p'} = 1$$

then the singular integral operator (23.2) *is bounded in* $L_p(E_m)$, $1 < p < \infty$.

† For an up to date summary of the theory see Chapter IX of P.P. Zabreyko et al., *Integral Equations - a reference text* (a translation from the Russian) Noordhoff International Publishing, Leyden (1975). (Eds.)

4. For simplicity we shall restrict attention to the case when $m = 2$. Let us expand the characteristic f as a Fourier series,

$$(23.5)\qquad f(x,\Theta) = \sum_{k=-\infty}^{\infty}{}' \; b_k(x)e^{ik\phi} \quad , \quad \Theta = e^{i\phi} ,$$

where the prime denotes the omission of the constant term which is necessary because of condition (23.3). Applying Mikhlin's formulae (see [104, §10])

$$(23.6)\qquad \begin{aligned} &\int_{E_2} \frac{e^{ik\phi}}{r^2} u(y)dy = \frac{2\pi i^{|k|}}{|k|} h^k u \quad , (k = \pm 1, \pm 2, \dots) , \\ &hu = \frac{1}{2\pi i} \int_{E_2} \frac{e^{i\phi}}{r^2} u(y)dy \quad , \quad h^{-1}u = \frac{1}{2\pi i} \int_{E_2} \frac{e^{-i\phi}}{r^2} u(y)dy \quad , \end{aligned}$$

we represent the singular integral (23.2)($m = 2$) as

$$(23.7)\qquad \begin{aligned} &\int_{E_2} \frac{f(x,\Theta)}{r^2} u(y)dy = \sum_{k=-\infty}^{\infty}{}' a_k(x)h^k u \quad , \\ &a_k(x) = \frac{2\pi i^{|k|}}{|k|} b_k(x) = \frac{i^{|k|}}{|k|} \int_0^{2\pi} f(x,\Theta)e^{-ik\phi}d\phi \quad . \end{aligned}$$

Let us introduce the singular operator

$$(23.8)\qquad Au \equiv a_0(x)u(x) + \int_{E_2} \frac{f(x,\Theta)}{r^2} u(y)dy + Tu \quad ,$$

where T denotes a completely continuous operator on $L_p(E_2)$. From equations (23.7) it follows that

$$(23.9)\qquad Au \equiv \sum_{k=-\infty}^{\infty} a_k h^k u + Tu \ .$$

The corresponding function

$$(23.10)\qquad \Phi(x,\Theta) = \sum_{k=-\infty}^{\infty} a_k(x)e^{ik\phi}$$

is called the *symbol* of the operator A. The singular operator $Au-Tu$ is sometimes called an operator in *simplest form*.

Theorem 23.3 (Mikhlin [104,§13]). *To the sum and product of singular operators of simplest form, there correspond the sum and product of their symbols.*

5. Let us examine singular integrals whose characteristics are independent of their poles, so that

$$V(x) = \frac{1}{2\pi} \int_{E_2} K(x-y)u(y)dy \ , \quad K(x-y) = f(\Theta)/r^2 \ . \tag{23.11}$$

Let F denote the Fourier transform. Formal application of the convolution theorem applied to equation (23.11) gives

$$FV = FK.Fu \ , \tag{23.12}$$

where FK is defined by

$$FK = \lim_{\substack{\epsilon\to 0 \\ N\to\infty}} \frac{1}{2\pi} \int_{\epsilon<|x|<N} e^{-i(x,z)}K(x)dx \ ,$$
$$(x,z) = \|x\| \, \|z\| \cos(\widehat{xz}) \ . \tag{23.13}$$

Theorem 23.4. (Mikhlin [104,§20]). *Let the characteristic of a singular integral be independent of the pole and square integrable on the unit sphere S. Then the symbol of this integral coincides with the Fourier transform of its kernel.*

As a consequence of this theorem, it follows that if, for the singular operator of simplest form

$$A_0u = a_0(x)u(x) + \frac{1}{2\pi} \int_{E_2} \frac{f(\Theta)}{r^2} u(y)dy \ , \tag{23.14}$$

the symbol $\Phi_A(\Theta)$ is independent of the pole and is bounded, then

$$A_0 = F^{-1}\Phi_A F \ , \tag{23.15}$$

and

$$\|A_0\|_{L_2(E_2)} = \text{ess sup } |\Phi_A(\Theta)| \ . \tag{23.16}$$

6. Let us examine the singular integral equation of simplest form

$$A_0u = F^{-1} \Phi(\Theta)Fu = g \ , \tag{23.17}$$

where $g \in L_2(E_2)$. We shall assume that the symbol $\Phi(\Theta)$ is finite almost everywhere on the unit sphere S, and that the greatest lower bound of its modulus is positive. Then it follows that equation (23.17) has a unique solution which can be written as

$$u = F^{-1}\ [\Phi(\Theta)]^{-1}\ Fg \ . \tag{23.18}$$

In the more general case of a singular integral equation of the form

$$Au \equiv a_0(x)u(x) + \int_{E_2} \frac{f(x,\Theta)}{r^2} u(y)dy + (Tu)(x) = g(x) \quad , \tag{23.19}$$

we obtain the solution u by a direct numerical solution of this equation.

The most appropriate approximate methods (as in the case of one-dimensional equations) are those of least squares (§3) and iteration (§5). We note that the main difficulty in the application of the method of least squares in a multi-dimensional domain is the necessity to use a significantly greater number of coordinate elements than in the one-dimensional case in order to ensure a pre-determined accuracy. Hence the necessity to investigate the economy of computer storage, the influence of round-off errors and the reliability of algorithms is increased (see §3 and §6.4). We also have, of course, the specific problem of selecting appropriate coordinate elements e_k and the evaluation of $y_k = Ae_k$.

Let us illustrate these points with an example. Vekua [19, 20] has shown that the solution of certain systems of elliptic equations can be reduced to the solution of the equation

$$K\omega \equiv \omega(z) - \frac{g(z)}{\pi} \iint_E \frac{\omega(\zeta)d\xi d\eta}{(\zeta-z)^2} = -\ g(z)\Phi'(z) \quad , \tag{23.20}$$

$z \in E$, where E is the complex plane $z = x+iy$, g is bounded almost everywhere on E, Φ is an entire function, and $\zeta = \xi+i\eta$. We shall consider equation (23.20) as a linear equation in the Hilbert space $L_2(E)$. Since $\{x^k \exp(-x^2)\}$ and $\{y^k \exp(-y^2)\}$ are complete systems in $L_2(-\infty,\infty)$, it follows that $\{z^k\bar{z}^j \exp(-z\bar{z})\}$ is a complete system in $L_2(E)$. Let us calculate the integral I, where

$$I = -\frac{1}{\pi} \iint_E \frac{\zeta^k\ \bar{\zeta}^j\ \exp(-\zeta\bar{\zeta})d\xi d\eta}{(\zeta-z)^2} \quad , \tag{23.21}$$

following the method suggested by Vekua [19]. We introduce the function G defined by

$$G = -\frac{1}{\pi} \iint_E \frac{\zeta^k\ \bar{\zeta}^j\ \exp(-\zeta\bar{\zeta})d\xi d\eta}{\zeta - z} \quad . \tag{23.22}$$

It is not difficult to show (see [19]) that $\partial G/\partial\bar{z} = z^k\bar{z}^j \exp(-z\bar{z})$. On the other hand the function F defined by

(23.23) $$F = \int^{\bar{z}} z^k \bar{z}^j \exp(-z\bar{z})d\bar{z}$$

obviously has the property $\partial F/\partial\bar{z} = z^k\bar{z}^j \exp(-z\bar{z}) = \partial G/\partial\bar{z}$. Hence

(23.24) $$G = F + \epsilon ,$$

where ϵ is an entire function. But for large $|z|$, G and F tend to zero, so that $\epsilon \equiv 0$. Thus $G = F$ and

(23.25) $$\partial G/\partial z = I(z) = \partial F/\partial z .$$

It should be noted that

(23.26) $$\left\| \frac{1}{\pi} \iint_E \frac{\omega(\zeta)d\xi d\eta}{(\zeta-z)^2} \right\|_{L_2(E)} = \|\omega(z)\|_{L_2(E)} ,$$

so that if $|g(z)| < 1$, then the operator K has an inverse in $L_2(E)$ and equation (23.20) can be solved by the method of simple iteration.

§24 NON-LINEAR EQUATIONS

Non-linear singular integral equations of the form

$$u(s) = \lambda\Phi\left[s;\ u(s);\ -\frac{1}{2\pi}\int_{-\pi}^{\pi} u(\sigma)\cot\left(\frac{\sigma-s}{2}\right)d\sigma\right], \tag{24.1}$$

where Φ is a given non-linear function and u is an unknown function, were first examined in detail by Guseynov [46]. Later, important results were obtained for (24.1) by other authors. In particular, questions concerning the numerical solution of (24.1) by the method of simple iteration (see Theorem 2.23) have been examined in [38] and [46].

As an illustration of the general approach to the approximate solution of non-linear operator equations (see §2.3 and §6.3), we shall consider a non-linear singular integral equation, which arises from the inverse boundary value problem for the determination of the lattice of profiles in two-dimensional steady flow of an incompressible fluid.

The problem is to obtain the form of the profile if we know the velocity v on the surface of the profile, the incident velocity $v_1 e^{i\alpha_1}$, the velocity $v_2 e^{i\alpha_2}$ downstream of the lattice, and the lattice spacing t. This and similar problems have been analyzed in considerable detail by Zhukovskiy [50]. Here we shall consider a method of solution which differs from those given by Zhukovskiy. We shall reduce the problem to that of the solution of a non-linear singular integral equation.

It is known that the complex fluid velocity due to a lattice of profiles in two dimensional, irrotational, incompressible flow is given by

$$W(z) = W_0 - \frac{1}{2ti}\int_{\Gamma} W(\zeta)\cot\left(\frac{\pi(\zeta-z)}{t}\right)d\zeta\ , \quad z \notin L\ , \tag{24.2}$$

where

$$W_0 = (v_1 e^{i\alpha_1} + v_2 e^{i\alpha_2})/2\ , \tag{24.3}$$

and Γ is the contour of the profile with length L. Let $z \to \zeta_0 \in \Gamma$. Applying the Sokhotski-Plemelj formulae we obtain

$$W(\zeta_0) = 2W_0 - \frac{1}{ti}\int_{\Gamma} W(\zeta)\cot\left(\frac{\pi(\zeta-\zeta_0)}{t}\right)d\zeta\ . \tag{24.4}$$

Since

(24.5) $$\cot\left(\frac{\pi(\zeta-\zeta_0)}{t}\right) = \frac{t}{\pi}\left[\frac{1}{(\zeta-\zeta_0)} + \sum_{k=1}^{\infty}\left(\frac{1}{\zeta-\zeta_0-kt} + \frac{1}{\zeta-\zeta_0+kt}\right)\right] ,$$

and the points $\zeta_0 \pm kt$ are outside of Γ, then $\int_\Gamma \frac{d\zeta}{\zeta-(\zeta_0\pm kt)} = 0$ and

(24.6) $$\frac{1}{ti}\int_\Gamma \cot\left(\frac{\pi(\zeta-\zeta_0)}{t}\right)d\zeta = 1 .$$

From (24.6) we have

(24.7) $$W(\zeta_0) = W_0 - \frac{1}{2ti}\int_\Gamma [W(\zeta) - W(\zeta_0)]\cot\left(\frac{\pi(\zeta-\zeta_0)}{t}\right)d\zeta .$$

The condition for the contour to be a streamline is

(24.8) $$W(\zeta)d\zeta = v(s)ds ,$$

or

(24.9) $$W(\zeta) = v(s)/u(s) \quad , \quad u(s) = \frac{d\zeta}{ds} ,$$
$$\zeta(s) - \zeta(s\ ) = \int_{s_0}^{s} u(x)dx .$$

On writing Ls for s, we can rewrite equation (24.7) in the form

(24.10) $$Au \equiv W_0u(s_0) - \frac{L}{2ti}\int_0^1 [v(s)u(s_0)-u(s)v(s_0)]\cot\left(\frac{\pi L}{t}\int_{s_0}^{s} u(x)dx\right)ds = v(s_0)$$

The unknown quantity L is determined from the equation

(24.11) $$t(v_1\cos\alpha_1 - v_2\cos\alpha_2) = L\int_0^1 v(s)ds .$$

It can be seen that the function u must satisfy the conditions

(24.12) $$\text{(i)}\ \int_0^1 u(s)ds = 0 ,\quad \text{(ii)}\ |u(s)| \equiv 1 ,\quad \text{(iii)}\ u(0) = u(1) .$$

We look for an approximate solution of equation (24.10) in the form

(24.13) $$u_n(s) = \sum_{k=-n}^{n}{}' \alpha_k \exp(ik2\pi s) ,$$

where the prime denotes that the term $k = 0$ is deleted. It is clear that conditions (i) and (iii) of equation (24.12) are automatically satisfied by u_n. Therefore we require values of α_k so as to minimise

$$Q(P) = \int_0^1 |Au_n - v(s_0)|^2 ds_0 + \int_0^1 (|u_n(s)| - 1)^2 ds \ , \tag{24.14}$$
$$P = P(\alpha_k^{(1)}, \alpha_k^{(2)}), \ \alpha_k = \alpha_k^{(1)} + i\alpha_k^{(2)} \ .$$

Usually we solve problems associated with the improvement of a given initial profile. In this case, it is possible to obtain initial values for α_k by solving approximately the equation $\tilde{u}(s) = \frac{d\tilde{\zeta}(s)}{ds}$ (where $\tilde{\zeta}(s) = \tilde{x}(s) + i\tilde{y}(s)$ is the equation of the original profile) using a polynomial of the form (24.13).

In the general case, it should be taken into account when minimizing $Q(P)$, that only certain profiles are admissible. This leads to a more general minimization problem for which some algorithms have been considered in §6.3.

BIBLIOGRAPHY

Editors' Introduction. Although most of the bibliography refers to articles in Russian, a larger number of these have been reviewed by Mathematical Reviews. Where possible we have included the Maths. Reviews reference. In addition we have also used Maths. Reviews abbreviations for the names of Journals, wherever possible. Maths. Reviews Indexes also contain lists of Journals which are translated into English; we have decided not to note explicitly where an English translation of a journal article exists, although we have done so for books.

[1] AKHIEZER N.I., *Izv. Akad. Nauk SSSR Ser. Mat.*, 9(1945)275-290, MR 7-439.

[2] AKHIEZER N.I., *Lectures in the Theory of Approximation*, Nauka, Moscow, 1965, MR 32#6108. (English translation of an earlier edition published by Ungar, New York, 1956, MR 20#1872).

[3] AKHIEZER N.I., GLAZMAN I.M., *Theory of Linear Operators in Hilbert Space*, Nauka, Moscow, 1966, MR 34#6527. (English translation of an earlier edition published by Ungar, New York, 1961, MR 41#9015 (a) and (b)).

[4] BABICH M.D., IVANOV V.V., *Ž. Vyčisl. Mat. i Mat. Fiz.*, 7(1967)988-1000, MR 38#4067.

[5] BELLMAN R., *Dynamic Programming*, Princeton Univ. Press, Princeton, 1957, MR 19-820.

[6] BELONOSOV S.M., *Plane Problems in the Theory of Elasticity*, Izd. SO AN SSSR, 1962.

[7] BEREZOVSKIY A.I., GAZ'KINA L.S., in *Proceedings of the First Conference of Young Scientists from a Sector of Technical Cybernetics*, Izd., IK Akad. Nauk Ukraine, Kiev, 1967, 22-25.

[8] BEREZIN I.S., ZHIDKOV N.P., *Computing Methods*, vols. 1 and 2, Fizmatgiz, Moscow, 1959, MR 22#12685. (English translation published by Pergamon Press, Oxford, 1965, MR 30#4372).

[9] BERNSTEIN S.N., *Collected Works*, Izd. Akad. Nauk SSSR, I, Moscow, 1952, 11-104, MR 14-2.

[10] BIRMAN M.Sh., *Vestnik Leningrad in-ta, Ser. Mat., Fiz. i Khim*, 9(1952)69-76.

[11] BITSADZE A.V., *Uspehi Mat. Nauk (N.S.)*,12(1957) no.5(77)185-190, MR 20#962.

[12] BOGOLYUBOV N.N., MEDVEDEV V.V., TAVKHELIDZE A.N., in *Problems of Continuum Mechanics*, English edition published by SIAM, Philadelphia, 1961, 39-55.

[13] BOCHAROV I.N., FEL'DBAUM A.A., *Avtomat. i Telemeh*, 23(1962)289-301, MR 28#4963.

[14] BOYARSKIY B.V., *Soobšč. Akad. Nauk Gruzin. SSR*, 21(1956)391-398.

[15] BUSLENKO N.P., SHREYDER Yu.A., *The Monte-Carlo Method and how it is carried out in Digital Computers*, Fizmatgiz, Moscow, 1961, MR 23#B1100.

[16] VEKUA I.N., *Dokl. Akad. Nauk SSSR*, 26(1940)335-338.

[17] VEKUA I.N., *Trudy Tbiliss. Mat. Inst.*,10(1941)45-72, MR 4-160.

[18] VEKUA I.N., *Soobšč. Akad. Nauk Gruzin.SSR*, 3(1942)869-876, MR 5-268.

[19] VEKUA I.N., *Mat. Sb. N.S.*, 31(1952)217-314, MR 15-230.

[20] VEKUA I.N., *Generalized Analytic Functions*, Fizmatgiz, Moscow, 1959, MR 21#7288. (English translation published by Pergamon Press, Oxford, 1962, MR 27#321).

[21] VEKUA N.P., *Systems of Singular Integral Equations*, Gostekhizdat, Moscow-Leningrad, 1950, MR 13-247. (English translation published by Noordhoff, Groningen, 1967, MR 35#2102).

[22] VEKUA N.P., *Akad. Nauk Gruzin. SSR Trudy Tbiliss. Mat. Inst. Razmadze*, 24(1957)135-147, MR 20#1898.

[23] VITUSHKIN A.G., *Estimation of the Complexity of the Tabulation Problem*, Fizmatgiz, Moscow, 1959, MR 22#8265.

[24] VOROB'EV Yu.V., *The Method of Moments in Applied Mathematics*, Fizmatgiz, Moscow, 1958, MR 21#7591.

[25] GABDULKHAEV B.G., *Functional Analysis and the Theory of Functions*, Izd. Kazansk. Univ., 3(1966)7-17.

[26] GAVURIN M.K., KANTOROVICH L.V., in *Forty Years of Mathematics in the U.S.S.R.*, vol. 1, Fizmatgiz, Moscow, 1959, 809-855.

[27] GARKAVI A.L., *Izv. Akad. Nauk SSSR Ser. Mat.*, 24(1960)103-128, MR 22#3921.

[28] GAKHOV F.D., *Mat. Sb.*, 2(44), No.4(1937)673-683.

[29] GAKHOV F.D., *Boundary Value Problems in the Theory of Analytic Functions and Singular Integral Equations*, Doctoral thesis, Tbilisi, 1941.

[30] GAKHOV F.D., *Izv. Kazan fiz-mat. obshch* , XIV ser.3 (1949)75-160.

[31] GAKHOV F.D., *Uspehi Mat. Nauk (N.S.)*, 7, No.4(50)(1952)3-54, MR 14-629.

[32] GAKHOV F.D., in *Problems of Continuum Mechanics*, English edition published by SIAM, Philadelphia, 1961, 118-132.

[33] GAKHOV F.D., *Boundary Value Problems*, Fizmatgiz, Moscow, 1963, MR 27#6094. (English translation published by Pergamon Press, Oxford, 1966, MR 33#6311).

[34] GAKHOV F.D., SMAGINA V.I., *Izv. Akad. Nauk SSSR Ser. Mat.*, 26(1962)361-390, MR 30#1374.

[35] GAKHOV F.D., CHERSKIY Yu.I., *Izv. Akad. Nauk SSSR Ser. Mat.*, 20(1956)33-52, MR 18-134.

[36] GEL'FAND I.M., SHILOV G.E., *Generalized Functions and Operations on Them*, Fizmatgiz, Moscow, 1958, MR 20#4182. (English translation published by Academic Press, New York, 1964, MR 29#3869).

[37] GEL'FAND I.M., TSETLIN M.L., *Uspehi Mat. Nauk*, 17(1962), No.1(103)3-25, MR 25#1081.

[38] GEKHT B.I., *Kazan Gos. Univ. Učen. Zap.*, 116(1956)111-139.

[39] GLUSHKOV V.M., *Introduction to Cybernetics*, Izd. Akad. Nauk Ukrain SSR, Kiev, 1964, MR 30#1009.

[40] GOLUZIN G.M., *Geometric Theory of Functions of a Complex Variable*, Gostekhizdat, Moscow, 1952, MR 15#112. (English translation of a later edition published by American Math. Soc., Providence, 1969, MR 40#308).

[41] GONCHAROV V.L., *Theory of Interpolation and Approximation of Functions*, Gostekhizdat, Moscow, 1954, MR 16-803.

[42] GOKHBERG I.Ts., *Kišinev Gos. Univ. Učen. Zap.*, 17(1955)35-44.

[43] GOKHBERG I.Ts., KREIN M.G., *Uspehi Mat. Nauk*, 12, No.2(1957)43-118, MR 20#3459.

[44] GOKHBERG I.Ts., FEL'DMAN I.A., *Dokl. Akad. Nauk SSSR*, 165(1965)268-271, MR 32#8085.

[45] GRINBERG G.A., FOK V.A., in *Survey of Radio Wave Propagation*, vol. 2, Izd. Akad. Nauk SSSR, Moscow-Leningrad, 1948, 69-96.

[46] GUSEYNOV A.I., *Azerbaĭdžan Gos. Univ. Učen. Zap. Ser. Fiz.-Mat. Nauk*, 10(1957).

[47] DANILYUK I.I., in *Problems of Continuum Mechanics*, English edition published by SIAM, Philadelphia, 1961, 93-103.

[48] DOMBROVSKAYA I.N., IVANOV V.K., *Izv. Vysš. Učebn Zaved Matematika*, No. 4(41) (1964)69-74, MR 29#3848.

[49] ERSHOV A.P., *Dokl. Akad. Nauk SSSR (N.S.)*, 100(1955)209-211, MR 16-1082.

[50] ZHUKOVSKIY M.I., *Calculation of the Flow in Lattices of Profiles of Turbomachines*, Mashgiz, Moscow, 1960.

[51] ZYGMUND A., *Trigonometric Series vols 1 and 2*, Cambridge Univ. Press, New York, 1959, MR 21#6498.

[52] ZUKHOVITSKIY S.I., AVDEEVA L.I., *Linear and Convex Programming*, Nauka, Moscow, 1964, MR 37#2521. (English translation published by W.B. Saunders, Philadelphia, 1966, MR 37#2522).

[53] IVANOV V.V., *Trudy Novocherkas Politekh in-ta*, 67(81)(1958)75-86.

[54] IVANOV V.V., *Dokl. Akad. Nauk SSSR (N.S.)*, 114(1957)945-948, MR 21#2883.

[55] IVANOV V.V., *Dokl. Akad. Nauk SSSR*, 121(1958)793-794, MR 20#4742.

[56] IVANOV V.V., *Izv. Vysš. Učebn. Zaved Matematika*, No. 4(5) (1958)89-95, MR 25#4111.

[57] IVANOV V.V., in *Investigations of Modern Problems in the Theory of Functions of a Complex Variable*, Fizmatgiz, Moscow, 1960, 439-450.

[58] IVANOV V.V., *Proceedings of the Scientific and Technical Conference,* Izd. Akad. Nauk Ukrain SSR, Kiev, 1960, 47-65.

[59] IVANOV V.V., *Ž. Vyčisl. Mat. i Mat. Fiz.,* 1(1961)963-971, MR 26#7134.

[60] IVANOV V.V., *Proceedings of the Seminar on Algorithm Formation for Industrial Processes,* Izd. KDNTP, Kiev, 1963.

[61] IVANOV V.V., in *Mathematical Analysis 1963,* Akad. Nauk SSSR Inst. Naučn. Informacii, Moscow, 1965, 125-177, MR 32#8086.

[62] IVANOV V.V., *Sbornik Standartnykh Podprogram na KMShP* (in Ukrainian), Tekhika, Kiev, 1966, 6-36.

[63] IVANOV V.V., ZADIRAKA V.K., *Vyčisl. Mat. (Kiev) Vyp.* 2, 1966, 3-30, MR 38#3658.

[64] IVANOV V.V., KARAGODOVA E.A., *Ukrain. Mat. Ž.,* 13(1961), No.1, 28-38, MR 25#1415.

[65] IVANOV V.V., KARAGODOVA E.A., YADRENKO E.K., *Ž. Vyčisl. Mat. i Mat. Fiz.,* 3(1961)523-541, MR 25#1969.

[66] IVANOV V.V., KUDRINSKIY V.Yu., *Ž. Vyčisl. Mat. i Mat. Fiz.,* 6(1966)831-841, MR 34#8625; and 7(1967)475-496, MR 36#1090.

[67] IVANOV V.V., SIMONENKO I.B., *Dokl. Akad. Nauk SSSR,* 126(1959)1172-1175, MR 21#7435.

[68] IVANOV V.K., *Dokl. Akad. Nauk SSSR,* 145(1962)270-272, MR 25#4357.

[69] IVANOV V.K., *Mat. Sb. (N.S.),* 61(103)(1963)211-223, MR 27#3932.

[70] KAGAN V.F., *The Theory of Determinants,* Gosizdat Ukraine, 1922.

[71] KALANDIYA A.I., *Mat. Sb. (N.S.),* 42(84)(1957)249-272, MR 21#6800.

[72] KALANDIYA A.I., *Dokl. Akad. Nauk SSSR,* 125(1959)715-718.

[73] KANTOROVICH L.V., *Mat. Sb.,* 41, 1(1934)235-245.

[74] KANTOROVICH L.V., *Uspehi Mat. Nauk (N.S.),* 3, No. 6(28), (1948) 89-185, MR 10-380.

[75] KANTOROVICH L.V., AKILOV G.P., *Functional Analysis in Normed Spaces,* Fizmatgiz, Moscow, 1959, MR 22#9837. (English translation published by Macmillan, New York, 1964, MR 35#4699).

[76] KANTOROVICH L.V., KRYLOV V.I., *Approximate Methods of Higher Analysis,* Fizmatgiz, Moscow-Leningrad, 1962, MR 27#4338. (English translation of an earlier edition published by Noordhoff, Groningen, 1958, MR 21#5268).

[77] KARAGODOVA E.A., *Vyčisl. Mat. (Kiev) Vyp. 1,* (1965)146-152, MR 34#8124.

[78] KAPITSA S.P., *Vyčisl. Mat.,* 1(1957)167-169, MR 19-776.

[79] KELDYSH M.V., SEDOV L.I., *Dokl. Akad. Nauk SSSR*, 16(1937)7-10.

[80] KIM ZE PKHEN, *Ž. Vyčisl. Mat. i Mat. Fiz.*, 3(1963)683-701, MR 27#2822.

[81] KIM ZE PKHEN, *Dokl. Akad. Nauk SSSR*, 150(1963)1249-1251, MR 28#442.

[82] KORNEYCHUK A.A., *Ž. Vyčisl. Mat. i Mat. Fiz.*, 4(1964), No. 4, suppl., 64-74, MR 30#5484.

[83] KRASNOSEL'SKIY M.A., *Topological Methods in the Theory of Non-Linear Integral Equations*, Gostekhizdat, Moscow, 1956, MR 20#3464. (English translation published by Macmillan, New York, 1964, MR 28#2414).

[84] KRASNOSEL'SKIY M.A., KREIN S.G., *Mat. Sb. N.S.*, 31(73)(1952) 315-334, MR 14-692.

[85] KREIN M.G., *Uspehi Mat. Nauk*, 13(1958), No. 5(83)3-120, MR 21#1507.

[86] KRYLOV V.I., *Approximate Calculation of Integrals*, Fizmatgiz, Moscow, 1959, MR 22#2002. (English translation published by Macmillan, New York, 1962, MR 26#2008).

[87] KUBLANOVSKAYA V.N., *Trudy Mat. Inst. Steklov*, 53(1959)145-185, MR 22#8641.

[88] KULIK V.T., *The Principles of Algorithm Formation and the Construction of Control Machines*, Gostekhizdat Ukrain SSR, Kiev, 1963.

[89] LAVRENTIEV M.A., *Trudy Ts.A.G.I.*, 118(1932)3-56.

[90] LAVRENTIEV M.A., SHABAT B.V., *Methods of the Theory of Functions of a Complex Variable*, Nauka, Moscow, 1965, MR 35#344.

[91] LAVRENTIEV M.M., *Some Improperly Posed Problems of Mathematical Physics*, Izdat Sibirsk. Otdel. Akad. Nauk SSSR, Novosibirsk, 1962, MR 36#5487 and 36#7363. (English translation published by Springer-Verlag, New York, 1967).

[92] LEVITIN E.S., POLYAK B.T., *Ž. Vyčisl. Mat. i Mat. Fiz.*, 6(1966) 787-823, MR 35#2468.

[93] LIKHT M.K., *Ž. Vyčisl. Mat. i Mat. Fiz.*, 3(1963)979-987, MR 28#3530.

[94] LUZIN N.N., *Integral and Trigonometric Series*, Gostekhizdat, Moscow-Leningrad, 1951, MR 14-2.

[95] LYUSTERNIK L.A., *Trudy Mat. Inst. Steklov*, 20(1947)49-64, MR 10-71.

[96] LYUSTERNIK L.A., SOBOLEV V.I., *Elements of Functional Analysis*, Nauka, Moscow, 1965, MR 35#698. (English translation of an earlier edition published by Ungar, New York, 1961, MR 25#5361).

[97] MAGNARADZE L.G., *Soobšč. Akad. Nauk Gruzin. SSR*, 4(1943)103-110, MR 6-158.

[98] MAL'TSEV A.I., *Foundations of Linear Algebra*, Gostekhizdat, Moscow-Leningrad, 1948, MR 11-412. (English translation published by Freeman, San Francisco, 1963, MR 29#3477).

[99] MANDZHAVIDZE G.F., in *Investigations of Modern Problems in the Theory of Functions of a Complex Variable*, Fizmatgiz, Moscow, 1960, 365-370, MR 22#6890.

[100] MARCHUK G.I., *Apl. Mat.*, 10(1965)255-267, MR 31#6390.

[101] MIKHALEVICH V.S., SHOR N.Z., *Nauchno-metodich materialy ekonomiko-matemat. seminara Akad. Nauk SSSR*, Moscow, 1(1962)15-41.

[102] MIKHLIN S.G., *Integral Equations*, Gostekhizdat, Moscow-Leningrad, 1949, MR 12-712. (English translation published by Pergamon Press, New York, 1957, MR 19-428).

[103] MIKHLIN S.G., *Variational Methods in Mathematical Physics*, Gostekhizdat, Moscow, 1957, MR 22#1981. (English translation by Macmillan, New York, 1964, MR 30#2712).

[104] MIKHLIN S.G., *Multidimensional Singular Integrals and Integral Equations*, Fizmatgiz, Moscow, 1962, MR 27#5105. (English translation published by Pergamon Press, Oxford, 1965, MR 32#2866).

[105] MIKHLIN S.G., SMOLITSKIY Kh.T., *Approximate Methods for the Solution of Differential and Integral Equations*, Izdat. Nauka, Moscow, 1965, MR 33#855. (English translation published by American Elsevier, New York, 1967, MR 36#1108).

[106] MOROZOV V.V., *Kazan. Gos. Univ. Učen. Zap.*, 112(1952)17-20.

[107] MUSKHELISHVILI N.I., *Some Basic Problems of the Mathematical Theory of Elasticity*, Izd. Akad. Nauk SSSR, Moscow, 1954, MR 16-1067. (English translation of an earlier edition published by Noordhoff, Groningen, 1953, MR 15-370).

[108] MUSKHELISHVILI N.I., *Singular Integral Equations*, Fizmatgiz, Moscow, 1962. (For review of an earlier edition, see MR 8-586). (English translation of an earlier edition published by Noordhoff, Groningen, 1953, MR 15-434).

[109] NATANSON I.P., *The Constructive Theory of Functions*, Gostekhizdat, Moscow-Leningrad, 1949, MR 11-591. (English translation published in 3 volumes by Ungar, New York, 1964/5, MR 33#4529(a), (b) and (c)).

[110] NATANSON I.P., *The Theory of Functions of a Real Variable*, Gostekhizdat, Moscow-Leningrad, 1957. (For review of an earlier edition see MR 12-598). (English translation of an earlier edition published in two parts by Ungar, New York, 1955 and 1961, MR 16-804 and MR 26#6309).

[111] NEMYTSKIY V.V., *Dokl. Akad. Nauk SSSR*, 131(1960)746-747, MR 22#12420.

[112] NIKOLAEVA M.V., *Trudy Mat. Inst. Steklov*, 28(1949)26-32, MR 13-496.

[113] NIKOLSKIY S.M., *Quadrature Formulae*, Fizmatgiz, Moscow, 1958.

[114] OSTROVSKIY G.M., *Avtomat. i Telemeh*, 23(1962)1284-1289, MR 28#2012.

[115] POLYAK B.T., *Ž. Vyčisl. Mat. i Mat. Fiz.*, 4(1964)995-1005, MR 32#4821.

[116] POL'SKIY N.I., *Ukrain. Mat. Ž.*, 7(1955)56-70, MR 17-64.

[117] POL'SKIY N.I., *Dokl. Akad. Nauk SSSR (N.S.)*, 111(1956)1181-1184, MR 18-802.

[118] PRIVALOV I.I., *Boundary Properties of Analytic Functions*, Gostekhizdat, Moscow-Leningrad, 1950, MR 13-926.

[119] PYKHTEEV G.N., *Prikl. Mat. Meh.*, 23(1959)1074-1082, MR 22#12698.

[120] PUGACHEV V.S., *Theory of Random Functions and its Application to Problems of Automatic Control*, Fizmatgiz, Moscow, 1960, MR 23#B2119.

[121] RYBAKOV L.M., *Matematicheskoe Prosvescenie*, 6(1961)262-263.

[122] SIMONENKO I.B., *Dokl. Akad. Nauk SSSR*, 124(1959)278-281, MR 22#3805.

[123] SMIRNOV V.I., *A Course of Higher Mathematics*, vol. 5, Fizmatgiz, Moscow, 1959. (For review of an earlier edition see MR 9-574). (English translation published by Pergamon Press, Oxford, 1964, MR 29#5964).

[124] SOBOLEV S.L., *Izv. Akad. Nauk SSSR Ser Mat.*, 20(1956)413-436, MR 18-322.

[125] SOLODOVNIKOV V.V., *Statistical Dynamics of Linear Automatic Control Systems*, Fizmatgiz, Moscow, 1960, MR 23#B3089.

[126] SOFRONOV I.D., *Dokl. Akad. Nauk SSSR (N.S.)*, 111(1956)37-39, MR 19-66.

[127] SUDAKOV V.N., KHALFIN L.A., *Dokl. Akad. Nauk SSSR*, 157(1964)1058-1060, MR 29#2858.

[128] TIKHONOV A.N., *Dokl. Akad. Nauk SSSR*, 39(1944)195-198.

[129] TIKHONOV A.N., *Dokl. Akad. Nauk SSSR*, 153(1963)49-52, MR 28#5577.

[130] TIKHONOV A.N., *Dokl. Akad. Nauk SSSR*, 161(1965)1023-1026, MR 34#3387.

[131] TIKHONOV A.N., GLASKO V.B., *Ž. Vyčisl. Mat. i Mat. Fiz.*, 4(1964)564-571, MR 29#6654.

[132] TIKHONOV A.N., GLASKO V.B., *Ž. Vyčisl. Mat. i Mat. Fiz.*, 5(1965)463-473.

[133] TITCHMARSH E., *Introduction to the Theory of Fourier Integrals*, Clarendon Press, Oxford, 1937.

[134] TOMPKINS C.B., in *Modern Mathematics for the Engineer*, First Series, ed. by E.F. Beckenbach, McGraw-Hill, New York, 1956, Chapter 18.

[135] WALSH J.L., *Interpolation and Approximation by Rational Functions in the Complex Domain*, Amer. Math. Soc. Colloq. Publ., XX, 3rd ed., 1960, MR 36#1672(a).

[136] FADEEV D.K., FADEEVA V.N., *Computational Methods of Linear Algebra*, Fizmatgiz, Moscow, 1960. (English translation published by Freeman, San Francisco, 1963, MR 28#1742).

[137] FAN VAN KHAP, *Ž. Vyčisl. Mat. i Mat. Fiz.*, 5(1965)171-184, MR 33#4623.

[138] FRIDMAN V.M., *Dokl. Akad. Nauk SSSR*, 128(1959)482-484, MR 22#1066.

[139] KHALILOV Z.I., *Izv. Akad. Nauk SSSR Ser. Mat.*, 13(1949)163-176, MR 11-39.

[140] KHARAZOV D.F., KHVEDELIDZE B.V., *Soobšč. Akad. Nauk Gruzin. SSR*, 28(1962)129-135, MR 25#1443.

[141] HOUSEHOLDER, A.S., *Principles of Numerical Analysis*, McGraw-Hill, New York, 1953, MR 15-470.

[142] KHVEDELIDZE B.V., *Akad. Nauk Gruzin. SSR. Trudy Tbiliss. Mat. Inst. Razmadze*, 23(1956)3-158, MR 21#5873.

[143] KHVEDELIDZE B.V., *Dokl. Akad. Nauk SSSR*, 140(1961)66-68, MR 26#563.

[144] TSELLNER V., *Z. Èksper. Teoret. Fiz.*, 36(1959)1103-1109, MR 22#1350.

[145] CHEBOTAREV G.N., *Kazan. Gos. Univ. Učen. Zap.*, 116(4)(1956)31-58.

[146] CHERNOFF H., MOSES L.E., *Elementary Decision Theory*, Wiley, New York, 1959, MR 21#3081.

[147] CHERSKIY Yu.I., *Kazan. Gos. Univ. Učen. Zap.*, 113(10)(1953)43-55, MR 18-135.

[148] CHERSKIY Yu.I., *Mat. Sb. N.S.*, 41(83)(1957)277-296, MR 19-559.

[149] CHIKIN L.A., *Kazan. Gos. Univ. Učen. Zap.*, 113(10)(1953)57-105.

[150] CHIKIN L.A., *Dokl. Akad. Nauk SSSR (N.S.)*, 111(1956)44-46, MR 19-540.

[151] SHAMANSKIY V.E., *Ukrain. Mat. Ž.*, 14(1962)100-109, MR 26#4464.

[152] BERG L., *Math. Nachr.*, 14(1955)193-212, MR 19-66.

[153] BERG L., *Wiss. Z. Univ. Rostock. Math. Nat. Reihe*, 4(1954/55)381-391, MR 22#4927.

[154] CALDERON A.P., ZYGMUND A., *Amer. J. Math.*, 78(1956)289-309, MR 18-894.

[155] DAVIS M.C., *IEEE Trans. Automatic Control*, AC-8 (1963)296-305.

[156] JONES C.E., TIKTOPOULOS G., *J. Mathematical Phys.*, 7(1966)311-315, MR 32#3294.

[157] ENSKOG D., *Acta Math.*, 54(1930)177-184.

[158] FRÉCHET M., *Rend. Circ. Mat. Palermo* (2), 1(1952)11-27, MR 14-237.

[159] HESTENES M.R., STIEFEL E., *J. Res. Nat. Bur. Standards*, 49(1952) 409-436, MR 15-651.

[160] LEVY P., *Compositio Math.*, 1(1934)1-14.

[161] MACCAMY R.C., *J. Math. Mech.*, 7(1958)355-375, MR 20#4163.

[162] MULTHOPP H., *Luftfahrtforschung*, 15(1938)153-169.

[163] OMNÈS R., *Nuovo Cimento*, 8(1958)316-326, MR 20#688.

[164] PLEMELJ J., *Monatsh. Math. und Phys.*, 19(1908)211-245.

[165] ROSEN J.B., *J. Soc. Indust. Appl. Math.*, 8(1960)181-217, MR 22#3601.

[166] SALZMAN G., SALZMAN F., *Phys. Rev.*, Second series, 8(1957)1619-1628.

[167] STEWART C.E., *J. Soc. Indust. Appl. Math.*, 8(1960)342-353, MR 22#4928.

[168] TRICOMI F., *Math. Z.*, 27(1928)87-133.

[169] WALTHER A., DEJON B., in *Symposium on the Numerical Treatment of Ordinary Differential Equations* etc., Birkhäuser, Basel, 1960, 645-671, MR 23#B602.

[170] WIENER N., *Extrapolation, Interpolation and Smoothing of Stationary Time Series*, Wiley, New York, 1949, MR 11-118.

[171] WIENER N., HOPF E., *S.-B. Deut. Akad. Wiss. Berlin Phys. Math. Kl.*, (1931)696-706.

[172] WILKINSON J.H., *The Algebraic Eigenvalue Problem*, Clarendon Press, Oxford, 1965, MR 32#1894.

[173] ZYGMUND A., *Duke Math. J.*, 12(1945)47-76, MR 7-60.

AUTHOR INDEX

SUBJECT INDEX

MIX
Papier aus verantwortungsvollen Quellen
Paper from responsible sources
FSC® C105338

If you have any concerns about our products,
you can contact us on
ProductSafety@springernature.com

In case Publisher is established outside the EU,
the EU authorized representative is:
Springer Nature Customer Service Center GmbH
Europaplatz 3, 69115 Heidelberg, Germany

Printed by Libri Plureos GmbH
in Hamburg, Germany